Revise AS

Chemistry

Rob Ritchie

Contents

Chapter 3 Energetics, rates and equilibrium

Chapter 4 Organic chemistry, analysis and the environment

Specification lists

AQA AS Chemistry

MODULE	SPECIFICATION TOPIC	CHAPTER REFERENCE	STUDIED IN CLASS	REVISED	PRACTICE QUESTIONS
AS Unit 1 (M1) Foundation chemistry	Atomic structure	1.1, 1.2, 1.3, 1.4			
	Amount of substance	1.4, 1.5, 1.6			
	Bonding	2.1, 2.2, 2.3, 2.4, 2.5			
	Periodicity	2.6			
	Introduction to organic chemistry	4.1			
	Alkanes	4.2, 4.3			
	Energetics	3.1, 3.2, 3.3			
	Kinetics	3.4, 3.5			
	Equilibria	3.6, 4.8			
	Redox reactions	1.7			
AS Unit 2 (M2) Chemistry in action	Group 7, the Halogens	2.8			
	Group 2, the alkaline earth metals	2.7			
	Extraction of metals	2.9			
	Haloalkanes	4.3, 4.6, 4.8			
	Alkenes	4.4			
	Alcohols	4.5, 4.8			
	Analytical techniques	4.7			

Examination analysis

AS Chemistry comprises two unit tests. All questions are compulsory.
Practical and investigative skills will also be assessed.

Unit 1	Structured questions: short and extended answers	1hr 15min test	$33^1/3\%$
Unit 2	Structured questions: short and extended answers	1hr 45min test	$46^2/3\%$
Unit 3	Internal assessment of practical and investigative skills		20%

OCR AS Chemistry

MODULE	SPECIFICATION TOPIC	CHAPTER REFERENCE	STUDIED IN CLASS	REVISED	PRACTICE QUESTIONS
AS Unit F321 **(M1)** Atoms, bonds and groups	Atoms	1.1, 1.4			
	Moles and equations	1.5, 1.6			
	Acids	1.8			
	Redox	1.7			
	Electron structure	1.2, 1.3			
	Bonding and structure	2.1, 2.2, 2.3, 2.4, 2.5			
	Periodicity	2.6			
	Group 2	2.7			
	Group 7	2.8			
AS Unit F322 **(M2)** Chains, energy and resources	Basic concepts	4.1, 4.4			
	Alkanes	4.2, 4.3			
	Alkenes	4.4			
	Alcohols	4.5			
	Halogenoalkanes	4.6			
	Modern analytical techniques	4.7			
	Enthalpy changes	3.1, 3.2, 3.3			
	Rates and equilibrium	3.4, 3.5, 3.6			
	Chemistry of the air	3.5, 4.7			
	Green chemistry	4.8			

Examination analysis

AS Chemistry comprises two unit tests. All questions are compulsory.
Practical and investigative skills will also be assessed.

Unit F321	Structured questions: short and extended answers	1hr test	30%
Unit F322	Structured questions: short and extended answers	1hr 45min test	50%
Unit F323	Internal assessment of practical and investigative skills		20%

Salters AS Chemistry

MODULE	SPECIFICATION TOPIC	CHAPTER REFERENCE	STUDIED IN CLASS	REVISED	PRACTICE QUESTIONS
AS Unit F331 (M1) *Chemistry for life*	Formulae, equations and amount of substance	1.1, 1.4, 1.5, 1.6, 2.1			
	Atomic structure	1.1, 1.2			
	Bonding and structure	2.1, 2.2, 2.5			
	Inorganic chemistry and the Periodic Table	2.6, 2.7, 4.3			
	Modern analytical techniques	1.4, 4.7			
	Energetics	3.1, 3.2, 3.3			
	Kinetics	3.5			
	Organic functional groups	4.1, 4.2			
	Organic reactions	4.3, 4.5			
	Isomerism	4.1			
	Applications	4.2, 4.8			
AS Unit F332 (M2) *Chemistry of natural resources*	Formulae, equations and amount of substance	1.5, 1.6			
	Atomic structure	1.2			
	Bonding and structure	2.1, 2.2, 2.3, 2.4, 2.5			
	Redox	1.7			
	Inorganic chemistry and the Periodic Table	1.3, 2.1, 2.6, 2.8			
	Organic functional groups	4.1, 4.4, 4.5			
	Organic reactions	4.4, 4.5, 4.8			
	Reaction mechanisms	4.1, 4.3, 4.4, 4.6			
	Isomerism	4.1, 4.4			
	Kinetics	3.4, 3.5			
	Equilibria	3.6			
	Applications of organic chemistry	4.8			
	Modern analytical techniques	4.7, 4.8			

Examination analysis

AS Chemistry comprises two unit tests. All questions are compulsory. Practical and investigative skills will also be assessed.

Unit F331	Structured questions: short and extended answers	1hr 15min test	30%
Unit F332	Structured questions: short and extended answers One question on an Advance Notice article	1hr 45min test	50%
Unit F333	Internal assessment of practical and investigative skills		20%

Edexcel AS Chemistry

MODULE	SPECIFICATION TOPIC	CHAPTER REFERENCE	STUDIED IN CLASS	REVISED	PRACTICE QUESTIONS
AS Unit 1 (M1) *The core principles of Chemistry*	Formulae, equations and amount of substance	*1.1, 1.4, 1.5, 1.6, 4.1*			
	Energetics	*3.1, 3.2, 3.3*			
	Atomic structure and the Periodic Table	*1.1, 1.2, 1.3, 1.4, 2.6*			
	Bonding	*2.1, 2.5*			
	Introductory organic chemistry	*4.1, 4.2, 4.3, 4.4*			
	Shapes of molecules and ions	*2.2*			
	Intermediate bonding and bond polarity	*2.3*			
	Intermolecular forces	*2.4, 2.5*			
	Redox	*1.7*			
AS Unit 2 (M2) *Application of core principles of Chemistry*	The Periodic Table – Groups 2 and 7	*2.7, 2.8*			
	Kinetics	*3.4, 3.5*			
	Chemical equilibria	*3.6*			
	Organic chemistry	*4.5, 4.6*			
	Mechanisms	*4.1, 4.3, 4.4, 4.6, 4.8*			
	Mass spectra and IR	*4.7, 4.8*			
	Green chemistry	*4.8*			

Examination analysis

AS Chemistry comprises two unit tests. All questions are compulsory. Practical and investigative skills will also be assessed.

Unit 1	Objective questions; Structured questions: short and extended answers	1hr 15min test	40%
Unit 2	Objective questions; Structured questions: short and extended answers contemporary context questions	1hr 15min test	40%
Unit 3	Internal assessment of practical and investigative skills		20%

WJEC AS Chemistry

MODULE	SPECIFICATION TOPIC	CHAPTER REFERENCE	STUDIED IN CLASS	REVISED	PRACTICE QUESTIONS
AS Unit CH1 (M1) Controlling and using chemical change	Basic ideas about atoms	1.1, 1.2, 1.3			
	Chemical calculations	1.4, 1.5, 1.6, 4.1			
	Chemical equilibrium and acid-base reactions	1.6, 1.8, 3.6			
	Energetics	3.1, 3.2, 3.3			
	Kinetics	3.4, 3.5			
	Application of principles	4.8			
AS Unit CH2 (M2) Properties, structure and bonding	Chemical bonding	2.1, 2.3			
	Forces between molecules	2.4			
	Shapes of molecules	2.2			
	Solubility of compounds in water	2.5			
	Solid structures	2.5			
	The Periodic Table	1.7, 2.6			
	Trends in properties of the elements of the s-block and Group 7	2.7, 2.8			
	Organic compounds and their reactions	1.6, 4.1, 4.2, 4.4, 4.5, 4.6			
	Hydrocarbons	4.1, 4.2, 4.3, 4.4			
	Halogenoalkanes	4.3, 4.6			
	Alcohols	4.5			
	Analytical techniques	4.7			

Examination analysis

AS Chemistry comprises two unit tests. All questions are compulsory. Practical and investigative skills will also be assessed.

Unit 1	Objective questions; Structured questions: short and extended answers	1hr 30min test	40%
Unit 2	Objective questions; Structured questions: short and extended answers	1hr 30min test	40%
Unit 3	Internal assessment of practical and investigative skills		20%

CCEA AS Chemistry

MODULE	SPECIFICATION TOPIC	CHAPTER REFERENCE	STUDIED IN CLASS	REVISED	PRACTICE QUESTIONS
AS Unit 1 (M1) *Basic concepts in physical and inorganic chemistry*	Formulae, equations and amounts of substance	1.5, 1.6, 1.8			
	Atomic structure	1.1, 1.2, 1.3, 1.4, 2.10			
	Bonding and structure	2.1, 2.3, 2.5			
	Shapes of molecules and ions	2.2			
	Intermolecular forces	2.4, 2.5			
	Redox	1.7			
	The Periodic Table	2.6			
	Group VII	2.8			
	Titrations	1.5			
AS Unit 2 (M2) *Further physical and inorganic chemistry and introduction to organic chemistry*	Formulae and formulae of a substance	1.5, 1.6, 4.1			
	Nomenclature and isomerism in organic compounds	4.1, 4.2			
	Hydrocarbons – alkanes	4.1, 4.2, 4.3, 3.5			
	Hydrocarbons – alkenes	4.4			
	Halogenoalkanes	4.6			
	Alcohols	4.5			
	Infra-red spectroscopy	4.7			
	Energetics	3.1, 3.2, 3.3			
	Equilibrium	3.6			
	Kinetics	3.4, 3.5			
	Group II elements and their compounds	2.7			
	Qualitative analysis	2.10			

Examination analysis

AS Chemistry comprises two unit tests. All questions are compulsory.
Practical and investigative skills will also be assessed.

Unit 1	Objective questions; Structured questions: short and extended answers	1hr 30min test	35%
Unit 2	Objective questions; Structured questions: short and extended answers contemporary context questions	1hr 30min test	35%
Unit 3	Internal assessment of practical and investigative skills		30%

AS/A2 Level Chemistry courses

AS and A2

All Chemistry A Level courses currently studied are in two parts, with three separate modules in each part. Students first study the AS (Advanced Subsidiary) course. Some will then go on to study the second part of the A Level course, called A2. Advanced Subsidiary is assessed at the standard expected halfway through an A Level course: i.e., between GCSE and Advanced GCE. This means that new AS and A2 courses are designed so that difficulty steadily increases:

- AS Chemistry builds from GCSE science
- A2 Chemistry builds from AS Chemistry.

How will you be tested?

Assessment units

For AS Chemistry, you will be tested by three assessment units. For the full A Level in Chemistry, you will take a further three units. AS Chemistry forms 50% of the assessment weighting for the full A Level.

One of the units in AS and in A2 is practically based. You will take two theory units in each of AS and A2. Each unit can normally be taken in either January or June. Alternatively, you can study the whole course before taking any of the unit tests. There is a lot of flexibility about when exams can be taken and the diagram below shows just some of the ways that the assessment units may be taken for AS and A Level Chemistry.

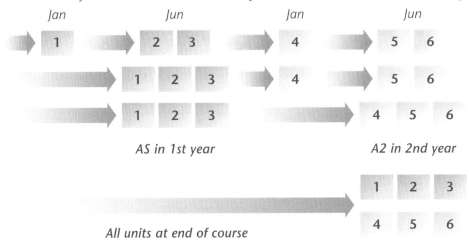

If you are disappointed with a module result, you can resit each module once. You will need to be very careful about when you take up a resit opportunity because you will have only one chance to improve your mark. The higher mark counts.

A2 and Synoptic assessment

After having studied AS Chemistry, you may wish to continue studying Chemistry to A Level. For this you will need to take three further units of Chemistry at A2. Similar assessment arrangements apply except some units, those that draw together different parts of the course in a synoptic assessment, have to be assessed at the end of the course.

Coursework

Coursework may form part of your A Level Chemistry course, depending on which specification you study. Where students have to undertake coursework, it is usually for the assessment of practical skills but this is not always the case. See pages 4–9.

Key Skills

To gain the key skills qualification, which is equivalent to an AS Level, you will need to collect evidence together in a 'portfolio' to show that you have attained a satisfactory level in Communication, Application of number and Information technology. You will also need to take a formal testing in each key skill. You will have many opportunities during AS Chemistry to develop your key skills.

What skills will I need?

The assessment objectives for AS Chemistry are shown below.

Knowledge with understanding

- recall of facts, terminology and relationships
- understanding of principles and concepts
- drawing on existing knowledge to show understanding of the responsible use of chemistry in society
- selecting, organising and presenting information clearly and logically

Application of knowledge and understanding, analysis and evaluation

- explaining and interpreting principles and concepts
- interpreting and translating, from one form into another, data presented as continuous prose or in tables, diagrams and graphs
- carrying out relevant calculations
- applying knowledge and understanding to familiar and unfamiliar situations
- assessing the validity of chemical information, experiments, inferences and statements

Experimental and investigative skills

Chemistry is a practical subject and part of the assessment of AS Chemistry will test your practical skills. You will be assessed on three main skills:

- implementing
- analysing evidence and drawing conclusions
- evaluating evidence and procedures.

The skills may be assessed in the context of separate practical exercises, although more than one skill could be assessed in any one exercise. They may also be assessed all together in the context of a single 'whole investigation'. An investigation may be set by your teacher or you may be able to pursue an investigation of your choice. Alternatively, you may take a practical examination.

You will receive guidance about how your practical skills will be assessed from your teacher. This study guide concentrates on preparing you for the written examinations testing the subject content of AS Chemistry.

Different types of questions in AS examinations

In AS Chemistry examinations, different types of question are used to assess your abilities and skills. Unit tests mainly use structured questions requiring both short answers and more extended answers.

Short-answer questions

A short-answer question may test recall or it may test understanding. Short-answer questions normally have space for the answers printed on the question paper.

Here are some examples (the answers are shown in *italics*):

What is meant by isotopes?

Atoms of the same element with different mass numbers.

Calculate the amount (in mol) of H_2O in 4.5 g of H_2O.

1 mol H_2O has a mass of 18 g. $\therefore$ 4.5 g of H_2O contains 4.5/18 = 0.25 mol H_2O.

Structured questions

Structured questions are in several parts. The parts usually have a common context and they often become progressively more difficult and more demanding as you work your way through the question. A structured question may start with simple recall, then test understanding of a familiar or an unfamiliar situation.

Most of the practice questions in this book are structured questions, as this is the main type of question used in the assessment of AS Chemistry.

When answering structured questions, do not feel that you have to complete one question before starting the next. The further you are into a question, the more difficult the marks are to obtain. If you run out of ideas, go on to the next question. You need to respond to as many parts of questions on an exam paper as possible. You will not score well if you spend so long trying to perfect the first questions that you do not have time to attempt later questions.

Here is an example of a structured question that becomes progressively more demanding.

(a) Write down the atomic structure of the two isotopes of potassium: ^{39}K and ^{41}K.

    (i) ^{39}K ...19... protons; ...20... neutrons; ...19... electrons. ✓

    (ii) ^{41}K ...19... protons; ...22... neutrons; ...19... electrons. ✓   [2]

(b) A sample of potassium has the following percentage composition by mass: ^{39}K: 92%; ^{41}K: 8% Calculate the relative atomic mass of the potassium sample.

    $92 \times 39/100 + 8 \times 41/100 = 39.16$ ✓   [1]

(c) What is the electronic configuration of a potassium atom?

    $1s^2 2s^2 2p^6 3s^2 3p^6 4s^1$ ✓   [1]

(d) The second ionisation energy of potassium is much larger than its first ionisation energy.

(i) Explain what is meant by the *first ionisation energy* of potassium.

The energy required to remove an electron ✓ from each atom in 1 mole ✓ of gaseous atoms ✓

(ii) Why is there a large difference between the values for the first and the second ionisation energies of potassium?

The 2nd electron removed is from a different shell ✓ which is closer to the nucleus and experiences more attraction from the nucleus. ✓ This outermost electron experiences less shielding from the nucleus because there are fewer inner electron shells than for the 1st ionisation energy. ✓   [6]

Extended answers

In AS Chemistry, questions requiring more extended answers may form part of structured questions or may form separate questions. They may appear anywhere on the paper and will typically have between 5 and 10 marks allocated to the answers as well as several lines of answer space. These questions are also often used to assess your abilities to communicate ideas and put together a logical argument.

The correct answers to extended questions are often less well-defined than to those requiring short answers. Examiners may have a list of points for which credit is awarded up to the maximum for the question.

An example of a question requiring an extended answer is shown below.

Magnesium oxide and sulfur trioxide are both solids at $10\,^\circ$C. Magnesium oxide has a melting point of $2852\,^\circ$C. Sulfur trioxide has a melting point of $17\,^\circ$C. Explain, in terms of structure and bonding, why these two compounds have such different melting points.   [7]

Points that the examiners might look for include:

Magnesium oxide has a giant lattice structure ✓ containing ionic bonds. ✓ These strong forces need to be broken during melting to allow ions to move free of the rigid lattice. ✓ This requires a large amount of energy supplied by a high temperature. ✓

Sulfur trioxide has a simple molecular structure ✓ held together by van der Waals' forces. ✓ These weak forces need to be broken during melting to allow molecules to move free of the rigid lattice. ✓ This requires a small amount of energy supplied by a low temperature. ✓

8 marking points → [7]

In this type of response, there may be an additional mark for a clear, well-organised answer, using specialist terms. In addition, marks may be allocated for legible text with accurate spelling, punctuation and grammar.

Other types of questions

Free-response and open-ended questions allow you to choose the context and to develop your own ideas. These are little used for assessing AS Chemistry but, if you decide to take the full A Level in Chemistry, you will encounter this type of question during synoptic assessment.

Multiple-choice or objective questions require you select the correct response to the question from a number of given alternatives. These are rarely used for assessing AS Chemistry, although it is possible that some short-answer questions will use a multiple-choice format.

Exam technique

Links from GCSE

Advanced Subsidiary Chemistry builds from grade CC in GCSE Science and GCSE Additional Science (combined), or GCSE Chemistry. This study guide has been written so that you will be able to tackle AS Chemistry from a GCSE Science background.

You should not need to search for important Chemistry from GCSE Science because this has been included where needed in each chapter. If you have not studied Science for some time, you should still be able to learn AS Chemistry using this text alone.

What are examiners looking for?

Examiners use instructions to help you to decide the length and depth of your answer.

If a question does not seem to make sense, you may have misread it – read it again!

State, define or list

This requires a short, concise answer, often recall of material that can be learnt by rote.

Explain, describe or discuss

Some reasoning or some reference to theory is required, depending on the context.

Outline

This implies a short response, almost a list of sentences or bullet points.

Predict or deduce

You are not expected to answer by recall but by making a connection between pieces of information.

Suggest

You are expected to apply your general knowledge to a 'novel' situation, one which you have not directly studied during the AS Chemistry course.

Calculate

This is used when a numerical answer is required. You should always use units in quantities and significant figures should be used with care.

Look to see how many significant figures have been used for quantities and give your answer to this degree of accuracy.

If the question uses three significant figures, then give your answer in three significant figures also.

Some dos and don'ts

Dos

Do answer the question

No credit can be given for good Chemistry that is irrelevant to the question.

Do use the mark allocation to guide how much you write

Two marks are awarded for two valid points – writing more will rarely gain more credit and could mean wasted time or even contradicting earlier valid points.

Do use diagrams, equations and tables in your responses

Even in 'essay-type' questions, these offer an excellent way of communicating chemistry.

Do write legibly

An examiner cannot give marks if the answer cannot be read.

Do write using correct spelling and grammar. Structure longer essays carefully

Marks are now awarded for the quality of your language in exams.

Don'ts

Don't fill up any blank space on a paper

In structured questions, the number of dotted lines should guide the length of your answer.

If you write too much, you waste time and may not finish the exam paper. You also risk contradicting yourself.

Don't write out the question again

This wastes time. The marks are for the answer!

Don't contradict yourself

The examiner cannot be expected to choose which answer is intended. You could lose a hard-earned mark, e.g.

A covalent bond is a shared pair of electrons ✓ bonded by the electrostatic attraction between ions. ✗

Don't spend too much time on a part that you find difficult

You may not have enough time to complete the exam. You can always return to a difficult part if you have time at the end of the exam.

What grade do you want?

Everyone would like to improve their grades but you will only manage this with a lot of hard work and determination. You should have a fair idea of your natural ability and likely grade in Chemistry and the hints below offer advice on improving that grade.

For a Grade A

You will need to be a very good all-rounder.

- You must go into every exam knowing the work extremely well.
- You must be able to apply your knowledge to new, unfamiliar situations.
- You need to have practised many, many exam questions so that you are ready for the type of question that will appear.

The exams test all areas of the syllabus and any weaknesses in your Chemistry will be found out. There must be no holes in your knowledge and understanding. For a Grade A, you must be competent in all areas.

At A level, you also have the opportunity to achieve an A* grade.

For a Grade C

You must have a reasonable grasp of Chemistry but you may have weaknesses in several areas and you may be unsure of some of the reasons for the Chemistry.

- Many Grade C candidates are just as good at answering questions as the Grade A students but holes and weaknesses often show up in just some topics.
- To improve, you will need to master your weaknesses and you must prepare thoroughly for the exam. You must become a better all-rounder.

For a Grade E

You cannot afford to miss the easy marks. Even if you find Chemistry difficult to understand and would be happy with a Grade E, there are plenty of questions in which you can gain marks.

- You must memorise all definitions.
- You must practise exam questions to give yourself confidence that you do know some Chemistry. In exams, answer the parts of questions that you know first. You must not waste time on the difficult parts. You can always go back to these later.
- The areas of Chemistry that you find most difficult are going to be hard to score on in exams. Even in the difficult questions, there are still marks to be gained. Show your working in calculations because credit is given for a sound method. You can always gain some marks if you get part of the way towards the solution.

What marks do you need?

The table below shows how your average mark is transferred into a grade.

average	80%	70%	60%	50%	40%
grade	A	B	C	D	E

A* grades are awarded for the A Level qualification only and not for the AS qualification or individual units.

For an A* grade, you need to achieve a...

- grade A overall (80% or more on uniform mark scale) for the whole A Level qualification
- grade A* (90% or more on the uniform mark scale) across your A2 units.

Four steps to successful revision

Step 1: Understand

- Study the topic to be learned slowly. Make sure you understand the logic or important concepts.
- Mark up the text if necessary – underline, highlight and make notes.
- Re-read each paragraph slowly.

GO TO STEP 2

Step 2: Summarise

- Now make your own revision note summary:
  What is the main idea, theme or concept to be learned?
  What are the main points? How does the logic develop?
  Ask questions: Why? How? What next?
- Use bullet points, mind maps, patterned notes.
- Link ideas with mnemonics, mind maps, crazy stories.
- Note the title and date of the revision notes
  (e.g. Chemistry: Atomic structure, 3rd March).
- Organise your notes carefully and keep them in a file.

This is now in **short-term memory**. You will forget 80% of it if you do not go to Step 3.
GO TO STEP 3, but first take a 10 minute break.

Step 3: Memorise

- Take 25 minute learning 'bites' with 5 minute breaks.
- After each 5 minute break test yourself:
  Cover the original revision note summary
  Write down the main points
  Speak out loud (record on tape)
  Tell someone else
  Repeat many times.

The material is well on its way to **long-term memory**.
You will forget 40% if you do not do step 4. **GO TO STEP 4**

Step 4: Track/Review

- Create a Revision Diary (one A4 page per day).
- Make a revision plan for the topic, e.g. 1 day later, 1 week later, 1 month later.
- Record your revision in your Revision Diary, e.g.
  Chemistry: Atomic Structure, 3rd March 25 minutes
  Chemistry: Atomic Structure, 5th March 15 minutes
  Chemistry: Atomic Structure, 3rd April 15 minutes
  ... and then at monthly intervals.

Chapter 1

Atoms, moles and reactions

The following topics are covered in this chapter:

- *Atoms and isotopes*
- *The electron structure of the atom*
- *Experimental evidence for electron structures*
- *Atomic mass*

- *The mole*
- *Formulae, equations and reacting quantities*
- *Redox reactions*
- *Acids, bases and salts*

1.1 Atoms and isotopes

After studying this section you should be able to:

- *recall the relative charge and mass of a proton, neutron and electron*
- *explain the existence of isotopes*
- *use atomic number and mass number to determine atomic structure*
- *describe characteristics of radioactivity*

LEARNING SUMMARY

What is an atom?

AQA	M1	OCR	
EDEXCEL		SALTERS	
CCEA	M1	WJEC	M1

An atom is the smallest part of an element that can exist on its own. Atoms are so tiny that there are more atoms in a full stop than there are people in the world. It is now possible to see individual atoms by using the most modern and powerful microscopes. However, nobody has yet seen *inside* an atom and we must devise models for the structure of an atom from experimental evidence.

Sub-atomic particles

Although there are various models for atomic structure, chemists use a model in which an atom is composed of three **sub-atomic particles**: protons, neutrons and electrons.

electron shells

nucleus (protons and neutrons)

In this model, protons and neutrons form the nucleus at the centre of the atom with electrons orbiting in shells. Compared with the total volume of an atom, the nucleus is tiny and extremely dense. Most of an atom is empty, made up of the space between the nucleus and the electron shells.

All matter is made up from the chemical elements and 116 are now known (although others will inevitably be discovered). The number of protons in the nucleus distinguishes the atoms of each element.

An important principle of chemistry is the link between an element and the number of protons in its atoms.

> Each atom of an element has the same number of protons.
> The number of protons in an atom of an element is called the **atomic number**, Z.

KEY POINT

Properties of protons, neutrons and electrons

Some properties of protons, neutrons and electrons are shown in the table below.

All atoms of hydrogen contain 1 proton.

All atoms of carbon contain 6 protons.

All atoms of oxygen contain 8 protons.

particle	relative mass	relative charge
proton, p	1	1+
neutron, n	1	0
electron, e	1/1840	1−

A proton has virtually the same mass as a neutron and, for most of chemistry, their masses can be assumed to be identical.

Atoms, moles and reactions

An electron has negligible mass compared with the mass of a proton or a neutron. In most of chemistry, the mass of an electron can be ignored.

The charge on a proton is opposite to that of an electron but each charge has the same magnitude. An atom is electrically neutral. To balance out the charges, an atom must have the same number of protons as electrons.

All atoms of hydrogen contain:
1 proton and 1 electron.

All atoms of carbon contain:
6 protons and 6 electrons.

All atoms of oxygen contain:
8 protons and 8 electrons.

> **KEY POINT**
>
> An atom is electrically neutral.
> An atom contains the same number of protons as electrons.

Isotopes

Without neutrons, the nucleus would just contain positively-charged protons. Like-charges repel, and a nucleus containing just protons would fly apart! Strong nuclear forces act between protons and neutrons and these hold the nucleus together. Neutrons can be thought of as 'nuclear glue'.

In each atom of an element, the number of protons in the nucleus is fixed. However, most elements contain atoms with different numbers of neutrons. These atoms are called **isotopes** and, because they have different numbers of neutrons, they also have different masses.

> **KEY POINT**
>
> Isotopes are atoms of the **same** element with **different** masses.
> Isotopes of an element have the **same** number of **protons and electrons**.
> Each isotope has a **different** number of **neutrons** in the nucleus.
> The combined number of protons and neutrons in an isotope of an element is called the **mass number**, A.

The isotopes of an element react in the same way. This is because chemical reactions involve electrons – neutrons make no difference.

Chemists represent the nucleus of an isotope in a special way.

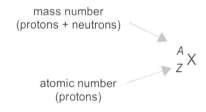

mass number
(protons + neutrons)

$^{A}_{Z}X$

atomic number
(protons)

The isotopes of carbon

Carbon exists as three isotopes, $^{12}_{6}C$, $^{13}_{6}C$, $^{14}_{6}C$. In all but the most accurate work, it is reasonable to assume that the relative mass of an isotope is equal to its mass number. It is easy to work out the atomic structure of an isotope from the representation above.

isotope	atomic number	mass number	protons	neutrons	electrons
$^{12}_{6}C$	6	12	6	6	6
$^{13}_{6}C$	6	13	6	7	6
$^{14}_{6}C$	6	14	6	8	6

Because all carbon isotopes have 6 protons, it is common practice to omit the atomic number, and $^{12}_{6}C$ is often shown as ^{12}C, or even as carbon-12.

Progress check

How many protons, neutrons and electrons are in the following isotopes?

1 $^{7}_{3}Li$; **2** $^{23}_{11}Na$; **3** $^{19}_{9}F$; **4** $^{27}_{13}Al$; **5** $^{55}_{26}Fe$.

5 $^{55}_{26}Fe$: 26p, 29n, 26e.
4 $^{27}_{13}Al$: 13p, 14n, 13e;
3 $^{19}_{9}F$: 9p, 10n, 9e;
2 $^{23}_{11}Na$: 11p, 12n, 11e
1 $^{7}_{3}Li$: 3p, 4n, 3e

Radioactivity

SALTERS

WJEC M1

The nuclei of some atoms are unstable and these atoms are **radioactive**.

Nuclear fission

During nuclear **fission**, the nuclei of radioactive atoms decay, emitting α and β-particles and γ-radiation.

- An α-**particle**, ^4_2He or $^4_2\alpha$, consists of 2p and 2n.
- A β-**particle**, $^0_{-1}\text{e}$ or $^0_{-1}\beta$, is an electron emitted from the nucleus.
- γ-**radiation** is high energy, electromagnetic radiation that often accompanies the emission of α and β-particles.

An α-particle is a helium nucleus.

Notice that α and β are particles; γ is electromagnetic radiation.

α-particles are the least penetrating form of radioactivity and can be stopped by paper.

γ-radiation is the most penetrating radioactivity.

	α	β	γ
relative charge	+2	−1	0
relative mass	4	0.00055	0
range in air	a few cm	a few m	long
deflection by electric field	small	large	none

During α-**decay**,

- the mass number goes **down** by 4
- the atomic number goes **down** by 2.

e.g. $$^{238}_{92}\text{U} \longrightarrow {}^{234}_{90}\text{Th} + {}^4_2\text{He}$$

During β-**decay**,

- a neutron changes into a proton: $^1_0\text{n} \longrightarrow {}^1_1\text{p} + {}^0_{-1}\text{e}$
- the mass number stays the **same**
- the atomic number goes **up** by 1.

On either side of a nuclear equation, the atomic numbers and mass numbers must balance.

e.g. $$^{14}_6\text{C} \longrightarrow {}^{14}_7\text{N} + {}^0_{-1}\text{e}$$

Half-life

Half-life is the time taken for half the radioactive nuclei in a sample to decay. A radioactive isotope has a fixed half-life.

The half-life of ^{90}Sr is 28 years.

Use of radioisotopes.

Radioisotopes can be used as 'tracers' in the body and dating of archaeological and geological material.

The half-life of 'tracers' must be of an appropriate length to allow detection but not cause undue damage.

- In one half-life (28 years), 4 g of ^{90}Sr decays to 2 g   ($\frac{1}{2}$ original)
- In two half-lives (56 years), 1 g of ^{90}Sr remains   ($\frac{1}{4}$ original)
- In three half-lives (84 years), 0.5 g of ^{90}Sr remains   ($\frac{1}{8}$ original)

Fusion

During nuclear **fusion**, nuclei of light atoms join to give heavier nuclei.

Nuclear fusion of hydrogen isotopes takes place in the nuclei of stars such as the Sun:

e.g. $$^1_1\text{H} + {}^2_1\text{H} \longrightarrow {}^3_2\text{He} + \gamma$$
$$^3_1\text{H} + {}^2_1\text{H} \longrightarrow {}^4_2\text{He} + {}^1_0\text{n}$$

1.2 The electron structure of the atom

Energy levels or 'shells'

AQA	M1	OCR	
EDEXCEL	M1	SALTERS	
CCEA	M1	WJEC	M1

energy

$n = 4$ ___32e⁻___

$n = 3$ ___18e⁻___

$n = 2$ ___8e⁻___

$n = 1$ ___2e⁻___

Electron energy levels are like a ladder and are filled from the bottom up.

Notice that the gap between successive energy levels becomes less with increasing energy.

You can imagine a model of an atom with electrons orbiting in shells around the nucleus. The electrons in each successive shell have an orbit further away from the nucleus. The further a shell is from the nucleus, the greater the shell's energy level. Each energy level is given a number called the principal quantum number, n. The shell closest to the nucleus has an energy level with $n = 1$, then $n = 2$ and so on.

The number of electrons that can occupy the first four energy levels is shown below:

n	shell	electrons
1	1st shell	2
2	2nd shell	8
3	3rd shell	18
4	4th shell	32

The principal quantum shell, n, is the shell number.

If you look closely, you can see a pattern. The number of electrons that can occupy a shell is $2n^2$.

Using this model, electrons can be placed into available shells, starting with the lowest energy level. Each shell must be full before the next starts to fill. The table below shows how the shells are filled for the first 11 elements in the Periodic Table.

element	atomic number	electrons $n = 1$	$n = 2$	$n = 3$
H	1	1		
He	2	2		
Li	3	2	1	
Be	4	2	2	
B	5	2	3	
C	6	2	4	
N	7	2	5	
O	8	2	6	
F	9	2	7	
Ne	10	2	8	
Na	11	2	8	1

This model breaks down as the $n = 3$ energy level is filled because each shell consists of sub-shells.

Sub-shells and orbitals

AQA	M1	OCR	
EDEXCEL		SALTERS	
CCEA	M1	WJEC	M1

A more advanced model of electron structure is used in which each shell is made up of **sub-shells**.

Sub-shells

There are different types of sub-shell: s, p, d and f. Each type of sub-shell can hold a different number of electrons:

sub-shell	electrons
s	2
p	6
d	10
f	14

The table below shows the shells and sub-shells for the first four principal quantum numbers.

Notice the labelling. The s sub-shell in the 2nd shell is labelled 2s.

n	shell	sub-shell				total number of electrons	
1	1st shell	1s				2	= 2
2	2nd shell	2s	2p			2 + 6	= 8
3	3rd shell	3s	3p	3d		2 + 6 + 10	= 18
4	4th shell	4s	4p	4d	4f	2 + 6 + 10 + 14	= 32

- Each successive shell contains a new type of sub-shell.
- The 1st shell contains 1 sub-shell, the second shell contains 2 sub-shells, and so on.

Orbitals

How do the electrons fit into the sub-shells? Mathematicians have worked out that electrons occupy negative charge clouds called **orbitals** and these make up each sub-shell.

> - An orbital can hold up to two electrons.
> - Each type of sub-shell has different orbitals: s, p, d and f.

KEY POINT

The table below shows how electrons fill the orbitals in each sub-shell.

sub-shell	orbitals	electrons
s	1	1 x 2 = 2
p	3	3 x 2 = 6
d	5	5 x 2 = 10
f	7	7 x 2 = 14

s-orbitals

One s-orbital

- An s-orbital has a spherical shape.

p-orbitals

Three p-orbitals

- A p-orbital has a 3-dimensional dumb-bell shape.
- There are three p-orbitals, p_x, p_y and p_z, at right angles to one another.

Three p-orbitals

Orbitals are regions around a nucleus that have electron density.

giving

d-orbitals and f-orbitals

The structures of d and f-orbitals are more complex.

Five d-orbitals

Seven f-orbitals

- There are five d-orbitals.
- There are seven f-orbitals.

How do two electrons fit into an orbital?

'Electrons in a box'

allowed

not allowed

Electrons are negatively charged and so they repel one another. An electron also has a property called **spin**. The two electrons in an orbital have opposite spins, helping to counteract the natural repulsion between their negative charges.

> **KEY POINT**
> - Chemists often represent an orbital as a box that can hold up to 2 electrons.
> - Each electron is shown as an arrow, indicating its spin: either ↑ or ↓.
> - Within an orbital, the electrons must have **opposite spins**.

Filling the sub-shells

AQA   M1   OCR

EDEXCEL   M1   SALTERS

CCEA   M1   WJEC   M1

Within a shell, the sub-shell energies are in the order: s, p, d and f.

Note that the 4s sub-shell is at a lower energy than the 3d sub-shell.

The 4s sub-shell fills before the 3d sub-shell.

Sub-shells have different energy levels. The diagram below shows the relative energies for the sub-shells in the first four shells.

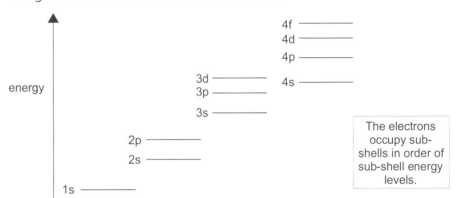

The electrons occupy sub-shells in order of sub-shell energy levels.

> **KEY POINT**
> - Shells and sub-shells are occupied in energy-level order.
> - Electrons occupy orbitals singly before pairing begins to prevent any repulsion caused by pairing.

The 2s orbital is occupied before the 2p orbitals because it is at a lower energy.

Note that the 2p orbitals are occupied singly before pairing begins.

The diagram below shows how electrons occupy orbitals from boron to oxygen.

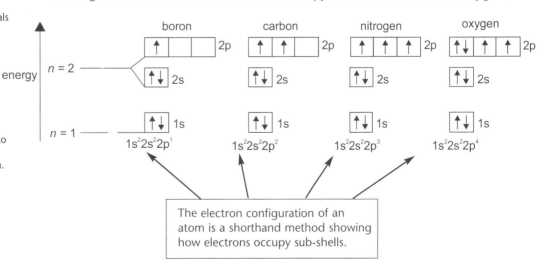

Use these 4 examples to see how electron configuration is written.

The electron configuration of an atom is a shorthand method showing how electrons occupy sub-shells.

The diagram below shows how the sub-shells are filled in an atom of potassium. Notice why the 4s sub-shell starts to fill before the 3d sub-shell.

The 19 electrons of K

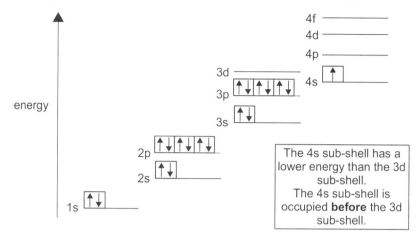

The 4s sub-shell has a lower energy than the 3d sub-shell.
The 4s sub-shell is occupied **before** the 3d sub-shell.

Sub-shells and the Periodic Table

AQA — M1   OCR
EDEXCEL M1   SALTERS
CCEA M1   WJEC — M1

You should be able to work out the electron configuration for any element in the first four periods, i.e. up to krypton (Z = 36).

The Periodic Table is structured in blocks of 2, 6, 10 and 14, linked to sub-shells. The order of sub-shell filling can be seen by dividing the Periodic Table into blocks.

	1s			1s
s-block				*p-block*
2s				2p
3s		*d-block*		3p
4s		3d		4p
5s		4d		5p
6s		5d		6p
7s				
		f-block		
		4f		
		5f		

Simplifying electron configurations

The similar electron configurations within a group of the Periodic Table can be emphasised with a simpler representation in terms of the previous noble gas.

The electron configurations for the elements in Group 1 are shown below:

Li: $1s^2 2s^1$ *or* $[He]2s^1$
Na: $1s^2 2s^2 2p^6 3s^1$ *or* $[Ne]3s^1$
K: $1s^2 2s^2 2p^6 3s^2 3p^6 4s^1$ *or* $[Ar]4s^1$

Progress check

1 Write the full electron configuration in terms of sub-shells for:
(a) C; (b) Al; (c) Ca; (d) Fe; (e) Br.

1 (a) C: $1s^2 2s^2 2p^2$
(b) Al: $1s^2 2s^2 2p^6 3s^2 3p^1$
(c) Ca: $1s^2 2s^2 2p^6 3s^2 3p^6 4s^2$
(d) Fe: $1s^2 2s^2 2p^6 3s^2 3p^6 3d^6 4s^2$
(e) Br: $1s^2 2s^2 2p^6 3s^2 3p^6 3d^{10} 4s^2 4p^5$

1.3 Experimental evidence for electron structures

After studying this section you should be able to:

- recall the definitions for first and successive ionisation energies
- understand the factors affecting the sizes of ionisation energies
- show how ionisation energies provide evidence for shells and sub-shells
- understand evidence for electron structure from electron emission and absorption spectra

LEARNING SUMMARY

Ions

AQA	M1	OCR	
EDEXCEL	M1	SALTERS	
CCEA	M1	WJEC	M1

Positive ions form when electrons are lost.

Negative ions form when electrons are gained.

An atom can either lose or gain electrons to form an **ion**: a charged atom.

A positive ion is formed when an atom loses electrons. For example, a lithium atom forms a **positive ion** with a 1+ charge by **losing** an electron:

$$Li \longrightarrow Li^+ + e^-$$
$$(3p^+, 4n, \mathbf{3e^-}) \quad (3p^+, 4n, \mathbf{2e^-})$$

A negative ion is formed when an atom gains electrons. For example, an oxygen atom forms a **negative ion** with a 2– charge by **gaining** two electrons:

$$O + 2e^- \longrightarrow O^{2-}$$
$$(6p^+, 6n, \mathbf{6e^-}) \qquad\qquad (6p^+, 6n, \mathbf{8e^-})$$

Ionisation energy

AQA	M1	OCR	
EDEXCEL	M1	SALTERS	
CCEA	M1	WJEC	M1

Ionisation energy measures the ease with which electrons are lost in the formation of positive ions. An element has as many ionisation energies as there are electrons.

> **KEY POINT**
>
> The **first** *ionisation energy* of an element is the energy required to remove **1 electron** from each atom in **1 mole** of **gaseous atoms** to form 1 mole of gaseous 1+ ions.

The equation representing the first ionisation energy of sodium is shown below.

$$Na(g) \longrightarrow Na^+(g) + e^- \qquad \text{1st ionisation energy} = +496 \text{ kJ mol}^{-1}$$

Factors affecting ionisation energy

Electrons are held in their shells by attraction from the nucleus. The first electron lost will be from the highest occupied energy level. This electron experiences least attraction from the nucleus.

Three factors affecting the size of this attraction are shown below.

These factors are very important and help to explain many chemical ideas throughout the course.

Learn them!

> **KEY POINT**
>
> **Atomic radius**
> The greater the distance between the nucleus and the outer electrons, the less the attractive force. Attraction falls rapidly with increasing distance and so this factor is very important and has a big effect.
>
> **Nuclear charge**
> The greater the number of protons in the nucleus, the greater the attractive force.
>
> **Electron shielding or 'screening'**
> The outer shell electrons are repelled by any inner shells between the electrons and the nucleus.
>
> This repelling effect, called electron shielding or screening, reduces the overall attractive force experienced by the outer electrons.

Evidence for shells

EDEXCEL M1   SALTERS M2
CCEA M1   WJEC M1
OCR M1

This pattern shows that a sodium atom has 3 shells:

the first shell ($n = 1$) contains 2 electrons

the second shell ($n = 2$) contains 8 electrons

the third shell ($n = 3$) contains 1 electron.

> **KEY POINT**
>
> The **second ionisation energy** of an element is the energy required to remove **1 electron** from each atom in **1 mole** of **gaseous 1+ ions** to form 1 mole of gaseous 2+ ions.

The equation representing the second ionisation energy of sodium is shown below.

$$Na^+(g) \longrightarrow Na^{2+}(g) + e^-$$ 2nd ionisation energy = +4563 kJ mol^{-1}

Sodium, with 11 electrons, has 11 successive ionisation energies, shown as the graph below. Successive ionisation energies provide evidence for the different energy levels of the principal quantum numbers.

ionisation energy /kJ mol^{-1}

These **2** electrons are in the 1st shell ($n = 1$), closest to the nucleus. These electrons experience very **strong** attractive forces from the nucleus.

These **8** electrons are in the 2nd shell ($n = 2$). These electrons experience reasonably strong attractive forces from the nucleus.

This **1** electron is in the 3rd shell ($n = 3$), furthest from the nucleus. This electron experiences comparatively **weak** attractive forces from the nucleus.

ionisation number

Trends in first ionisation energies

AQA M1   OCR M1
EDEXCEL M1   SALTERS M2
CCEA M1   WJEC M2

The graph below shows the variation of first ionisation energy with increasing atomic number from hydrogen to calcium:

first ionisation energy

Each 'peak' marks a noble gas (He, Ne, Ar) at the end of each period.

Each 'trough' marks an alkali metal (Li, Na, K) at the start of each period

atomic number

The graph shows:

Across a period, increased nuclear charge is most important.

- a general **increase** in first ionisation energy across a period
  (see H→He; Li→Ne; Na→Ar).
  This results from the **increase** in nuclear charge as electrons are added to the **same shell** across each period.
- a sharp **decrease** in first ionisation energy between the end of one period and the start of the next period (see He→Li; Ne→Na; Ar→K).
  This reflects the addition of a new outer shell with the resulting increase in **distance** and **shielding**.

Down a group, increased distance and shielding are most important.

- a **decrease** in first ionisation energy down a group
  (see He→Ne→Ar and other groups).
  This reflects the presence of **extra shells** down the group.

The reasons for the general trend in ionisation energies are similar to those explaining the trend in atomic radii (page 73).

Evidence for sub-shells

AQA   M1
EDEXCEL   M1
CCEA   M1

The variation in first ionisation energies across a period in the Periodic Table provides evidence for the existence of sub-shells. The graph below shows this variation across Period 2 (Li → Ne)

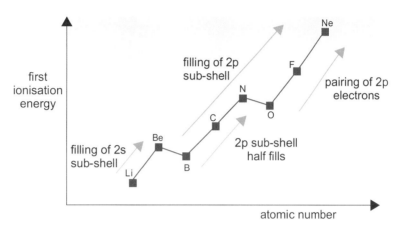

Across the period, the graph shows:

- a rise from lithium to beryllium,
- a fall to boron followed by a rise to nitrogen,
- a fall to oxygen followed by a rise to neon.

Comparing beryllium and boron

The fall in 1st ionisation energy from beryllium to boron marks the start of filling the 2p sub-shell which is at higher energy than the 2s sub-shell.

Beryllium versus boron: higher energy level.

- Beryllium's outermost electron is in the 2s sub-shell.
- Boron's outermost electron is in the 2p sub-shell.
- Boron's outermost electron is easier to remove because the 2p sub-shell has a **higher energy level** than the 2s sub-shell.

Comparing nitrogen and oxygen

The fall in 1st ionisation energy from nitrogen to oxygen marks the start of electron pairing in the p-orbitals of the 2p sub-shell.

Nitrogen versus oxygen: electron pairing.

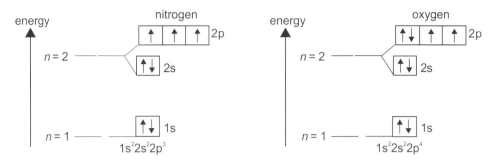

- Both nitrogen and oxygen have their outer electrons in the same 2p sub-shell with the same energy level.
- Oxygen has one 2p orbital with an electron pair.
- Nitrogen, with one electron in each 2p orbital, has unpaired 2p electrons only.
- Because of **electron repulsion**, it is easier to remove one of oxygen's paired 2p electrons.

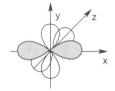

Nitrogen
2p sub-shell half-full
1 electron in each 2p orbital
parallel spins at right angles

Oxygen
2p electrons start to pair
Paired electrons repel

Atomic spectra of hydrogen

SALTERS
WJEC   M1
CCEA   M1

Emission and absorption

The emission and absorption spectra of atomic hydrogen are the result of electron transitions between atomic energy levels.

Each element has its own unique set of lines in its atomic spectra. This allows elements to be identified. We know that the Sun contains hydrogen and helium gases because their atomic spectra have been observed.

- Energy is absorbed, exciting an electron from its ground state to a higher energy level.
- The electron returns to lower energy levels, emitting energy.

An **absorption** spectrum shows as black lines on a coloured or bright background.

- The **black lines** show that **radiation is absorbed** as electrons are excited from a **lower to higher** energy level.

An **emission** spectrum shows as bright or coloured lines on a black background.

- The **bright lines** show that **radiation is emitted** as electrons move from **higher to lower** energy levels.

Energy and frequency

The energy emitted or absorbed is proportional to the frequency of a line produced in the spectrum.

frequency, *f*, is related to the speed of light, *c*, and wavelength, λ:

$f = c/\lambda$

    energy change, $\Delta E = hf$

    h = the Planck constant; *f* = frequency

The hydrogen spectrum

The emission spectrum of atomic hydrogen has several series of lines.

Emission and absorption spectra both have lines in the same position for a given element:

sets of lines represent transitions to or from a particular level

the lines become closer at higher frequencies.

- The lines in a series get closer together with increasing frequency.
- Eventually the lines converge as the energy levels also converge.

This provides evidence for discrete (separate and distinct) energy levels.

- Transitions from a higher energy level to the $n = 1$ energy level (the ground state of hydrogen) are in the ultraviolet region of the spectrum. This is the Lyman series.
- Transitions from a higher energy level to the $n = 2$ energy level are in the visible region of the spectrum. This is the Balmer series.

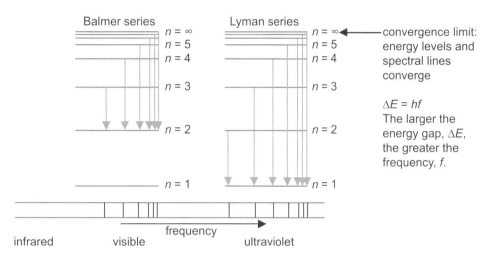

convergence limit: energy levels and spectral lines converge

$\Delta E = hf$
The larger the energy gap, ΔE, the greater the frequency, f.

$c = 3.00 \times 10^8$ m s^{-1}

$h = 6.62 \times 10^{-34}$ J s

The calculation gives the energy per H atom. We need to multiply by the Avogadro constant (6.02×10^{23} mol^{-1}) to get the energy per mole of H atoms.

Calculating the ionisation energy

The ionisation energy of hydrogen can be calculated from the frequency of the convergence limit in the Lyman series.

In the Lyman series, the lines converge at $\lambda = 91.1$ nm.

$f = c/\lambda = 3.00 \times 10^8/91.1 \times 10^{-9} = 3.29 \times 10^{15}$ s^{-1}

$\Delta E = hf = 6.62 \times 10^{-34} \times 3.29 \times 10^{15} = 2.18 \times 10^{-18}$ J $= 2.18 \times 10^{-21}$ kJ

The first ionisation energy $= 6.02 \times 10^{23} \times 2.18 \times 10^{-21} = 1310$ kJ mol^{-1}.

Progress check

1 State and explain the general trend in 1st ionisation energy across a period.

2 Explain why the 1st ionisation energy falls slightly:
   (a) between beryllium and boron
   (b) between nitrogen and oxygen.

2 (a) In B, the highest energy electron is in the 2p sub-shell, higher in energy than the 2s sub-shell in Be.
   (b) In O, 2 electrons pair up in a 2p sub-shell. In N, the 2p sub-shell is half filled with a single electron in each orbital. The paired electrons in O repel one another slightly making it easier to remove an electron.

1 The 1st ionisation energy decreases across a period. Across a period a period the number of protons increases and electrons are being added to the same shell. Therefore attraction from the nucleus increases.

1.4 Atomic mass

After studying this section you should be able to:

- *define relative masses, based on the ^{12}C scale*
- *describe the basic principles of the mass spectrometer*
- *calculate the relative atomic mass of an element from its isotopic abundance or mass spectrum*
- *calculate the relative molecular mass of a compound from relative atomic masses*

Relative atomic mass

AQA	M1	OCR	M1
EDEXCEL	M1	SALTERS	M1
CCEA	M1	WJEC	M1

Carbon-12: an international standard

The mass of an atom is too small to be measured on even the most sensitive balance. Chemists use **relative** masses to compare the atomic masses of different elements. First, we must have an atom with which to compare other atoms and an atom of the **carbon-12 isotope** is chosen as the international standard for the measurement of atomic mass.

Unified atomic mass unit (u)

1 unified atomic mass unit (u) is the mass of one-twelfth the mass of an atom of the carbon-12 isotope. This provides the base measurement for atomic masses:

$1\ u = 1.661 \times 10^{-23}\ g$

- The mass of an atom of carbon-12 is exactly 12 unified atomic mass units (u).
- The mass of one-twelfth of an atom of carbon-12 is exactly 1 u.

All atoms in a pure isotope have the same atomic structure and the same mass. An isotope's *relative* mass is found by comparison with carbon-12.

> **Relative isotopic mass** is the mass of an atom of an isotope compared with one-twelfth the mass of an atom of carbon-12.

In most chemistry work, it is reasonable to:
- neglect the tiny contribution to atomic mass from electrons
- take both the mass of a proton and a neutron as 1 u.

Relative isotopic mass is then simply the mass number of the isotope, e.g. the relative isotopic mass of ^{16}O is 16.

Relative atomic mass

Note that a relative mass has no units. It is simply a ratio of masses and any mass units will cancel.

Most elements consist of a mixture of isotopes, each with a different mass number. To work out the relative atomic mass of an element we must find the **weighted average mass** of the isotopes present from:

- the natural abundances of the isotopes
- the relative isotopic masses of the isotopes.

Examples of relative atomic masses:

H: 1.008
Cl: 35.45
Pb: 207.19

> **Relative atomic mass**, A_r, is the weighted average mass of an atom of an element compared with one-twelfth of the mass of an atom of carbon-12.

Calculating a relative atomic mass

In most chemistry work, the relative isotopic mass can be taken as a whole number.

Naturally occurring chlorine consists of 75% ^{35}Cl and 25% ^{37}Cl.

$\frac{75}{100}$ of chlorine is the ^{35}Cl isotope;

$\frac{25}{100}$ of chlorine is the ^{37}Cl isotope.

The relative atomic mass of chlorine $= \frac{75}{100} \times 35 + \frac{25}{100} \times 37 = 35.5$

Measuring relative atomic masses

A relative atomic mass can be determined using a **mass spectrometer**.

- A sample of the element is placed into the mass spectrometer and vaporised.
- The sample is bombarded with electrons forming positive ions.
- The positive ions are accelerated using an electric field.
- The positive ions are deflected using a magnetic field.
- Ions of lighter isotopes are deflected more than ions of heavier isotopes. This separates different isotopes.
- The ions are detected to produce a **mass spectrum**.

You can follow the stages in the diagram below:

Relative atomic mass from a mass spectrum

A mass spectrum provides:

- the relative isotopic masses of the isotopes in an element
- isotopic abundances.

The diagram below shows how a relative atomic mass can be calculated from a mass spectrum.

Mass spectrum of a copper sample

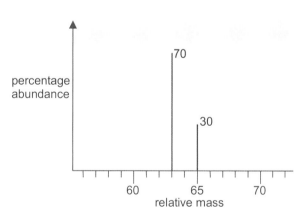

ion	relative mass	percentage abundance
$^{63}Cu^+$	63	70%
$^{65}Cu^+$	65	30%

The relative atomic mass of copper

$$= \frac{70}{100} \times 63 + \frac{30}{100} \times 65$$

$$= 63.6$$

Remember VIADD:

Vaporisation
Ionisation
Acceleration
Deflection
Detection

Notice that ions are detected in the mass spectrometer. The ions are formed following electron bombardment.

(**Salters only**) Time-of-flight mass spectrometry uses a different method to identify ions.

The ions are accelerated by an electric field of known strength.

The time taken for the ions to reach a detector is measured.

Lighter ions travel faster than heavier ions.

An ion is identified from the time it takes to reach the detector.

Mass spectrometers have been sent into space to identify elements, e.g. Mars space probe. On Earth, mass spectrometers are used in environmental monitoring and for forensic analysis.

More relative masses

AQA	M1	OCR	M1
EDEXCEL	M1	SALTERS	M1
CCEA	M1	WJEC	M1

You will find out more about chemical bonding and structure in Chapter 2, p.51.

Relative masses can be used to compare the masses of any chemical species. Much of matter is made up of compounds.

Relative molecular mass

A simple molecule of Cl_2, H_2O or CO_2 comprises small groups of atoms, held together by chemical bonds.

> **KEY POINT**
>
> Relative molecular mass, M_r, is the weighted average mass of a molecule of a compound compared with one-twelfth the mass of an atom of carbon-12.

Mass spectrometry can be used with compounds to determine the relative molecular mass.

The *relative molecular mass* of a compound is found by adding together the relative atomic masses of each atom in a molecule.

Examples of relative molecular masses
Cl_2: $M_r = 35.5 \times 2 = 71.0$;
H_2O: $M_r = 1.0 \times 2 + 16.0 = 18.0$

For giant structures, the **formula unit** of the compound is the simplest ratio of atoms or ions in the structure.

Some compounds, such as sand SiO_2, exist as 'giant' molecules comprising hundreds of thousands of atoms bonded together. Each molecule in these structures is the size of each crystal and, in a sense, the whole crystal is a molecule.

- Each 'molecule' of SiO_2 has twice as many oxygen atoms as silicon atoms.

Many compounds, such as common salt NaCl, are made up of ions and not molecules. These ionic compounds do not have single molecules and form giant structures of oppositely charged ions.

- NaCl has the same number of sodium ions as chloride ions.

Relative formula mass can be used to describe the relative mass of any compound.

Although relative **molecular** mass is often used with these compounds, a better term to use is relative **formula** mass.

> **KEY POINT**
>
> Relative formula mass is the average mass of the formula unit of a compound compared with one-twelfth the mass of an atom of carbon-12.

The *relative formula mass* of a compound is found by adding together the relative atomic masses of each atom in the formula.

Examples of relative formula masses
$CaBr_2$: relative formula mass $= 40.1 + 79.9 \times 2 = 199.9$;
Na_3PO_4: relative formula mass $= 23.0 \times 3 + 31.0 + 16.0 \times 4 = 164.0$

Progress check

1 Calculate the relative atomic mass, A_r, of the following elements from their isotopic abundances:
   (a) Gallium containing 60% [69]Ga and 40% [71]Ga.
   (b) Silicon containing 92.18% [28]Si, 4.71% [29]Si and 3.11% [30]Si.

2 Use A_r values from the Periodic Table to calculate the relative formula mass of:
   (a) CO_2
   (b) NH_3
   (c) Fe_2O_3
   (d) $C_6H_{12}O_6$
   (e) $Pb(NO_3)_2$.

1 (a) Gallium: 69.8
  (b) Silicon: 28.11.
2 (a) 44.0
  (b) 17.0
  (c) 159.6
  (d) 180.0
  (e) 331.2.

1.5 The mole

After studying this section you should be able to:

- understand the concept of the mole as the amount of substance
- calculate molar quantities using masses, solutions and gas volumes

The mole

AQA	M1	OCR	
EDEXCEL		SALTERS	
CCEA	M1	WJEC	M1

Counting and weighing atoms

To make a compound such as H_2O, it would be useful to be able to count out atoms. For H_2O, we would need twice as many hydrogen atoms as oxygen atoms. Atoms are far too small to be counted individually but they can be counted using the idea of relative mass:

element	H	C	O

| relative mass | 1 | 12 | 16 |

For 1 atom of H, C and O, the relative masses are in the ratio 1 : 12 : 16.
For 10 atoms of each, the ratio is 10 : 120 : 160 which is still 1 : 12 : 16.

> Provided we have the same number of atoms of each, the masses are in the ratio of the relative masses.

By measuring masses, chemists are able to actually count atoms.
For 1 g of H, 12 g of C and 16 g of O, the masses are still in the ratio 1 : 12 : 16.
1 g of H contains the same number of atoms as 12 g of C and 16 g of O.

Amount of substance

Chemists use a quantity called *amount of substance* for counting atoms.
Amount of substance is:
- given the symbol n
- measured using a unit called the **mole** (abbreviated as **mol**).

As with relative masses, amount of substance uses the carbon-12 isotope as the standard:

The amount n of carbon-12 atoms in 12 g of ^{12}C is 1 *mol*.

The actual **number** of atoms in 1 mol of carbon-12 is 6.02×10^{23}.

This can be expressed by saying:

'There are 6.02×10^{23} atoms per mole of carbon-12 atoms (or 6.02×10^{23} mol^{-1})'.

6.02×10^{23} mol^{-1} is a constant known as the **Avogadro constant**, L, or N_A.

*Note that carbon-12 is the **substance** here. So **amount of** substance translates to **amount of carbon-12**.*

> This is a very powerful and useful idea. To make a water molecule from hydrogen and oxygen, we cannot **count** out the atoms but we can **weigh** them out in the correct ratio. To make H_2O, we need 2 atoms of hydrogen for every 1 atom of oxygen, which we can get by weighing out 2 g of hydrogen and 16 g of oxygen.
>
> **2 g of H contains twice as many atoms as 16 g of O.**
> **2 g of H atoms contains 2 mol of H atoms.**
> **16 g of O atoms contains 1 mol of O atoms.**

KEY POINT

Particle

A single particle of a substance may refer to an atom, a molecule, an ion, an electron or to any identifiable particle. Chemists refer to a collection of particles as a **chemical species**.

Amount of substance is not restricted just to atoms. We can have an amount of any chemical species: 'amount of atoms', 'amount of molecules', 'amount of ions', etc.

> A **mole** is the amount of substance that contains as many single particles as there are atoms in exactly 12 g of the carbon-12 isotope.

Using the definition of the mole above, this means that:

- 1 mole of hydrogen atoms, H, contains 6.02×10^{23} hydrogen atoms, H.

- 1 mole of oxygen molecules, O_2, contains 6.02×10^{23} oxygen molecules, O_2.

- 1 mole of electrons, e^-, contains 6.02×10^{23} electrons, e^-.

It is important to always refer to what the particle is.

For example:

'1 mole of oxygen' could refer to oxygen atoms, O, or to oxygen molecules, O_2.

It is always safest to quote the formula to which the amount of substance refers.

Moles from masses

AQA	M1	OCR	M
EDEXCEL	M1	SALTERS	N
CCEA	M1	WJEC	M1

Molar mass, *M*

> Molar mass, *M*, is the mass per mole of a substance.
> The units of molar mass are g mol⁻¹.

For the atoms of an element, the molar mass is simply the relative atomic mass in g mol⁻¹.

- Molar mass of Mg = 24.3 g mol⁻¹
- This simply means that each mole of Mg atoms has a mass of 24.3 g.

For a compound, the molar mass is found by adding together the relative atomic masses of each atom in the formula:

- Molar mass of CO_2 = 12 + (16 × 2) = 44 g mol⁻¹
- 1 mole of CO_2 has a mass of 44 g.

Molar mass is an extremely useful term that can be applied to any chemical species.

Examples of molar masses

H	1 g mol⁻¹
H_2O	18 g mol⁻¹
CO_2	44 g mol⁻¹
HNO_3	63 g mol⁻¹

> Amount of substance *n*, mass and molar mass are linked by the expression below:
> $$n = \frac{mass\ (in\ g)}{molar\ mass}$$

In 24 g of C, *amount of C* $= \dfrac{mass\ (in\ g)}{molar\ mass} = \dfrac{24}{12} = 2$ mol

In 11 g of CO_2, *amount of CO_2* $= \dfrac{mass\ (in\ g)}{molar\ mass} = \dfrac{11}{44} = 0.25$ mol

Progress check

1 (a) What is the amount of each substance (in mol) in the following:
  (i) 124 g P; (ii) 64 g O; (iii) 64 g O_2; (iv) 10.01 g $CaCO_3$;
  (v) 3.04 g Cr_2O_3?
(b) What is the mass of each substance (in g) in the following:
  (i) 0.1 mol Na; (ii) 2.5 mol NO_2; (iii) 0.3 mol Na_2SO_4;
  (iv) 0.05 mol H_2SO_4; (v) 0.025 mol Ag_2CO_3?

1 (a) (i) 4 mol; (ii) 4 mol; (iii) 2 mol; (iv) 0.1 mol; (v) 0.02 mol.
(b) (i) 2.3 g; (ii) 115 g; (iii) 42.63 g; (iv) 4.905 g; (v) 6.895 g.

Moles from solutions

AQA	M1	OCR	M1
EDEXCEL	M3	SALTERS	M2
CCEA	M1	WJEC	M1

> **KEY POINT**
>
> The concentration of a solution is the amount of solute, in mol, dissolved in each dm³ (1000 cm³) of solution.

'Dilute' solutions contain few moles of solute in a volume.

If the volume of the solution is in dm³:

$$n = c \times V \text{ (in dm}^3)$$

n = amount of substance, in mol.

c = concentration of solution, in mol dm⁻³.

V = volume of solution, in dm³.

Dividing a volume measured in cm³ by 1000 converts the volume automatically to dm³.

It is more convenient to measure smaller volumes of solutions in cm³ and the expression becomes:

$$n = c \times \frac{V \text{ (in cm}^3)}{1000}$$

Molarity

Molarity refers to the concentration in mol dm⁻³.

Thus 2 mol dm³ and 2 Molar (or 2M) mean the same: 2 moles of solute in 1 dm³ of solution.

Example

What is the amount of NaCl (in mol) in 25.0 cm³ of an aqueous solution of concentration 2.00 mol dm⁻³?

$$n(\text{NaCl}) = c \times \frac{V}{1000} = 2.00 \times \frac{25.0}{1000} = 0.0500 \text{ mol}$$

Concentrations are sometimes referred to in g dm⁻³.

The Na_2CO_3 here has a **molar** concentration of 0.100 mol dm⁻³ and a **mass** concentration of 10.6 g dm⁻³ of Na_2CO_3.

Standard solutions

Chemists often need to prepare **standard solutions** with an exact concentration. Using an understanding of the mole, the mass required to prepare such a solution can easily be worked out.

Example

Find the mass of sodium carbonate required to prepare 250 cm³ of a 0.100 mol dm⁻³ solution.

Find the amount of Na_2CO_3 (in mol) required in solution:

amount, n, of $Na_2CO_3 = c \times \frac{V}{1000} = 0.100 \times \frac{250}{1000} = 0.0250$ mol

Very dilute solutions are measured in parts per million (ppm), e.g. ppm is 1 milligram (0.001 g) in 1 dm³ of solution.

Convert moles to grammes

$$n = \frac{\text{mass}}{\text{molar mass}} \quad \therefore \quad \text{mass} = n \times \text{molar mass}$$

molar mass of Na_2CO_3 = 23.0 × 2 + 12.0 + 16.0 × 3 = 106.0 g mol⁻¹

mass of Na_2CO_3 required = 0.0250 × 106.0 = 2.65 g

Progress check

1 (a) What is the amount of each substance, in mol, in:
    (i) 250 cm³ of a 1.00 mol dm⁻³ solution;
    (ii) 10 cm³ of a 2.0 mol dm⁻³ solution?

  (b) Find the concentration, in mol dm⁻³, for:
    (i) 4 mol in 2 dm³ of solution;
    (ii) 0.0100 mol in 100 cm³ of solution.

  (c) Find the concentration, in g dm⁻³, for:
    (i) 2 mol of NaOH in 4 dm³ of solution;
    (ii) 0.500 mol of HNO_3 in 200 cm³ of solution.

1 (a) (i) 0.25 mol; (ii) 0.02 mol.
(b) (i) 2 mol dm⁻³; (ii) 0.1 mol dm⁻³.
(c) (i) 20 g dm⁻³; (ii) 157.5 g dm⁻³.

Moles from gas volumes

AQA	M1	OCR	M1
EDEXCEL	M1	SALTERS	M2
CCEA	M2	WJEC	M1

This is sometimes summarised by Avogadro's hypothesis:

'Equal volumes of gases contain the same number of molecules under the same conditions of temperature and pressure.'

For a gas, the amount of gas molecules (in mol) is most conveniently obtained by measuring the gas volume.

Provided that the pressure and temperature are the same, equal volumes of gases contain the same number of molecules.

This means that it does not matter which gas is being measured. By measuring the volume, we are indirectly also counting the number of molecules.

> **KEY POINT**
>
> At room temperature and pressure (r.t.p), 298 K (25°C) and 100 kPa, 1 mol of a gas occupies approximately 24 dm^3 = 24 000 cm^3.

At r.t.p., if the volume of the gas is in **dm^3**: $n = \dfrac{V \text{ (in dm}^3)}{24}$

if the volume of the gas is in **cm^3**: $n = \dfrac{V \text{ (in cm}^3)}{24000}$

Example

How many moles of gas molecules are in 480 cm^3 of a gas at r.t.p.?

$$n = \frac{V \text{ (in cm}^3)}{24000} = \frac{480}{24000} = 0.0200 \text{ mol of gas molecules.}$$

The Ideal Gas Equation (AQA only)

The conditions will not always be room temperature and pressure. A gas volume depends on temperature and pressure.

The Ideal Gas Equation can be used to find the amount of gas molecules n (in mol) in a volume V (in m^3) at any temperature T (in K) and pressure p (in Pa).

The Ideal Gas Equation: $pV = nRT$ ($R = 8.31$ J K^{-1} mol^{-1})

Before using $pV = nRT$, you must remember to convert any values to Pa, K and m^3.

Conversion rules

cm^3 to m^3	× 10^{-6}
dm^3 to m^3	× 10^{-3}
°C to K	+ 273
kPa to Pa	× 10^3

Note that M (molar mass) has units: g mol^{-1}

M_r (relative molecular mass) has no units.

Example

A 0.215 g sample of a volatile liquid, **X**, produces 77.5 cm^3 of gas at 100°C and 100 kPa. Calculate the relative molecular mass of **X**.

Use The Ideal Gas Equation:

$$pV = nRT \quad \therefore n = \frac{pV}{RT}$$

$$\therefore n = \frac{1.00 \times 10^5 \times 7.75 \times 10^{-5}}{8.31 \times 373} = 0.00250 \text{ mol}$$

Find the molar mass:

$$n = \frac{\text{mass}}{\text{molar mass}}$$

$$\therefore \text{ molar mass} = \frac{\text{mass}}{n} = \frac{0.215}{0.00250} = 86.0 \text{ g mol}^{-1}$$

$$\therefore \text{ Relative molecular mass } M_r = 86.0$$

Progress check

1 (a) What is the volume, at room temperature and pressure (r.t.p.) of
(i) 44 g CO$_2$(g); (ii) 7 g N$_2$(g); (iii) 5.1 g NH$_3$(g)?
(b) What is the mass, at r.t.p. of
(i) 1.2 dm^3 O$_2$; (ii) 720 cm^3 CO$_2$(g); (iii) 48 cm^3 CH$_4$(g)?

1 (a) (i) 24 dm³; (ii) 6 dm³; (iii) 7.2 dm³.
(b) (i) 1.6 g; (ii) 1.32 g; (iii) 0.032 g.

1.6 Formulae, equations and reacting quantities

After studying this section you should be able to:

- *understand what is meant by empirical and molecular formula*
- *calculate the empirical and molecular formula of a substance*
- *write balanced equations*
- *use chemical equations to calculate reacting masses and reacting volumes (and vice versa)*
- *perform calculations involving volumes and concentrations of solutions in simple acid-base titrations and back titrations*

LEARNING SUMMARY

Types of chemical formula

AQA	M1	OCR	
EDEXCEL	M1	SALTERS	
CCEA	M2	WJEC	M1

Empirical and molecular formula

The **empirical formula** of ethane is CH_3.

Ethane has 1 carbon atom for each of 3 hydrogen atoms.

The molecular formula of ethane is C_2H_6

Each molecule of ethane contains 2 carbon atoms and 6 hydrogen atoms.

> **Empirical formula** represents the simplest, whole-number ratio of atoms of each element in a compound.
>
> **Molecular formula** represents the actual number of atoms of each element in a molecule of a compound.

KEY POINT

Formula determination

A formula can be calculated from experimental results using the Mole Concept.

Example 1

Find the empirical formula of a compound formed when 6.75 g of aluminium reacts with 26.63 g of chlorine. [A_r: Al, 27.0; Cl, 35.5.]

Find the molar ratio of atoms:

$$\text{Al} \quad : \quad \text{Cl}$$
$$= \frac{6.75}{27.0} \quad : \quad \frac{26.63}{35.5}$$
$$0.25 \quad : \quad 0.75$$

Divide by smallest number (0.25):   1   :   3

∴ Empirical formula = $AlCl_3$

Example 2

Find the molecular formula of a compound containing carbon, hydrogen and oxygen only with the composition by mass of carbon: 40.0%; hydrogen: 6.7%. [$M_r = 180$. A_r: H, 1.00; C, 12.0; O, 16.0.]

Analysis often doesn't give the oxygen content directly but it can easily be calculated.

Percentage of oxygen in compound = 100 – (40.0 + 6.7) = 53.3%
100.0 g of the compound contains 40.0 g C, 6.7 g H and 53.3 g O.

Find the molar ratio of atoms:

$$\text{C} \quad : \quad \text{H} \quad : \quad \text{O}$$
$$= \frac{40.0}{12.0} : \frac{6.7}{1.00} : \frac{53.3}{16.0}$$
$$= 3.33 : 6.7 : 3.33$$

Divide by smallest number (3.33):   1 : 2 : 1

∴ Empirical formula = CH_2O

Relate to molecular mass

Each CH_2O unit has a relative mass of 12 + (1 × 2) + 16 = 30
M_r, of the compound is 180 containing 180/30 = 6 CH_2O units
∴ molecular formula is $C_6H_{12}O_6$

Progress check

1 Find the empirical formula of the following:
   (a) sodium oxide (2.3 g of sodium reacts to form 3.1 g of sodium oxide).
   (b) iron oxide (11.16 g of iron reacts to form 15.96 g of iron oxide).

2 Use the following percentage compositions by mass to find the empirical formula of the compound.
   (a) A compound of sulfur and oxygen. [S, 40.0%; O, 60.0%]
   (b) A compound of iron, sulfur and oxygen. [Fe, 36.8%; S, 21.1%; O, 42.1%]

3 2.8 g of a compound of carbon and hydrogen is formed when 2.4 g of carbon combines with hydrogen [M_r: 56]. What is the empirical and molecular formula of the compound formed?

1 (a) Na_2O; (b) Fe_2O_3.
2 (a) SO_3; (b) $FeSO_4$.
3 CH_2; C_4H_8.

Equations

AQA   M1   OCR   M1
EDEXCEL M1   SALTERS M1
CCEA   M1   WJEC   M1

Chemical reactions involve the **rearrangement** of atoms and ions.

Chemical equations provide two types of information about a reaction:

- qualitative – *which* atoms or ions are rearranging;
- quantitative – *how* many atoms or ions are rearranging.

Balancing equations

The formula of each substance shows the chemicals involved in the reaction.

State symbols can be added to show the physical states of each species under the conditions of the reaction.

State symbols give the physical states of each species in a reaction:

gaseous state, (g);
liquid state, (l);
solid state, (s);
aqueous solution, (aq).

The **qualitative** equation for the reaction of hydrogen with oxygen is shown below:

$$\text{hydrogen} \quad + \quad \text{oxygen} \quad \longrightarrow \quad \text{water}$$
$$H_2(g) \quad + \quad O_2(g) \quad \longrightarrow \quad H_2O(l)$$

In this equation, hydrogen is balanced, oxygen is **not** balanced.

	$H_2(g)$	+	$O_2(g)$	$\longrightarrow$	$H_2O(l)$	
hydrogen	2				2	✓
oxygen			2		1	✗

The equation must be **balanced** to give the **same number** of particles of each element on each side of the equation.

Balancing an equation often takes several stages.

- Oxygen can be balanced by placing a '2' in front of H_2O.
- However, this unbalances hydrogen:

	$H_2(g)$	+	$O_2(g)$	$\longrightarrow$	$2 H_2O(l)$	
hydrogen	2				4	✗
oxygen			2		2	✓

- H can be re-balanced by placing a '2' in front of H_2.
- The equation is now balanced.

Notice that no formula has been changed. You balance an equation by adding numbers in front of a formula only.

	$2 H_2(g)$	+	$O_2(g)$	$\longrightarrow$	$2 H_2O(l)$	
hydrogen	4				4	✓
oxygen			2		2	✓

More balancing numbers.

$3 Na_2SO_4$ means
Na: $3 \times 2 = 6$;
S: $3 \times 1 = 3$;
O: $3 \times 4 = 12$.

With brackets, the subscript applies to everything in the preceding bracket:

$2 Ca(NO_3)_2$ means
Ca: $2 \times 1 = 2$;
N: $2 \times (1 \times 2) = 4$;
O: $2 \times (3 \times 2) = 12$.

> **KEY POINT**
>
> In a formula, a subscript applies **only** to the symbol immediately preceding,
>
> e.g. NO_2 comprises **1** N and **2** O.
>
> When balancing an equation, you must **not** change any formula. You can only add a balancing number in front of a formula.
>
> The balancing number multiplies everything in the formula by the balancing number,
>
> e.g. $2AlCl_3$ comprises Al: $2 \times 1 = 2$; Cl: $2 \times 3 = 6$.

Progress check

Balance the following equations:
(a) $Li(s) + O_2(g) \longrightarrow Li_2O(s)$
(b) $CH_4(g) + O_2(g) \longrightarrow CO_2(g) + H_2O(l)$
(c) $Al(s) + HCl(aq) \longrightarrow AlCl_3(aq) + H_2(g)$

(a) $4Li(s) + O_2(g) \longrightarrow 2Li_2O(s)$
(b) $CH_4(g) + 2O_2(g) \longrightarrow CO_2(g) + 2H_2O(l)$
(c) $2Al(s) + 6HCl(aq) \longrightarrow 2AlCl_3(aq) + 3H_2(g)$.

Reacting quantities

AQA	M1	OCR
EDEXCEL		SALTERS
CCEA	M1, M2	WJEC M1

The balancing numbers give the ratio of the **amount** of each substance, in mol. Using the example above:

equation	$2 H_2(g)$	$+$	$O_2(g)$	$\longrightarrow$	$2 H_2O(l)$
moles	2 mol		1 mol	$\longrightarrow$	2 mol

Chemists use this **quantitative** information to find:
- the *reacting quantities* required to prepare a particular quantity of a product
- the *quantities of products* formed by reacting particular quantities of reactants.

An understanding of molar reacting quantities provides chemists with the required recipes to make quantities of chemicals to order.

By working out the reacting quantities from a balanced chemical equation, quantities can be adjusted to take into account the required scale of preparation.

Example 1

These examples assume a 100% yield of products. See also percentage yield, page 123.

Calculate the masses of nitrogen and oxygen required to form 3 g of nitrogen oxide, NO. [A_r: N, 14; O, 16].

equation	$N_2(g)$	$+$	$O_2(g)$	$\longrightarrow$	$2 NO(g)$
moles	1 mol	$+$	1 mol	$\longrightarrow$	2 mol
reacting masses	(14×2) g	$+$	(16×2) g	$\longrightarrow$	$2 (14 + 16)$ g
	28 g	$+$	32 g	$\longrightarrow$	60 g
for 3 g NO(g), $\div 20$:	1.4 g	$+$	1.6 g	$\longrightarrow$	3 g

$\therefore$ 1.4 g of $N_2(g)$ react with 1.6 g of $O_2(g)$ to form 3 g NO(g).

Example 2

You must keep the proportions the same as the reacting quantities.

Any scaling must be applied to all reacting quantities.

Calculate the volume of oxygen, at r.t.p., formed by the decomposition of an aqueous solution containing 5 g of hydrogen peroxide, H_2O_2. [A_r: H, 1; O, 16.]

equation	$2 H_2O_2(aq)$	$\longrightarrow$	$2 H_2O(l)$	$+$	$O_2(g)$
moles	2 mol	$\longrightarrow$	2 mol	$+$	1 mol
reacting quantities	$2 [(1 \times 2) + (16 \times 2)]$ g	$\longrightarrow$			1×24 dm³
	68 g	$\longrightarrow$			24 dm³
for 1 g H_2O_2, $\div 68$,	1 g	$\longrightarrow$			$\frac{24}{68}$ dm³
for 5 g H_2O_2, $\times 5$,	5 g	$\longrightarrow$			$\frac{24}{68} \times 5$ dm³

$\therefore$ 5 g of H_2O_2 decomposes to form 1.8 dm³ of $O_2(g)$

Progress check

1 For the reaction: $S(s) + O_2(g) \longrightarrow SO_2(g)$
   (a) How many moles of S and O_2 react to form 0.5 mole of SO_2?
   (b) What mass of SO_2 is formed by reacting 64.2 g of S with O_2?
   (c) What volume of SO_2 forms when 0.963 g of S reacts with O_2 at r.t.p.?

2 Balance the equation: $Mg(s) + O_2(g) \longrightarrow MgO(s)$
   (a) What mass of MgO forms by burning 8.1 g of Mg?
   (b) What volume of O_2 at r.t.p. will react with this mass of Mg?
   (c) What masses of Mg and O_2 are needed to prepare 1 g of MgO?

2 (a) 13.4 g; (b) 4 dm³; (c) 0.6 g Mg, 0.4 g O_2.
1 (a) 0.5 mol S, 0.5 mol O_2; (b) 96.2 g; (c) 0.72 dm³.

Calculations in acid-base titrations

AQA   M1   OCR
EDEXCEL M2   SALTERS
CCEA   M1   WJEC   M1

The acid is added from the burette to the alkali.

An indicator is required to show the 'end-point' when all alkali has reacted with the added acid.

The concentration and volume of NaOH are known.

Using the equation determine the number of moles of the second reagent

Work out the concentration of $H_2SO_4(aq)$, in mol dm⁻³.

Calculations for acid-base titrations are best shown with an example:

In a titration, 25.0 cm³ of 0.100 mol dm⁻³ sodium hydroxide NaOH(aq) were found to react exactly with 20.80 cm³ of sulfuric acid, $H_2SO_4(aq)$. Find the concentration of the sulfuric acid.

To solve the problem, we must use the following pieces of information:

• the balanced equation
$$2NaOH(aq) + H_2SO_4(aq) \longrightarrow Na_2SO_4(aq) + 2H_2O\ (l)$$
• the concentration c_1 and reacting volume V_1 of NaOH(aq)
• the concentration c_2 and reacting volume V_2 of $H_2SO_4(aq)$.

From the titration results, the amount of NaOH (in mol) can be calculated:
$$\text{amount of NaOH} = c \times \frac{V}{1000} = 0.100 \times \frac{25.0}{1000} = 0.00250 \text{ mol}$$

From the equation, the amount of H_2SO_4 (in mol) can be determined:
$$2\ NaOH\ (aq) + H_2SO_4(aq) \longrightarrow Na_2SO_4(aq) + 2H_2O\ (l)$$
  2 mol   1 mol *(balancing numbers)*
  ∴ 0.00250 mol NaOH reacts with 0.00125 mol H_2SO_4
  amount of H_2SO_4 that reacted = 0.00125 mol

The concentration (in mol dm⁻³) of H_2SO_4 can be calculated by scaling to 1000 cm³:
  20.80 cm³ $H_2SO_4(aq)$ contains 0.00125 mol H_2SO_4
  1 cm³ $H_2SO_4(aq)$ contains $\frac{0.00125}{20.80}$ mol H_2SO_4

  1 dm³ (1000 cm³) $H_2SO_4(aq)$ contains $\frac{0.00125}{20.80} \times 1000 = 0.0601$ mol H_2SO_4

  ∴ concentration of $H_2SO_4(aq)$ is 0.0601 mol dm⁻³

Back titrations (CCEA only)

A back titration is a technique that can be used to analyse a sample that is insoluble in water.

In a **back titration**:

• a sample is first reacted with an **excess** of a solution of which the volume and concentration are both known
• the excess amount added is determined by carrying out a titration.

From the results, we can find out information about the sample.

AS exam questions would help you through titrations such as this by asking each step in the calculation as a question.

Worked Example

75.0 cm^3 of 0.500 mol dm^{-3} hydrochloric acid is added to 1.25 g of an impure sample of magnesium carbonate, $MgCO_3$.

The equation for the reaction between HCl and $MgCO_3$ is shown below:

equation 1 $2HCl(aq) + MgCO_3(s) \longrightarrow MgCl_2(aq) + H_2O(l) + CO_2(g)$

This is an excess of HCl.

- all of the $MgCO_3$ reacts
- some hydrochloric acid is left over unreacted.

The excess of HCl(aq) is titrated with 0.500 mol dm^{-3} sodium hydroxide and 21.60 cm^3 is required for complete reaction.

The equation for the reaction is shown below:

equation 2 $NaOH(aq) + HCl(aq) \longrightarrow NaCl(aq) + H_2O(l)$

Calculate the percentage of $MgCO_3$ in the impure sample.

From the titration results, the amount (in mol) of NaOH can be calculated:

The concentration and volume of NaOH are known.

amount of NaOH $= c \times \dfrac{V}{1000} = 0.500 \times \dfrac{21.60}{1000} = 0.0108$ mol

From equation 2, the amount (in mol) of HCl that reacts with the NaOH can be determined:

Using the equations, determine the number of moles of HCl

From equation 2,
1 mol NaOH reacts with 1 mol HCl
$\therefore$ 0.0108 mol NaOH reacts with 0.0108 mol HCl
This is the excess of HCl left over unreacted.

The amount (in mol) of HCl that reacted with the sample can now be determined:

At the start, 75.0 cm^3 of 0.500 mol dm^{-3} hydrochloric acid is used.

The amount of HCl that reacted in equation 1 is now found.

amount (in mol) of HCl used $= 0.500 \times \dfrac{75.0}{1000} = 0.0375$ mol

amount of HCl that reacts = original amount HCl – excess amount HCl
$= 0.0375 - 0.0108 = 0.0267$ mol

Calculate the mass of $MgCO_3$ in the rock sample:

From equation 1,
2 mol HCl reacts with 1 mol $MgCO_3$.

Using the equations, determine the number of moles of $MgCO_3$.

$\therefore$ 0.0267 mol HCl reacts with $\dfrac{0.0267}{2} = 0.01335$ mol $MgCO_3$.

mass of $MgCO_3 = 0.01335 \times (24.3 + 12 + 16 \times 3) = 1.125$ g

Calculate the percentage by mass of $MgCO_3$ in the impure sample:

Percentage of $MgCO_3$ in impure sample $= \dfrac{\text{mass of } MgCO_3}{\text{mass of impure } MgCO_3} \times 100$

$= \dfrac{1.125}{1.25} \times 100 = 90.0\%$

Progress check

1 25.0 cm^3 of 0.500 mol dm^{-3} NaOH(aq) reacts with 23.2 cm^3 of HNO_3(aq).
$$HNO_3(aq) + NaOH(aq) \longrightarrow NaNO_3(aq) + H_2O(l)$$
Find the concentration of the nitric acid, HNO_3.

2 25.0 cm^3 of 0.100 mol dm^{-3} KOH(aq) reacts with 26.6 cm^3 of H_2SO_4(aq).
$$H_2SO_4(aq) + 2KOH(aq) \longrightarrow K_2SO_4(aq) + 2H_2O(l)$$
Find the concentration of the sulfuric acid, H_2SO_4.

2 0.0470 mol dm^{-3}.

1 0.539 mol dm^{-3}.

1.7 Redox reactions

After studying this section you should be able to:

- *explain the terms reduction and oxidation in terms of electron transfer*
- *explain the terms oxidising agent and reducing agent*
- *apply the rules for assigning oxidation states*
- *identify changes in oxidation numbers from an equation*
- *construct an overall equation for a redox reaction from half-equations*

LEARNING SUMMARY

Oxidation and Reduction

AQA	M2	OCR	M1
EDEXCEL	M2	SALTERS	M2
CCEA	M1	WJEC	M2

Redox reactions

Oxidation and *reduction* were originally used for reactions involving oxygen.

> Oxidation is the gain of oxygen.
> Reduction is the loss of oxygen.

KEY POINT

Nowadays, oxidation and reduction have a much broader definition in terms of electron transfer in a *redox* reaction (**red**uction and **ox**idation).

> **Reduction** is the **gain** of electrons.
> **Oxidation** is the **loss** of electrons.

KEY POINT

Reduction and oxidation must take place together:
- if one species **gains** electrons
- another species **loses** the same number of electrons

OIL

RIG

↓

Oxidation

Is

Loss of electrons

Reduction

Is

Gain of electrons

Half-equations

The formation of magnesium chloride from its elements is a redox reaction:

overall reaction $Mg + Cl_2 \longrightarrow MgCl_2$

The overall equation conceals the electron transfer that has taken place. This can be shown by writing half-equations:

Half-equations are useful to identify the species being oxidised or reduced. Notice that the number of electrons lost and gained must balance.

electron transfer $Mg \longrightarrow Mg^{2+} + 2e^-$ oxidation (loss of electrons)
$Cl_2 + 2e^- \longrightarrow 2Cl^-$ reduction (gain of electrons)

Oxidising and reducing agents

Non-metals are oxidising agents.

Metals are reducing agents.

> An **oxidising agent accepts** electrons from another reactant.
> Non-metals are oxidising agents, e.g. F_2, Cl_2, O_2.
>
> A **reducing agent donates** electrons to another reactant.
> Metals are reducing agents, e.g. Na, Fe, Zn.

KEY POINT

In the example above:
- Mg is the reducing agent – it has *reduced* the Cl_2 to $2Cl^-$ by *adding* electrons
- Cl_2 is the oxidising agent – it has *oxidised* Mg to Mg^{2+} by *removing* electrons.

Using oxidation numbers

AQA	M2	OCR	M1
EDEXCEL	M2	SALTERS	M2
CCEA	M1	WJEC	M2

Chemists use the concept of **oxidation number** as a means of accounting for electrons.

The oxidation number of a species is assigned by applying a set of rules.

Oxidation number rules

Exceptions to rules

In compounds with fluorine and in peroxides, the oxidation number of oxygen is not –2 and must be calculated from other oxidation numbers.

In metal hydrides, the oxidation number of hydrogen is –1.

species	oxidation number	examples
uncombined element	0	C, zero; Na, zero; O_2, zero.
combined oxygen	–2	H_2O; CaO.
combined hydrogen	+1	NH_3, H_2S.
simple ion	charge on ion	Na^+, +1; Mg^{2+}, +2; Cl^-, –1
combined fluorine	–1	NaF, CaF_2.

> When applying oxidation numbers to elements, compounds and ions, the sum of the oxidation numbers must equal the overall charge. **KEY POINT**

Examples of applying oxidation numbers

Oxidation Number rules can be applied to any compound, whether ionic or covalent.

In CO_2, the overall charge is zero.

- There are **2** oxygen atoms, each with an oxidation number of **–2**, giving a total contribution of **–4**.
- The oxidation number of carbon must be **+4** to give the overall charge of zero.

In NO_3^-, the overall charge is 1–.

- There are **3** oxygen atoms, each with an oxidation number of **–2**, giving a total contribution of **–6**.
- The oxidation number of nitrogen must be **+5** to give the overall charge of –1.

Oxidation number applies to each atom in a species.

Notice how this has been shown in the examples.

Sometimes an element can form compounds or ions in which its atoms can have different oxidation states. The oxidation number is included in the name as a Roman numeral.

For example, there are two nitrate ions:

nitrate(III), NO_2^- N: +3

nitrate(V), NO_3^- N: +5

In a compound:	CO_2		In an ion:	NO_3^-
• the total of all the oxidation numbers is zero:	+4 –2 –2		• the total of all the oxidation numbers is equal to the overall charge on the ion:	+5 –2 –2 –2
	+4 –4			+5 –6
CO_2: overall charge = 0	0		NO_3^-: overall charge = –1	–1

Using oxidation number with equations

Oxidation numbers can be used to identify redox reactions in which electron loss and electron gain are not easy to see.

By applying oxidation numbers to an equation:

- the species being oxidised and reduced can be identified
- the number of electrons on both sides of the equation can be checked.

	$Cr_2 O_3$ (s)	+	2 Al (s)	⟶	$Al_2 O_3$ (s)	+	2 Cr (s)
oxidation	+3 –2		0		+3 –2		0
numbers	+3 –2		0		+3 –2		0
	–2				–2		
sum of oxidation numbers:	0		0		0		0

The sum of the oxidation numbers on both sides of a chemical equation must be the same.

In this reaction, the changes in oxidation number are:

Cr: +3 ⟶ 0   reduction   *oxidation number decreases*
Al: 0 ⟶ +3   oxidation   *oxidation number increases*

> **Oxidation** is an **increase** in oxidation number.
> **Reduction** is a **decrease** in oxidation number.

Combining half-equations

An overall equation for a redox reaction can be constructed by combining the half-equations showing the transfer of electrons in the reaction.

Example: The reaction of Ag⁺ ions react with Zn metal.

electron gain: reduction

electron loss: oxidation

The half-equations are:

$$Ag^+(aq) \ + \ e^- \ \longrightarrow \ Ag(s)$$
$$Zn(s) \ \longrightarrow \ Zn^{2+}(aq) \ + \ 2e^-$$

To combine the half-equations in an overall equation, the number of electrons transferred must be the same in each half-equation. This will ensure that every electron lost by Zn(s) is gained by Ag⁺.

Balance the electrons:

Ag⁺ reaction x 2.

- The Ag⁺ half-equation is multiplied by '2' to balance the electrons.
$$2Ag^+(aq) \ + \ 2e^- \ \longrightarrow \ 2Ag(s)$$

- The half-equations are now added:
$$2Ag^+(aq) \ + \ 2e^- \ + \ Zn(s) \ \longrightarrow \ 2Ag(s) + Zn^{2+}(aq) + 2e^-$$

Cancel the electrons to give the overall equation.

- Any species appearing on both sides are cancelled to give the overall equation.
$$2Ag^+(aq) \ + \ Zn(s) \ \longrightarrow \ 2Ag(s) + Zn^{2+}(aq)$$

- Finally, check that the oxidation numbers balance on either side of the equation:

	2Ag⁺(aq)	+	Zn(s)	⟶	2Ag(s)	+	Zn²⁺(aq)
oxidation numbers:	+1		0		0		+2
	+1				0		

oxidation number check: +2   +2

Progress check

1   What is the oxidation state of the elements in the following:
    (a) Ag⁺; (b) F_2; (c) N^{3-}; (d) Fe; (e) MgF_2; (f) $NaClO_3$?

2   Write down the oxidation number of sulfur in the following:
    (a) H_2S; (b) SO_2; (c) SO_4^{2-}; (d) SO_3^{2-}; (e) $Na_2S_2O_3$.

3   The following reaction is a redox process:
    $$Mg + 2HCl \longrightarrow MgCl_2 + H_2$$
    (a) Identify the changes in oxidation number.
    (b) Which species is being oxidised and which is being reduced?
    (c) Identify the oxidising agent and the reducing agent.

1 (a) +1; (b) 0; (c) −3; (d) 0; (e) Mg: +2, F: −1; (f) Na: +1, Cl: +5, O: −2.
2 (a) −2; (b) +4; (c) +6; (d) +4; (e) +2.
3 (a) Mg, 0 ⟶ +2, H, +1 ⟶ 0; (b) Mg oxidised, H reduced;
(c) Oxidising agent: HCl, Reducing agent: Mg.

1.8 Acids, bases and salts

After studying this section you should be able to:

- *know what is meant by acid, base, alkali and salt*
- *know what is meant by neutralisation*
- *describe the typical reactions of acids*
- *understand what is meant by hydrated, anhydrous and water of crystallisation*
- *know how to determine the water of crystallisation in a hydrated salt*

LEARNING SUMMARY

Acids, bases and alkalis

OCR

WJEC · M1

A hydrogen ion, H^+ is simply a hydrogen atom with an electron removed.

Because a hydrogen atom comprises one proton and one electron, removing the electron leaves just a proton.

'H^+', 'Hydrogen ion' and 'proton' refer to the same thing.

The common acids are: H_2SO_4, HCl and HNO_3

Acids

> All acids release $H^+(aq)$ ions in aqueous solution.

KEY POINT

Common acids include:

- strong acids: sulfuric acid, H_2SO_4; hydrochloric acid, HCl; nitric acid, HNO_3
- weak organic acids: ethanoic acid, CH_3COOH; methanoic acid, HCOOH

When added to water, HCl ionises and releases protons into the solution.

$$HCl(g) \quad + \quad aq \longrightarrow H^+(aq) \quad + \quad Cl^-(aq)$$

Bases

> All bases accept $H^+(aq)$ ions.

KEY POINT

Common bases include:

- metal oxides: magnesium oxide, MgO; copper oxide, CuO
- metal hydroxides: sodium hydroxide, NaOH; iron(III) hydroxide, $Fe(OH)_3$
- ammonia and amines: ammonia, NH_3; methylamine, CH_3NH_2

Alkalis

> An alkali is a soluble base that dissolves in water, releasing $OH^-(aq)$ ions.

KEY POINT

Common alkalis are: NaOH, KOH and NH_3

You need to learn these formulae as you will use these compounds very often.

There are only a few common alkalis:

- strong alkalis: potassium hydroxide, KOH; lithium hydroxide, LiOH
- weak alkalis: ammonia, NH_3

When added to water, the alkali NaOH ionises and releases hydroxide ions into the solution.

$$NaOH(s) \quad + \quad aq \longrightarrow Na^+(aq) \quad + \quad OH^-(aq)$$

Neutralisation

A base neutralises the H^+ of the acid to form water. The equation for neutralisation of an acid by an alkali is shown below:

$$H^+(aq) \quad + \quad OH^-(aq) \longrightarrow H_2O(l)$$

Salts

When acids react, they often form salts.

A salt is a compound formed from an acid, when a H^+ ion from the acid has been replaced by a metal ion or another positive ion, such as the ammonium ion, NH_4^+.

Examples of some common salts are shown below:

acid		salt	example
sulfuric acid	H_2SO_4	sulfate	sodium sulfate: Na_2SO_4
hydrochloric acid	HCl	chloride	sodium chloride: NaCl
nitric acid	HNO_3	nitrate	potassium nitrate: KNO_3

Typical reactions of acids

OCR

You will often come across these reactions of acids.

You need to learn how acids react with carbonates, bases, alkalis and metals.

Salts are always formed in solution from these typical acid reactions. The salt can be isolated by evaporation of water.

Acids react with carbonates, bases, alkalis and metals.

Neutralisation reactions

Acids react with carbonates, bases and alkalis in **neutralisation** reactions.

- The H^+ of the acid is neutralised to form water.

Acids react with carbonates, forming a salt, carbon dioxide gas and water:

$$2HCl(aq) \; + \; CaCO_3(s) \; \longrightarrow \; CaCl_2(aq) \; + \; CO_2(g) \; + \; H_2O(l)$$
$$\textbf{acid} \; + \; \textbf{carbonate} \; \longrightarrow \; \textbf{salt} \; + \; \textbf{carbon dioxide} \; + \; \textbf{water}$$

Acids react with bases, forming a salt and water:

$$2HCl(aq) \; + \; CaO(s) \; \longrightarrow \; CaCl_2(aq) \; + \; H_2O(l)$$
$$\textbf{acid} \; + \; \textbf{base} \; \longrightarrow \; \textbf{salt} \; + \; \textbf{water}$$

Acids react with alkalis, forming a salt and water:

$$HCl(aq) \; + \; NaOH(aq) \; \longrightarrow \; NaCl(aq) \; + \; H_2O(l)$$
$$\textbf{acid} \; + \; \textbf{alkali} \; \longrightarrow \; \textbf{salt} \; + \; \textbf{water}$$

Redox reactions

Redox reactions are discussed in detail on pages 42–44.

Acids react with many metals in **redox** reactions:

- the metal is oxidised
- hydrogen is reduced.

Acids react with metals, forming a salt and hydrogen gas:

$$2HCl(aq) \; + \; Mg(s) \; \longrightarrow \; MgCl_2(aq) \; + \; H_2(g)$$
$$H_2SO_4(aq) \; + \; Mg(s) \; \longrightarrow \; MgSO_4(aq) \; + \; H_2(g)$$

+1 —————————————————→ 0   H reduced

0 ————→ +2   Mg oxidised

$$\textbf{acid} \; + \; \textbf{metal} \; \longrightarrow \; \textbf{salt} \; + \; \textbf{hydrogen}$$

Ammonia as a base

OCR

Ammonia, NH_3 is a colourless gas that is very soluble in water. It forms a weak alkaline solution when dissolved in water.

$$NH_3(aq) \; + \; H_2O(l) \; \rightleftharpoons \; NH_4^+(aq) \; + \; OH^-(aq)$$

Much of the ammonia formed in the Haber process (see pages 113–114) is neutralised by acids to form ammonium salts, such as ammonium nitrate:

$$NH_3(aq) + HNO_3(aq) \longrightarrow NH_4NO_3(aq)$$

Ammonium salts such as ammonium nitrate are very soluble in water and have a high nitrogen content. They are extensively used as artificial fertilisers.

Water of crystallisation

CCEA   M1
OCR   M1

Hydrated crystals

Many salts form crystals, which contain water in their structure. The water can be easily driven off by heat, often forming a non-crystalline powder.

- The crystalline form containing water molecules is **hydrated**.
- The powdered form without water is **anhydrous**.
- The molar proportion of water in the hydrated crystals is **water of crystallisation**.

Notice how water of crystallisation is shown with a '•' in the formula of the hydrated form.

In hydrated copper sulfate:

- the molar proportion of copper sulfate : water is 1 : 5
- the formula is shown as $CuSO_4 \bullet 5H_2O$

Calculating the formula of a hydrated salt

The formula of a hydrated salt can be determined by heating the hydrated salt to drive off the water of crystallisation.

Example

From an experiment to determine the formula of hydrated sodium sulfate,

mass of the **hydrated** salt, $Na_2SO_4 \bullet xH_2O$ = 7.08 g
mass of the **anhydrous** salt, Na_2SO_4 = 3.12 g
∴ mass of H_2O in $Na_2SO_4 \bullet xH_2O$ = 3.96 g

Calculate the amount (in mol) of anhydrous Na_2SO_4:

Molar mass, $M(Na_2SO_4)$ = 23.0 x 2 + 32.1 + (16 x 4) = 142.1 g mol^{-1}

Remember $n = \dfrac{mass}{Molar\ mass}$

$n(Na_2SO_4) = \dfrac{3.12}{142.1} = 0.0220$ mol

Calculate the amount (in mol) of water:

$n(H_2O) = \dfrac{3.96}{18.0} = 0.220$ mol

Determine the formula of the hydrated salt:

Molar ratio $Na_2SO_4 : H_2O = 0.0220 : 0.220 = 1 : 10$

∴ formula of the hydrated salt = $Na_2SO_4 \bullet 10H_2O$

Progress check

1 Write equations for the following reactions:
(a) sulfuric acid and potassium hydroxide
(b) nitric acid and copper(II) carbonate
(c) hydrochloric acid and zinc

2 From the experimental results, work out the formula of the hydrated salt.
Mass of $FeSO_4 \bullet xH_2O$ = 8.34 g
Mass of $FeSO_4$ = 4.56 g

2 $FeSO_4 \bullet 7H_2O$
(c) $2HCl(aq) + Zn(s) \longrightarrow ZnCl_2(aq) + H_2(g)$
(b) $2HNO_3(aq) + CuCO_3(s) \longrightarrow Cu(NO_3)_2(aq) + H_2O(l) + CO_2(g)$
1 (a) $H_2SO_4(aq) + 2KOH(aq) \longrightarrow K_2SO_4(aq) + 2H_2O(l)$

Sample question and model answer

The determination of reacting quantities using the Mole Concept is one of the most important concepts in chemistry. This can be tested in exam questions almost anywhere and it underpins much of the content of a chemistry course at this level. To succeed at AS Chemistry, it is essential that you grasp this concept.

A student carried out an investigation using a sample of hydrochloric acid, HCl(aq).

(a) The concentration of the hydrochloric acid was first determined by titration of a 25.0 cm³ sample against 0.148 mol dm⁻³ sodium hydroxide of which 28.40 cm³ were required.

Calculate the concentration, in mol dm⁻³, of the hydrochloric acid.

no. of moles of NaOH = $\frac{0.148 \times 28.40}{1000}$ = 4.20 × 10⁻³ ✓

NaOH + HCl ⟶ NaCl + H₂O
1 mol NaOH reacts with 1 mol HCl

∴ 4.20 × 10⁻³ mol NaOH reacts with 4.20 × 10⁻³ mol HCl ✓
25.0 cm³ HCl(aq) contains 4.20 × 10⁻³ mol HCl

1 cm³ HCl(aq) contains $\frac{4.20 \times 10^{-3}}{25}$ mol HCl

1 dm³ (1000 cm³) HCl(aq) contains $\frac{4.20 \times 10^{-3}}{25}$ × 1000 = 0.168 mol HCl.

∴ concentration of HCl(aq) is 0.168 mol dm⁻³ ✓

[3]

(b) The student took a 300 cm³ sample of the same hydrochloric acid. The student intended to neutralise the hydrochloric acid with calcium hydroxide, Ca(OH)₂. The equation is shown below:

$$Ca(OH)_2 + 2HCl \longrightarrow CaCl_2 + 2H_2O$$

Calculate the mass, in g, of calcium hydroxide, Ca(OH)₂ required to neutralise this hydrochloric acid.

no. of moles of 0.168 mol dm⁻³ HCl in 300 cm³ = $\frac{0.168 \times 300}{1000}$ = 0.0504 ✓
1 mol Ca(OH)₂ reacts with 2 mol HCl

Notice that the same ideas are used here.

The correct reacting quantities from the equation are essential.

∴ $\frac{0.0504}{2}$ mol Ca(OH)₂ reacts with 0.0504 mol HCl ✓

Molar mass of Ca(OH)₂ = 40 + (16 + 1)2 = 74.1 g mol⁻¹ ✓

Mass of Ca(OH)₂ = $\frac{0.0504}{2}$ × 74.1 = 1.87 g (to 3 significant figures) ✓

[4]

(c) The student made some calcium hydroxide by heating calcium carbonate and then adding water:

$$CaCO_3 \longrightarrow CaO + CO_2$$
$$CaO + H_2O \longrightarrow Ca(OH)_2$$

Calculate the mass of calcium carbonate that the student would need to produce 10.0 g of calcium hydroxide.

It is important to always show full working. If you do make a mistake early on in a calculation, you will only be penalised once if the method that follows is sound.

In 10.0 g Ca(OH)₂, number of moles of Ca(OH)₂ = $\frac{10.0}{74.1}$ = 0.135 ✓

From the equations, 1 mol CaCO₃ ⟶ 1 mol CaO ⟶ 1 mol Ca(OH)₂
∴ number of moles of CaCO₃ = 0.135
∴ Mass of CaCO₃ = 0.135 × 100.1 = 13.5 g (to 3 significant figures) ✓

[2]
[Total: 9]

Practice examination questions

1 Chemists use a model of an atom that consists of sub-atomic particles: protons, neutrons and electrons.

(a) Complete the table below to show the properties of these sub-atomic particles.

particle	relative mass	relative charge
proton		
neutron		
electron		

[3]

(b) The particles in each pair below differ **only** in the number of protons **or** neutrons **or** electrons. Explain what the difference is within each pair.
(i) ^{12}C and ^{13}C
(ii) ^{16}O and $^{16}O^{2-}$
(iii) $^{23}Na^+$ and $^{24}Mg^{2+}$.

[6]

[Total: 9]

2 (a) In terms of the numbers of sub-atomic particles, state **one** difference and **two** similarities between two isotopes of the same element. [3]

(b) Give the chemical symbol, including its mass number, for an atom which has 5 electrons and 6 neutrons. [1]

(c) An element has an atomic number of 15 and it forms an ion with a charge of 3–.
(i) Deduce the electron configuration of this ion.
(ii) Which block in the Periodic Table is this element in? [2]

(d) (i) Write an equation for the process involved in the first ionisation energy of silicon.
(ii) Explain why the fifth ionisation energy of silicon is much greater than the fourth. [6]

[Total: 12]

3 (a) A student added 0.117 g of lithium to 200 cm³ of water. The following reaction took place.

$$2Li(s) + 2H_2O(l) \longrightarrow 2LiOH(aq) + H_2(g)$$

(i)   Calculate the number of moles of lithium that reacted.

(ii) Calculate the concentration, in mol dm⁻³, of the 200 cm³ of aqueous lithium hydroxide solution that formed.

(iii) Calculate the volume of hydrogen gas produced at room temperature and pressure (298 K and 100 kPa).

[6]

(b) In another experiment 25.0 cm³ of 0.126 mol dm⁻³ lithium hydroxide were neutralised by 21.25 cm³ of sulfuric acid as shown below.

$$2LiOH(aq) + H_2SO_4(aq) \longrightarrow Li_2SO_4(aq) + 2H_2O(l)$$

Calculate the concentration, in mol dm⁻³, of the sulfuric acid.

[3]

[Total: 9]

Practice examination questions

4 A student investigated different methods to neutralise 0.580 mol dm^{-3} sulfuric acid, H_2SO_4.

(a) The student decided to add 1.25 mol dm^{-3} NaOH to a 25.0 cm^3 sample of the 0.580 mol dm^{-3} sulfuric acid. The reaction that takes place is shown below.

$$2NaOH(aq) + H_2SO_4(aq) \longrightarrow Na_2SO_4(aq) + 2H_2O(l)$$

(i) Calculate the amount, in moles, of H_2SO_4 in 25.0 cm^3 of 0.580 mol dm^{-3} of sulfuric acid.

(ii) What is the amount, in moles, of NaOH that is needed to neutralise this amount of acid?

(iii) Calculate the volume, in cm^3, of 1.25 mol dm^{-3} NaOH that would be needed to neutralise this amount of acid.

[3]

(b) The student then decided to add sodium carbonate, Na_2CO_3, to a second 25.0 cm^3 sample of the 0.580 mol dm^{-3} sulfuric acid. The equation for this reaction is shown below.

$$Na_2CO_3(s) + H_2SO_4(aq) \longrightarrow Na_2SO_4(s) + H_2O(l) + CO_2(g)$$

Calculate the mass of Na_2CO_3 that would be needed to neutralise this amount of sulfuric acid.

[2]

(c) In a third experiment the student added magnesium metal to the sulfuric acid. The equation for the reaction that takes place is shown below.

$$Mg(s) + H_2SO_4(aq) \longrightarrow MgSO_4(aq) + H_2(g)$$

Use oxidation numbers to show that this is a redox reaction.

[2]
[Total: 7]

5 Sodium azide, NaN_3, decomposes when heated to release nitrogen gas.

$$2NaN_3(s) \longrightarrow 2Na(l) + 3N_2(g)$$

(a) A student prepared 1.50 dm^3 of nitrogen in the laboratory by this method. This gas volume was measured at room temperature and pressure (r.t.p.).

(i) How many moles of N_2 did the student prepare?
[Assume that 1 mole of gas molecules occupies 24.0 dm^3 at r.t.p.]

(ii) What mass of NaN_3 did the student heat? [4]

(b) In this reaction, 0.96 g of sodium metal was also formed. The student carefully reacted the sodium with water to form 50.0 cm^3 of aqueous sodium hydroxide:

$$2Na(s) + 2H_2O(l) \longrightarrow 2NaOH(aq) + H_2(g)$$

(i) Calculate the concentration, in mol dm^{-3}, of the aqueous sodium hydroxide.

(ii) Use oxidation numbers to show that this is a redox reaction. [4]
[Total: 8]

Chapter 2

Bonding, structure and the Periodic Table

The following topics are covered in this chapter:

- *Chemical bonding*
- *Shapes of molecules*
- *Electronegativity, polarity and polarisation*
- *Intermolecular forces*
- *Bonding, structure and properties*

- *The modern Periodic Table*
- *The s-block elements: Group 1 and Group 2*
- *The Group 7 elements and their compounds*
- *Extraction of metals*
- *Qualitative analysis*

2.1 Chemical bonding

After studying this section you should be able to:

- *appreciate that compounds often contain atoms or ions with electron structures of a noble gas*
- *understand the nature of an ionic bond in terms of electrostatic attraction between ions*
- *determine the formula of an ionic compound from the ionic charges*
- *understand evidence for ionic bonds from lattice enthalpies*
- *understand the nature of polarisation in ionic compounds*
- *understand the nature of a covalent bond and a dative covalent (coordinate) bond in terms of the sharing of an electron pair*
- *construct 'dot-and-cross' diagrams to demonstrate ionic and covalent bonding*
- *describe metallic bonding*

LEARNING SUMMARY

Why do atoms bond together?

AQA	M1	OCR	M1
EDEXCEL		SALTERS	M
CCEA	M1	WJEC	M2

The Octet Rule

Under normal conditions, the only elements that can exist as single atoms are the Noble Gases in Group 0 of the Periodic Table. The atoms of all other elements are bonded together. To find out why, we need to look at the electron configurations of the atoms of the Noble Gases.

The Noble Gases get their name from their unreactivity.

In the atoms of a noble gas:

- all electrons are paired,
- the bonding shells are full.

Atoms of the Noble Gases are stable because all their electrons are paired.

He atom
$1s^2$

Ne atom
$1s^2 2s^2 2p^6$

Ar atom
$1s^2 2s^2 2p^6 3s^2 3p^6$

This arrangement of electrons is particularly stable and is often referred to as a stable 'octet' (as in neon and argon).

This tendency to acquire a noble gas electron structure is often referred to as the 'Octet Rule'.

Apart from the Noble Gases, atoms are bonded together. Thus the elements oxygen and nitrogen exists as O_2 and N_2. Compounds exist when atoms of different elements bond together as in CO_2, CO and NaCl. In all these examples, electrons have been shared, transferred or rearranged in such a way that the atoms of the elements have acquired a noble gas electron structure. If this creates an outer shell of 8 electrons (as in all these cases), the '*Octet Rule*' is obeyed. All these themes are discussed in the sections that follow.

Types of bonding

Many mistakes are made in chemistry by confusing the type of bonding.

Covalent and ionic compounds are bonded differently and they behave differently. You cannot apply 'ionic bonding ideas' to a compound that is covalent.

Chemical bonds are classified into three main types: ionic, covalent and metallic.

- **Ionic** bonding occurs between the atoms of a **metal** and a **non-metal**. For example, NaCl, MgO, Fe_2O_3.
- **Covalent** bonding occurs between the atoms of **non-metals**. For example, O_2, H_2, H_2O, diamond and graphite.
- **Metallic** bonding occurs between the atoms of **metals**. For example, all metals such as iron, zinc, aluminium, etc.; alloys as in brass (copper and zinc), bronze (copper and tin).

> Always decide the type of bonding before:
> - drawing a 'dot-and-cross' diagram
> - predicting how a compound reacts
> - predicting the properties of a compound.
>
> **KEY POINT**

Ionic bonding

AQA	M1	OCR	
EDEXCEL	M1	SALTERS	
CCEA	M1	WJEC	M2

Ionic bonds

Ionic bonds are present in a compound of a metal and a non-metal.

An ionic bond is formed when electrons are **transferred** from a metal atom to a non-metal atom forming **oppositely-charged ions**.

> An ionic bond is the electrical attraction between oppositely charged ions.
>
> **KEY POINT**

A positive ion is called a **cation** because it is attracted to a **cathode** (a negative electrode). A negative ion is called an **anion** because it is attracted to an **anode** (a positive electrode).

Example: sodium chloride, NaCl

Sodium chloride, NaCl, is formed by the transfer of one electron from a sodium atom ($[Ne]3s^1$) to a chlorine atom ($[Ne]3s^23p^5$) with the formation of ions:

For showing chemical bonding, 'dot-and-cross' diagrams often only show the outer shell.

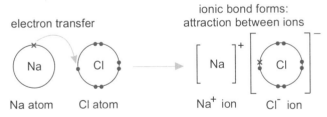

electron transfer

ionic bond forms: attraction between ions

Na atom   Cl atom Na^+ ion Cl^- ion

A sodium atom has lost an electron to acquire the electron structure of neon:

$$Na \longrightarrow Na^+ + e^-$$
$$[Ne]3s^1 \longrightarrow [Ne] + e^-$$

The ions that are formed have stable noble gas electron structures with full outer electron shells.

A chlorine atom has gained an electron to acquire the electron structure of argon:

$$Cl + e^- \longrightarrow Cl^-$$
$$[Ne]3s^23p^5 + e^- \longrightarrow [Ne]3s^23p^6 \text{ or } [Ar]$$

Example: magnesium chloride, $MgCl_2$

A magnesium atom ($[Ne]3s^2$) has two electrons in its outer shell. One electron is transferred to each of two chlorine atoms ($[Ne]3s^23p^5$) forming ions.

The Mg^{2+} and Cl^- ions have noble gas electron configurations.

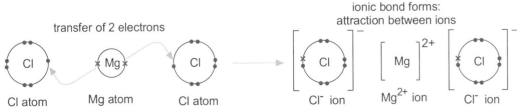

transfer of 2 electrons

ionic bond forms: attraction between ions

Cl atom   Mg atom   Cl atom Cl^- ion Mg^{2+} ion Cl^- ion

Giant ionic lattices

Each ion is surrounded by oppositely-charged ions, forming a giant ionic lattice.

Edexcel studies evidence for the model of ionic bonding in detail. See pages 57–58.

- Although it is convenient to look at ionic bonding between two ions only, each ion is able to attract oppositely charged ions in all directions.
- This results in a **giant ionic lattice** structure comprising hundreds of thousands of ions (depending upon the size of the crystal).
- This arrangement is characteristic of all ionic compounds.

Part of the sodium chloride lattice

- Na^+
- Cl^-

- Each Na^+ ion surrounds 6 Cl^- ions
- Each Cl^- ion surrounds 6 Na^+ ions

Ionic charges and the Periodic Table

An ionic charge can be predicted from an element's position in the Periodic Table.

The diagram below shows elements in Periods 2 and 3 of the Periodic Table.

group	1	2	3	4	5	6	7	0
number of outer shell electrons	1	2	3	4	5	6	7	8
element	Li	Be	B	C	N	O	F	Ne
ion	Li^+				N^{3-}	O^{2-}	F^-	
element	Na	Mg	Al	Si	P	S	Cl	Ar
ion	Na^+	Mg^{2+}	Al^{3+}		P^{3-}	S^{2-}	Cl^-	

Remember that any metal ions will be positive and non-metal ions negative.

- Metals in Groups 1–3 **lose** sufficient electrons from their atoms to form ions with the electron configuration of the **previous** noble gas;
- Non-metals in Groups 5–7 **gain** sufficient electrons to their atoms to form ions with the electron configuration of the **next** noble gas.
- Be, B, C and Si do not normally form ionic compounds because of the large amount of energy required to transfer electrons forming their ions.

Groups of covalent-bonded atoms can also lose or gain electrons to give ions such as those shown below.

Learn these

1+	1–		2–		3–	
ammonium NH_4^+	hydroxide	OH^-	carbonate	CO_3^{2-}	phosphate	PO_4^{3-}
	nitrate	NO_3^-	sulfate	SO_4^{2-}		
	nitrite	NO_2^-	sulfite	SO_3^{2-}		
	hydrogen-carbonate	HCO_3^-				

Predicting ionic formulae

Although an ionic compound comprises charged ions, its overall charge is zero.

In an ionic compound,
total number of positive charges from positive ions = total number of negative charges from negative ions.

KEY POINT

Working out an ionic formula from ionic charges

calcium chloride:	ion	charge	aluminium sulfate:	ion	charge
equalise charges:	Ca^{2+}	2+	equalise charges:	Al^{3+}	3+
				Al^{3+}	3+
	Cl^-	1–		SO_4^{2-}	2–
	Cl^-	1–		SO_4^{2-}	2–
				SO_4^{2-}	2–
total charge must be zero:		0	total charge must be zero:		0
formula:		$CaCl_2$	formula:		$Al_2(SO_4)_3$

Covalent bonding

AQA	M1	OCR	M1
EDEXCEL	M1	SALTERS	M1
CCEA	M1	WJEC	M2

Covalent bonds

A covalent compound comprises molecules: groups of atoms held together by covalent bonds.

> **KEY POINT**
> A molecule is the smallest part of a covalent compound that can take part in a chemical reaction.

Covalent bonds are present in a compound of two non-metals.

A covalent bond is formed when electrons are **shared** rather than transferred.

> **KEY POINT**
> A covalent bond is a shared pair of electrons.

Example: hydrogen, H_2

The covalent bond in a hydrogen molecule H_2 forms when two hydrogen atoms ($1s^1$) bond together, each contributing 1 electron to the bond:

In forming the covalent bond, each hydrogen atom in the H_2 molecule has:

A single covalent bond is written as A–B.

- control of 2 electrons – one of its own and the second from the other atom
- the electron configuration of the noble gas helium.

When a covalent bond forms, each atom *often* acquires a stable noble gas electron configuration with a full outer bonding shell.

This diagram shows how two atoms are bonded together by a covalent bond.

Unlike an ionic bond, a covalent bond is *directional* and acts solely between the two atoms involved in the bond. Thus hydrogen exists simply as H_2 molecules.

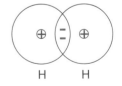

In a hydrogen molecule, the nucleus of each hydrogen atom is attracted towards the electron pair of the covalent bond.

The diagrams below show examples of covalent bonding in some simple molecules.

Notice that each atom contributes one electron to the covalent bond.

Cl_2

H_2O

NH$_3$

CH_4

Multiple covalent bonds

Other common examples include:

C_2H_4: $H_2C=CH_2$
CO_2: $O=C=O$
C_2H_2: $H–C\equiv C–H$

A covalent bond in which **one** pair of electrons is shared is known as a **single bond**, e.g. Cl_2, written as Cl–Cl.

Atoms can also share more than one pair of electrons to form a **multiple bond**.

* Sharing of **two** pairs of electrons forms a **double bond**, e.g. O_2, written as O=O.
* Sharing of **three** pairs of electrons forms a **triple bond**, e.g. N_2, written as $N\equiv N$.

O_2 N_2

Dative covalent or coordinate bonds

A dative covalent bond forms when the shared pair of electrons comes from just **one** of the bonded atoms.

A dative covalent or coordinate bond is one in which **one** of the atoms supplies **both** the shared electrons to the covalent bond. A dative covalent bond is written as A→B, where the direction of the arrow shows the direction in which the electron pair is donated.

Example: ammonium ion, NH_4^+

* The ammonium ion, NH_4^+, forms when ammonia NH_3 bonds with a proton H^+, with both the bonding electrons coming from NH_3.
* Note that the ammonium ion has three covalent bonds and one dative covalent bond:

The involvement of the electron pair as a dative covalent bond is often shown as an arrow:

Note that once NH_4^+ has formed, all 4 bonds are equivalent and you cannot tell which was formed by a dative covalent bond.

NH$_3$ NH$_4$$^+$

In exams, these are easy marks provided that the facts have been learnt.

The Octet Rule doesn't always work

The Octet Rule is useful in chemistry but it can be broken. It is probably more important that all unpaired electrons are paired.

When covalent bonds form, unpaired electrons **often** pair up to form a noble gas electron configuration according to the Octet Rule (see page 51). Many molecules, however, form electron configurations that do not form an octet. Some molecules have 6, 10, 12 or even 14 electrons in the outer bonding shell. These compounds form by the pairing of unpaired electrons.

Phosphorus forms two chlorides, PCl_3 and PCl_5.

PCl_3 obeys the Octet Rule.

In PCl_5,

* the outer shell of phosphorus has expanded allowing all 5 electrons to be used in bonding
* the Octet Rule is broken and phosphorus now has 10 electrons in its outer shell:

In PCl_3, phosphorus is using 3 valence electrons.

In PCl_5, phosphorus is using 5 valence electrons (the same as the Group number of phosphorus).

     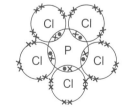

PCl_3: only 3 electrons from the outer bonding shell of phosphorus are used in bonding.

PCl_5: all 5 electrons from the outer bonding shell of phosphorus are used in covalent bonds.

An electron that takes part in forming chemical bonds is called a **valence electron**.

For elements in the s-block and p-block of the Periodic Table, the maximum number of valence electrons available is usually equal to the group number.

Metallic bonding

AQA	M1	OCR	
EDEXCEL	M1	SALTERS	
CCEA	M1	WJEC	M2

A metallic bond holds atoms together in a solid metal or alloy.

In solid metals, the atoms are ionised.

- The positive ions occupy fixed positions in a lattice.
- The outer shell electrons are **delocalised** – they are spread throughout the metallic structure and are able to move freely throughout the lattice.

> **KEY POINT**
>
> A **metallic bond** is the electrostatic attraction between the positive metal ions and delocalised electrons.

A metallic lattice is held together by electrostatic attraction between positive metal ions and electrons.

In the metallic lattice, each metal atom exists as + ions by releasing its outer valence electrons to the delocalised pool of electrons (often called a 'sea of electrons').

The 'sea of electrons'

Delocalised and localised electrons

The **delocalised** electrons in metals are spread throughout the metal structure. The delocalisation is such that electrons are able to move throughout the structure and it is impossible to assign any electron to a particular positive ion.

In a covalent bond, the **localised** pair of electrons is always positioned between the two atoms involved in the bond. The electron charge is much more concentrated.

Progress check

1 Draw 'dot-and-cross' diagrams of (a) MgO; (b) Na_2O.

2 Using the information on page 53, predict formulae for the following ionic compounds:

(a) lithium chloride

(b) potassium sulfide

(c) lithium nitride

(d) aluminium oxide

(e) sodium carbonate

(f) calcium hydroxide

(g) aluminium sulfate

(h) ammonium phosphate.

3 Draw 'dot-and-cross' diagrams for
(a) C_2H_6;   (b) HCN;   (c) H_3O^+.

2 (a) LiCl; (b) K_2S; (c) Li_3N; (d) Al_2O_3; (e) Na_2CO_3; (f) $Ca(OH)_2$; (g) $Al_2(SO_4)_3$; (h) $(NH_4)_3PO_4$.

Testing the model of ionic bonding

EDEXCEL

Lattice enthalpy

Lattice enthalpy indicates the strength of the ionic bonds in an ionic lattice.

> **KEY POINT**
>
> The lattice enthalpy ($\Delta H^{\ominus}_{\text{L.E.}}$) of an ionic compound is the enthalpy change that accompanies the formation of 1 mole of an ionic compound from its constituent gaseous ions. ($\Delta H^{\ominus}_{\text{L.E.}}$ is exothermic.)
>
> e.g. $Na^+(g) + Cl^-(g) \longrightarrow Na^+Cl^-(s)$

Determination of lattice enthalpies

For Hess' Law see page 99.

Lattice enthalpies cannot be determined directly and must be calculated indirectly using Hess' Law from other enthalpy changes that can be found experimentally. The energy cycle used to calculate a lattice enthalpy is the **Born-Haber cycle**.

The basis of the Born-Haber cycle is the formation of an ionic lattice from its elements. For sodium chloride, one route is the single stage process corresponding to its enthalpy change of formation.

ROUTE 1
$$Na(s) + \tfrac{1}{2}Cl_2(g) \longrightarrow Na^+Cl^-(s)$$

A second route can be broken down into separate steps resulting in:

ROUTE 2

- production of gaseous atoms:
$$Na(s) \longrightarrow Na(g)$$
$$\tfrac{1}{2}Cl_2(g) \longrightarrow Cl(g)$$
- production of gaseous ions:
$$Na(g) \longrightarrow Na^+(g) + e^-$$
$$Cl(g) + e^- \longrightarrow Cl^-(g)$$
- production of the solid ionic lattice:
$$Na^+(g) + Cl^-(g) \longrightarrow Na^+Cl^-(s)$$

The Born-Haber cycle

The complete Born-Haber cycle for sodium chloride is shown below. Definitions for these enthalpy changes are shown on page 58.

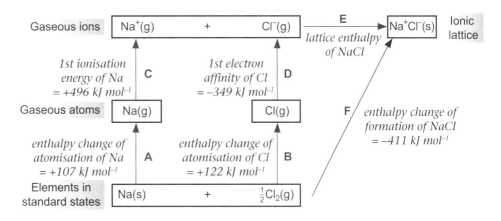

For NaCl, the production of gaseous ions from gaseous atoms comprises two changes:

the 1st ionisation energy of Na is **endothermic**

the 1st electron affinity of Cl is **exothermic**.

Route 1: **A + B + C + D + E**

Route 2: **F**

Using Hess' Law: **A + B + C + D + E = F**

$\therefore$ +107 + 122 + 496 + (–349) + E = –411

Hence, the lattice energy of $Na^+Cl^-(s)$, **E = –787** kJ mol^{-1}

Definitions for enthalpy changes

The enthalpy changes involved in these two routes are shown below.

> **The standard enthalpy change of formation $\Delta H^{\ominus}_f$** is the enthalpy change that takes place when one mole of a compound in its standard state is formed from its constituent elements in their standard states under standard conditions.
>
> **KEY POINT**

Be careful when using $\Delta H^{\ominus}_{at}$ involving diatomic elements.

$\Delta H^{\ominus}_{at}$ relates to the formation of 1 mole of atoms. For chlorine, this involves $\frac{1}{2}Cl_2$ only.

$$Na(s) + \frac{1}{2}Cl_2(g) \longrightarrow Na^+Cl^-(s) \qquad \Delta H^{\ominus}_f = -411 \text{ kJ mol}^{-1}$$

> **The standard enthalpy change of atomisation, $\Delta H^{\ominus}_{at}$** of an element is the enthalpy change that accompanies the formation of 1 mole of gaseous atoms from the element in its standard state.
>
> **KEY POINT**

$$Na(s) \longrightarrow Na(g) \qquad \Delta H^{\ominus}_{at} = +107 \text{ kJ mol}^{-1}$$
$$\frac{1}{2}Cl_2(g) \longrightarrow Cl(g) \qquad \Delta H^{\ominus}_{at} = +122 \text{ kJ mol}^{-1}$$

> **The first ionisation energy $\Delta H^{\ominus}_{I.E.}$** of an element is the enthalpy change that accompanies the removal of 1 electron from each atom in 1 mole of gaseous atoms to form 1 mole of gaseous 1+ ions.
>
> **KEY POINT**

$$Na(g) \longrightarrow Na^+(g) + e^- \qquad \Delta H^{\ominus}_{I.E.} = +496 \text{ kJ mol}^{-1}$$

Be careful using electron affinities. $\Delta H^{\ominus}_{E.A.}$ relates to the formation of 1 mole of 1⁻ ions. For chlorine, this involves Cl(g) only

> **The first electron affinity $\Delta H^{\ominus}_{E.A.}$** of an element is the enthalpy change that accompanies the addition of 1 electron to each atom in 1 mole of gaseous atoms to form 1 mole of gaseous 1⁻ ions.
>
> **KEY POINT**

$$Cl(g) + e^- \longrightarrow Cl^-(g) \qquad \Delta H^{\ominus}_{E.A.} = -349 \text{ kJ mol}^{-1}$$

Limitations of lattice enthalpies

The ionic model for ionic bonding can be tested by comparing 'Born-Haber' lattice enthalpies with theoretical values, calculated by considering each ion as a charged sphere. Any difference between a **theoretical** lattice enthalpy and lattice enthalpy calculated using a Born-Haber cycle indicates a proportion of covalent bonding caused by **polarisation**. This is explained below.

Polarisation in ionic compounds

EDEXCEL   M1
CCEA   M2
WJEC   M2

A 100% ionic bond would only form by complete transfer of electrons from a metallic atom to a non-metallic atom. Many ionic compounds have a degree of covalency, caused by polarisation and resulting in an incomplete transfer of electrons.

In Period 3, from Na to Al, the radius of the positive ions decreases.

- The cations Na^+, Mg^{2+} and Al^{3+} are isoelectronic (have the same number of electrons).
- The number of protons increases.

The electrons are therefore attracted closer to the nucleus, increasing the charge density.

Polarisation is most apparent with a small, densely charged 3+ cation.

Na^+	Mg^{2+}	Al^{3+}
$10e^-$	$10e^-$	$10e^-$
$11p^+$	$12p^+$	$13p^+$

Charge/size ratio increases

Cations and Anions

During electrolysis, the negative electrode or *cathode* attracts the positive ions, called **cations**.

The positive electrode or *anode* attracts the negative ions, called **anions**.

Cations are smaller than their atoms.

Anions are larger than their atoms.

Down a group, the ionic radius increases. For example, the ionic radius of the halide ions, from Cl^- to I^-, increases in size as shells are added, producing less charge density.

Polarisation is greatest between:

- a small + ion (cation) with multiple + charges (i.e. a high charge density) and
- a large – ion (anion) with a low charge density.

The electric field around a **small cation** distorts the electron shells around a **large anion**. As a result, electrons are attracted towards the cation, giving a degree of covalency (electron sharing). This is called **polarisation**.

Small Al^{3+} ion is highly charged.
O^{2-} ion is distorted and polarised resulting in partial covalent character.

Stability of ionic compounds

EDEXCEL   M1

The stability of ionic compounds can be compared using values calculated for enthalpy changes of formation based on Born-Haber cycles.

e.g.   MgCl: -94 kJ mol^{-1}; $MgCl_2$ -643 kJ mol^{-1}; $MgCl_3$ $+3943$ kJ mol^{-1}

$MgCl_2$ is the compound that actually forms: it is the most stable magnesium chloride because it has the most exothermic enthalpy change of formation.

Progress check

1 Using the information below:

(a) name each enthalpy change **A** to **E**

(b) construct a Born-Haber cycle for sodium bromide and calculate the lattice enthalpy of sodium bromide. (Use the Born-Haber cycle for NaCl on page 57 as a guide.)

enthalpy change	equation	$\Delta H^{\ominus}$ / kJ mol^{-1}
A	$Na(s) + \frac{1}{2}Br_2(l) \longrightarrow NaBr(s)$	-361
B	$Br(g) + e^- \longrightarrow Br^-(g)$	-325
C	$Na(s) \longrightarrow Na(g)$	$+107$
D	$Na(g) \longrightarrow Na^+(g) + e^-$	$+496$
E	$\frac{1}{2}Br_2(l) \longrightarrow Br(g)$	$+112$

1 (a) A enthalpy change of formation of sodium bromide
B 1st electron affinity of bromine
C enthalpy change of atomisation of sodium
D 1st ionisation energy of sodium
E enthalpy change of atomisation of bromine
(b) -751 kJ mol^{-1}

2.2 Shapes of molecules

After studying this section you should be able to:

- use electron-pair repulsion to explain the shapes and bond angles of simple molecules
- understand that lone pairs have a larger repulsive effect than bonded pairs, leading to distortion of the molecular shape
- understand how multiple bonds affect the shape of a molecule

Electron-pair repulsion theory

AQA	M1	OCR	M1
EDEXCEL	M2	SALTERS	M1
CCEA	M1	WJEC	M2

Electron-pair repulsion theory states that:

- the shape of a molecule depends upon the number of electron pairs surrounding the central atom
- electron pairs repel one another and move as far apart as possible.

The shape of any molecule can be predicted by applying this theory.

Bonds and lone pairs are negative charge clouds that repel one another.

Molecules with bonded pairs

Learn these shapes and bond-angles.

You can predict the shape of a molecule from its **'dot-and-cross'** diagram. Try to relate the number of electron pairs in the examples below with the three-dimensional shape of each molecule.

molecule	$BeCl_2$	BF_3	CH_4	PCl_5	SF_6
dot-and-cross diagram					
electron pairs around central atom	2	3	4	5	6
shape and bond angle					
name of shape	linear	trigonal planar	tetrahedral	trigonal bipyramid	octahedral

Molecules with lone pairs

Each of the examples on page 61 are molecules with four electron pairs surrounding the central atom. The molecular shape will therefore be based upon a tetrahedron. However, a lone pair is closer to an atom than a bonded pair of electrons and has a larger repulsive effect than a bonded pair.

> **KEY POINT**
>
> The relative magnitudes of electron-pair repulsions are:
>
> **lone**-pair/**lone**-pair > **bonded**-pair/**lone**-pair > **bonded**-pair/**bonded**-pair.

Lone pairs distort the shape of a molecule and reduce the bond angle. Look at the examples below and notice how each lone pair decreases the bond-angle by 2.5°.

Always draw 3-D shapes – you may be penalised in exams otherwise.

molecule	CH_4	NH_3	H_2O
'dot-and-cross' diagram			
number of lone pairs	0	1	2
shape and bond angle	109.5°	107°	104.5°
name of shape	tetrahedral	pyramidal	non-linear

Molecules with double bonds

In molecules containing multiple bonds, each double bond is treated in the same way as a bonded pair. In the diagram of carbon dioxide below, each double bond is treated as a *'bonding region'*.

molecule	*'dot-and-cross' diagram*	*number of bonding regions*	*shape and bond angle*	*name of shape*
CO_2		2	180° $O=C=O$	linear

Progress check

1 For each of the following molecules, predict the shape and bond angles.

(a) BeF_2
(b) $AlCl_3$
(c) SiH_4
(d) H_2S
(e) PH_3
(f) CS_2
(g) SO_3
(h) SO_2

1 (a) linear, 180°
(b) trigonal planar, 120°
(c) tetrahedral, 109.5°
(d) non-linear, 104.5°
(e) pyramidal, 107°
(f) linear, 180°
(g) trigonal planar, 120°
(h) non-linear, 120°.

2.3 Electronegativity, polarity and polarisation

After studying this section you should be able to:

- *appreciate that many chemical bonds have bonding intermediate between ionic and covalent bonding*
- *describe electronegativity in terms of attraction for bonding electrons*
- *understand the nature of polarity in molecules of covalent compounds*

Ionic or covalent?

AQA	M1	OCR	M1
EDEXCEL	M2	SALTERS	M2
CCEA	M1	WJEC	M2

Between the extremes of ionic and covalent bonding, there is a whole range of intermediate bonds, which have both ionic and covalent contributions.

An ionic bond with 100% ionic character would require the complete transfer of an electron from a metal atom to a non-metal atom. In practice, this never completely happens, although some compounds have close to 100% ionic character.

In a hydrogen molecule, H_2, the two hydrogen atoms are identical and form a covalent bond with 100% covalent character by equally sharing the bonded pair of electrons. However, in molecules such as HCl, the bonded pair of electrons is not shared equally.

An ionic bond is often formed with some covalent character:
- the electron transfer is incomplete
- there is a degree of electron sharing.

A covalent bond is often formed with some ionic character:
- the electrons are not equally shared
- there is a degree of electron transfer.

The sections that follow discuss intermediate bonding in covalent compounds.

Electronegativity

AQA	M1	OCR	M1
EDEXCEL	M2	SALTERS	M2
CCEA	M1	WJEC	M2

The nuclei of the atoms in a molecule attract the electron pair in a covalent bond.

> **KEY POINT**
>
> Electronegativity is a measure of the attraction of an atom in a molecule for the pair of electrons in a covalent bond.

The most electronegative atoms attract bonding electrons most strongly.

- In general, small atoms are electronegative atoms.
- The most electronegative atoms are those of highly reactive non-metallic elements (such as O, F and Cl).
- Reactive metals (such as Na and K) have the least electronegative atoms.

How is electronegativity measured?

Notice that the Noble Gases are not included. Although neon and helium atoms are smaller than those of fluorine, they do not form bonds and so they have no affinity for bonded electrons.

The 'Pauling scale' is often used to compare the electronegativities of different elements. The diagram below shows how the electronegativity of an element relates to its position in the Periodic Table. The numbers give the Pauling electronegativity values.

See how the Pauling values of electronegativity relate to the element's position in the Periodic Table.

electronegativity increases						
Li	Be	B	C	N	O	F
1.0	1.5	2.0	2.5	3.0	3.5	4.0
Na						Cl
0.9						3.0
K						Br
0.8						2.8

Fluorine, the most electronegative element, has small atoms with a greater attraction for the pair of electrons in a covalent bond than larger atoms.

> The greater the **difference** between electronegativities, the greater the **ionic** character of the bond.
>
> The greater the **similarity** in electronegativities, the greater the **covalent** character of the bond.

Polar and non-polar molecules

AQA	M1	OCR	MI
EDEXCEL	M2	SALTERS	M2
CCEA	M1	WJEC	M2

When bonding atoms are the same, the attraction for the bonded pair of electrons is the same and the bond is non-polar.

Hydrocarbons are non-polar because C and H have very similar electronegativities.

Non-polar bonds

A covalent bond is non-polar when:

- the bonded electrons are shared **equally** between both atoms
- the bonded atoms have similar electronegativities.

A covalent bond **must** be non-polar if the bonded atoms are the same, as in molecules of H_2 and Cl_2 shown here:

H_2 molecule

Cl_2 molecule

Polar bonds

In a polar covalent bond, the bonded electrons are attracted towards the more electronegative of the atoms.

The electronegativities of hydrogen halides are discussed in more detail on page 80.

A covalent bond is polar when:

- the electrons in the bond are shared **unequally** making a *polar bond*
- the bonded atoms are different, each has a different electronegativity.

In hydrogen chloride, HCl, the Cl atom is more **electronegative** than the H atom, making the H–Cl covalent bond polar.

The molecule is *polarised* with a small positive charge δ+ on the hydrogen atom and a small negative charge δ− on the chlorine atom.

$\delta+$ $\delta-$
H Cl
bonded pair of electrons attracted closer to chlorine

The hydrogen chloride molecule is *polar* with a *permanent dipole*.

Symmetrical and unsymmetrical molecules

If the molecule is symmetrical, any dipoles will cancel and the molecule will not have a permanent dipole. The diagram below shows why the symmetrical molecule, CCl_4, is non-polar.

- Each C–Cl bond is polar but the dipoles act in different directions.
- The overall effect is for the dipoles to cancel each other.
- ∴ CCl_4 is a non-polar molecule.

Dipoles in symmetrical molecules cancel.

CCl_4 is non-polar

Although each C–Cl bond is polar, the dipoles cancel

Progress check

1 Decide whether each molecule is polar and state which atom, if any, has the δ− charge.
(a) Br_2; (b) H_2O; (c) O_2; (d) HBr; (e) NH_3.

2 (a) Predict the shape of a molecule of BF_3 and of PF_3.
(b) Explain why BF_3 is non-polar whereas PF_3 is polar.

atoms are all on the same side of the molecule resulting in a permanent dipole.
(b) BF_3 has a symmetrical shape. Although each bond is polar, the dipoles cancel. PF_3 is polar. PF_3 is not symmetrical. The F

2 (a) BF_3: trigonal planar; PF_3: pyramidal.

1 (a) non-polar; (b) polar, $O^{\delta-}$; (c) non-polar; (d) polar, $Br^{\delta-}$; (e) polar, $N^{\delta-}$.

2.4 Intermolecular forces

After studying this section you should be able to:

- *understand the nature of van der Waals' forces (induced dipole-dipole interactions)*
- *understand the nature of permanent dipole-dipole interactions and hydrogen bonds*
- *describe the anomalous properties of water arising from hydrogen bonding.*

LEARNING SUMMARY

Van der Waals' forces or induced dipole-dipole forces

AQA	M1	OCR	M1
EDEXCEL	M2	SALTERS	M2
CCEA	M1	WJEC	M2

Van der Waals' forces (induced dipole-dipole forces) exist between all molecules whether polar or non-polar.

Induced dipole-dipole interactions are often referred to as van der Waals' forces.

What causes van der Waals' forces?

Van der Waals' forces are:

- weak intermolecular forces
- caused by attractions between very small dipoles in molecules.

The diagram below shows how these forces arise between single atoms in a noble gas.

Ionic and covalent bonds are of comparable strength.

Intermolecular forces are far weaker.

Movement of electrons produces an oscillating dipole

dipole oscillates and continually changes with time

An oscillating or instantaneous dipole is caused by an uneven distribution of electrons at an instant of time.

Oscillating dipole induces a dipole in a neighbouring molecule which is induced onto further molecules

induced dipoles attract one another

What affects the strength of van der Waals' forces?

Van der Waals' forces result from interactions of electrons between molecules.

The greater the number of electrons in each molecule:

- the larger the oscillating and induced dipoles
- the greater the attractive forces between molecules
- the greater the van der Waals' forces.

Van der Waals' forces increase in strength with increasing number of electrons.

This can be seen by comparing the boiling points of the Noble Gases.

Noble gas	b. pt./°C	number of electrons	trend
He	−269	2	• easier to distort electron clouds
Ne	−246	10	
Ar	−186	18	• induced dipoles increase
Kr	−152	36	
Xe	−107	54	• boiling point increases
Rn	− 62	86	

Permanent dipole-dipole interactions

AQA   M1   OCR
EDEXCEL   M3   SALTERS
CCEA   M1   WJEC   M2

Permanent dipole-dipole interactions are 'non-directional'.

They are simply weak attractions between dipole charges on different molecules.

The small δ+ and δ- charges on a polar molecule attract oppositely charged dipoles on another polar molecule.

This gives a weak *intermolecular force* called a permanent dipole-dipole interaction.

Example: Intermolecular forces between HCl molecules

Between hydrogen chloride molecules, there will be both:

- van der Waals' forces and
- permanent dipole-dipole interactions.

Although both are weak forces, the permanent dipole-dipole interactions are still stronger than the van der Waals' forces.

Hydrogen bonds

AQA   M1   OCR
EDEXCEL   SALTERS
CCEA   M1   WJEC   M2

A hydrogen bond is a special type of permanent dipole-dipole interaction found between molecules containing the following groups:

$$\overset{\delta+}{H}—\overset{\delta-}{O} \qquad \overset{\delta+}{H}—\overset{\delta-}{N} \qquad \overset{\delta+}{H}—\overset{\delta-}{F}$$

A hydrogen bond is a comparatively strong intermolecular attraction between:

- an electron deficient hydrogen atom, $H^{\delta+}$, on one molecule and
- a lone pair of electrons on a highly electronegative atom of F, O or N on another molecule.

Hydrogen bonding occurs between molecules such as H_2O:

Note the role of the lone pair – this is essential in hydrogen bonding.

A hydrogen bond is shown between molecules as a dashed line.

hydrogen bond formed by attraction between dipole charges on different molecules

Hydrogen bonding is especially important in organic compounds containing –OH or –NH bonds: e.g. alcohols, carboxylic acids, amines, amino acids.

Special properties of water arising from hydrogen bonding

A hydrogen bond has only one-tenth the strength of a covalent bond. However, hydrogen bonding is strong enough to have significant effects on physical properties, resulting in some unexpected properties for water.

The solid (ice) is less dense than the liquid (water)

Solids are usually denser than liquids – but ice is less dense than water.

- Particles in solids are **usually** packed closer together than in liquids.
- Hydrogen bonds hold water molecules apart in an open lattice structure.

Therefore ice is less dense than water.

When water changes state, the covalent bonds between the H and O atoms in an H_2O molecule are strong and do **not** break.

It is the intermolecular forces that break.

The diagram below shows how the open lattice of ice collapses on melting.

tetrahedral open lattice in ice

hydrogen bonds break

MELTING OF ICE

ice lattice collapses: molecules move closer together

Ice has a relatively high melting point, and water a relatively high boiling point

- There are relatively strong hydrogen bonds between H_2O molecules.
- The hydrogen bonds are extra forces over and above van der Waals' forces.
- These extra forces result in higher melting and boiling points than would be expected from just van der Waals' forces.
- When the ice lattice breaks, hydrogen bonds are broken.

Other properties

The extra intermolecular bonding from hydrogen bonds also explains the relatively high surface tension and viscosity in water.

Evidence for hydrogen bonds

The boiling points of the Group 6 and Group 7 hydrides are shown below.

	Group 6		Group 7	
hydride	boiling point / K		hydride	boiling point / °C
H_2O	373		HF	293
H_2S	212		HCl	188
H_2Se	232		HBr	206
H_2Te	271		HI	238

The boiling points increase from $H_2S \longrightarrow H_2Te$ and from $HCl \longrightarrow HI$:

- the number of electrons in molecules increase
- van der Waals' forces increase
- more energy is required to break the intermolecular bonds to vaporise the hydrides.

Without hydrogen bonding, H_2O would be a gas at room temperature and pressure.

The boiling point of the first hydride in each group is higher than expected. This provides evidence that there are some extra forces acting between the molecules that must be broken to boil each hydride. These extra forces are hydrogen bonds.

Progress check

1 Which of the following molecules have hydrogen bonding: H_2O, H_2S, CH_4, CH_3OH, NO_2?

2 Draw diagrams showing hydrogen bonding between:
(a) 2 molecules of ammonia
(b) 1 molecule of water and 1 molecule of ethanol.

1 H_2O, CH_3OH.

2 (a)

(b)

2.5 Bonding, structure and properties

After studying this section you should be able to:

- *describe the typical properties of an ionic compound in terms of its structure*
- *describe the typical properties of a covalent compound in terms of simple molecular and giant molecular structures*
- *describe the structure and associated properties of diamond and graphite*
- *understand the properties of metals in terms of metallic bonding*

Bonds and forces

AQA	M1	OCR	M1
EDEXCEL		SALTERS	
CCEA	M1	WJEC	M2

Bonding, structure and properties are all related.

A structure is held together by bonds and forces. The different types of bonds introduced in the last section are shown below.

These are very important and are needed to understand the links between bonding, structure and properties.

> **KEY POINT**
>
> Covalent bonds act between atoms.
> Ionic bonds act between ions.
> Metallic bonds act between positive ions and electrons.
> Hydrogen bonds act between polar molecules.
> Dipole-dipole interactions act between polar molecules.
> Van der Waals' forces act between molecules.

Ionic, covalent and metallic bonds are of comparable strength.

Intermolecular forces are much weaker, and their relative strengths are compared with the strength of covalent bonds in the table below.

type of bond	bond enthalpy / kJ mol⁻¹
covalent bond	200–500
hydrogen bond	5–40
van der Waals' forces	~2

The properties of a substance depend upon its **bonding** and **structure**.

Properties of ionic compounds

AQA	M1	OCR	M1
EDEXCEL		SALTERS	M1, M2
CCEA	M1	WJEC	M2

Ionic compounds form giant ionic lattices with each ion surrounded by ions of the opposite charge. The ions are held together by **strong** electrostatic attraction between positive and negative ions. (See page 52.)

Ionic compounds are solids at room temperature.

The strong forces between ions result in high melting and boiling points.

High melting point and boiling point

- High temperatures are needed to break the strong electrostatic forces holding the ions rigidly in the solid lattice.

Therefore ionic compounds have high melting and boiling points.

Electrical conductivity

Ionic compounds conduct only when ions are free to move – when molten or in aqueous solution.

*In the **solid** lattice,*

- the ions are in a fixed position and there are no mobile charge carriers.

Therefore an ionic compound is a **non-conductor** of electricity in the solid state.

*When **melted** or **dissolved** in water,*

- the solid lattice breaks down
- the ions are now free to move as mobile charge carriers.

Therefore an ionic compound is a **conductor** of electricity in liquid and aqueous states.

Solubility

- The ionic lattice often dissolves in **polar** solvents (e.g. water).
- Polar water molecules break down the lattice and surround each ion in solution as shown below for sodium chloride.

The solubility of ionic compounds in water increases with increasing temperature.

 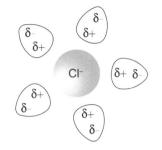

- Water molecules attract Na^+ and Cl^- ions.
- Lattice breaks down as it dissolves.
- Water molecules surround ions.

Na^+ attracts $\delta-$ partial charges on the O atoms of water molecules.

Cl^- attracts $\delta+$ partial charges on the H atoms of water molecules.

Solubility curves

WJEC   M2

The solubility of a solute in a solvent typically depends on temperature.

- For most solid solutes, the solubility increases with temperature.

A solubility curve shows how the solubility of a substance varies with temperature.

A solution is saturated when no more of the solute salt dissolves in a solvent at that particular temperature.

Solubility is usually measured as the mass of solute that saturates 100 grams of water at a particular temperature.

The diagram below shows solubility curves of NaCl and KNO_3.

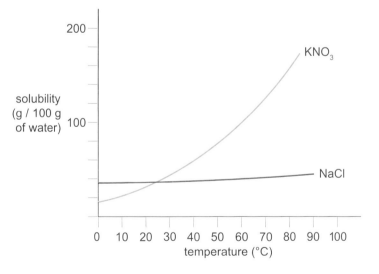

Soluble salts can be recovered from an aqueous solution by crystallisation.

- A solution is saturated with the salt by evaporation of the solvent using heat.
- On cooling, the water can no longer dissolve all of the salt at the lower temperature and the excess salt crystallises out.

For the two salts above, far more KNO_3 can be recovered on cooling because the difference in solubility with temperature is much greater than for NaCl.

Properties of covalent compounds

AQA	M1	OCR	
EDEXCEL	M2	SALTERS	
CCEA	M1	WJEC	M2

Elements and compounds with covalent bonds have either of two structures:
- a simple molecular structure
- a giant covalent structure.

Simple molecular structures, e.g. iodine, I_2

Simple molecular structures form solid lattices with small molecules, such as Ne, H_2, O_2, N_2, held together by **weak** intermolecular forces.

Intermolecular forces are weak.

Only the intermolecular forces break when a simple molecule melts or boils.

A common mistake in exams is to confuse intermolecular forces with covalent bonds.

Low melting point and boiling point

- Low temperatures provide sufficient energy to break the weak intermolecular forces between molecules.

Therefore simple molecular structures have low melting and boiling points.

The simple molecular structure of solid I_2

Giant structure:

high melting point.

Simple molecular structure:

low melting point.

- When the I_2 lattice breaks down, only the weak van der Waals' forces between the I_2 molecules break.
- In the I_2 molecule, the covalent bond, I–I, is strong and does **not** break when the lattice breaks down.

Strong covalent bonds hold together each I_2 molecule

Weak van der Waals' forces between I_2 molecules

Electrical conductivity

- There are no mobile charged particles.

Therefore simple molecular structures are **non-conductors** of electricity.

Solubility

- Van der Waals' forces form between a simple molecular structure and a non-polar solvent, such as hexane. These weaken the structure.

Therefore simple molecular structures are often soluble in **non-polar** solvents (e.g. hexane).

Giant covalent structures, e.g. carbon (diamond and graphite)

Diamond, graphite and SiO_2 are examples of giant covalent lattices.

Giant covalent structures have thousands of atoms bonded together by **strong** covalent bonds. This type of structure is known by a variety of names: a giant **covalent** lattice, a giant **molecular** lattice or a giant **atomic** lattice.

Covalent bonds are strong.

The covalent bonds break when a giant molecular structure melts or boils – this only happens at high temperatures.

High melting point and boiling point

- High temperatures are needed to break the strong covalent bonds in the lattice.

Therefore giant covalent structures have high melting and boiling points.

Electrical conductivity

- Except for graphite (see page 70), there are no mobile charged particles.

Therefore giant covalent structures are **non-conductors** of electricity.

Carbon can also form other structures: fullerenes and nanotubes.

Solubility

- The strong covalent bonds in the lattice are too strong to be broken by either polar **or** non-polar solvents.

Therefore giant covalent structures are insoluble in polar *and* non-polar solvents.

Silicon (IV) oxide, SiO_2, has a similar tetrahedral structure to diamond.

Comparison between the properties of diamond and graphite

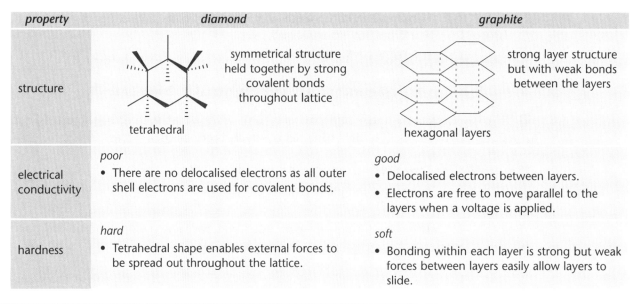

property	diamond	graphite
structure	*tetrahedral* — symmetrical structure held together by strong covalent bonds throughout lattice	*hexagonal layers* — strong layer structure but with weak bonds between the layers
electrical conductivity	*poor* • There are no delocalised electrons as all outer shell electrons are used for covalent bonds.	*good* • Delocalised electrons between layers. • Electrons are free to move parallel to the layers when a voltage is applied.
hardness	*hard* • Tetrahedral shape enables external forces to be spread out throughout the lattice.	*soft* • Bonding within each layer is strong but weak forces between layers easily allow layers to slide.

Properties of metals

AQA M1 OCR M1
EDEXCEL S1 SALTERS M1, M2
CCEA M1 WJEC M2

Metals have a giant metallic lattice structures held together by **strong** electrostatic attractions between positive ions and negative electrons.

High melting point and boiling point

• Generally high temperatures are needed to separate the ions from their rigid positions within the lattice.

Therefore most metals have high melting and boiling points.

When a metal conducts electricity, only the electrons move.

Good thermal and electrical conductivity

• The existence of mobile, delocalised electrons allows metals to conduct heat and electricity well, even in the solid state.
• The electrons are free to flow between positive ions.
• The positive ions do **not** move.

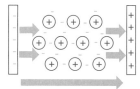

drift of delocalised electrons across potential difference

Modern materials

EDEXCEL M2
WJEC M2

Modern research is investigating new types of materials for a whole range of uses. These include nanomaterials and smart materials.

Nanomaterials

Nanomaterials have:

1 nanometre,
1 nm = 10^{-9} m.

• dimensions on a molecular scale: 1 – 100 nm
• unique properties arising from their nanoscale.

A single-walled carbon nanotube is a single layer of graphite rolled up into a cylinder with a diameter of approximately 1 nm (10^{-9} m). Nanotubes have the potential to carry drugs into cells but extensive uses of nanomaterials have yet to be realised.

Smart materials (WJEC only)

A smart material changes properties with a change in conditions (e.g. temperature, moisture, pH, or electric and magnetic fields) and this is often caused by a change in structure.

Summary of properties from structure and bonding

structure	bonding	melting pt / boiling pt	reason	electrical conductivity	reason	solubility	reason
giant ionic	ionic bonds throughout structure	high	strong electrostatic attraction between oppositely charged ions	poor when solid good when aqueous or molten	ions in a fixed position in lattice lattice has broken down: mobile ions	good in polar solvents, e.g. water	attraction between ionic lattice and polar solvent
simple molecular	covalent bonds **within** molecules, van der Waals' forces **between** molecules	low	weak van der Waals' forces between molecules	poor	no mobile charged particles (electrons or ions)	good in non-polar solvents	van der Waals' forces between molecular structure and solvent
giant covalent	covalent bonds throughout structure	high	strong covalent bonds between atoms	poor	no mobile charged particles (electrons or ions)	poor	forces within lattice too strong to be broken by solvents
hydrogen bonded	hydrogen bonds between molecules	low *but* higher than expected	dipole-dipole attractions between molecules	poor	no mobile charged particles (electrons or ions)	good in polar solvents, e.g. water	attraction between dipoles
giant metallic	metallic bonds throughout structure	usually high	strong electrostatic attractions between ions and electrons	good	mobile electrons, even in solid state	poor	forces within lattice too strong to be broken by solvents

> A high melting point is the result of any giant structure, bonded together with strong forces.
>
> Giant structures can be ionic, covalent or metallic.

KEY POINT

Progress check

1 For (a) MgO; (b) CH_4; (c) SiO_2, state the structure and explain the following physical properties: melting and boiling points, electrical conductivity, solubility.

1 (a) giant ionic: high m.pt/b.pt. – strong forces between ions.
Non-conductor when solid – ions fixed in lattice. Conductor when molten or dissolved in water – ions are able to move.
Soluble in water – dipole on water attracted to ions.
(b) simple molecular: low m. pt/b. pt. – weak van der Waals' forces between molecules.
Non-conductor – no mobile charge carriers, electrons localised in covalent bonds.
Soluble in non-polar solvents – van der Waals' forces in solvent interact with van der Waals' forces in CH_4.
(c) giant molecular: high m. pt. – strong covalent bonds between atoms.
Non-conductor – no mobile charge carriers, electrons localised in covalent bonds.
Insoluble in all solvents – strong covalent bonds are too strong to be broken by either polar or non-polar solvents.

2.6 The modern Periodic Table

After studying this section you should be able to:

- *describe the Periodic Table in terms of atomic number, periods and groups*
- *classify the Periodic Table into s, p, d and f blocks*
- *describe and explain periodic trends in atomic radii, ionisation energies, melting and boiling points, and electrical conductivities*

Arranging the elements

AQA	M1	OCR	
EDEXCEL	M1	SALTERS	
CCEA	M1	WJEC	M2

In the Periodic Table:

- Elements are arranged in order of increasing atomic number
- Groups are vertical columns including elements with similar properties
- Periods are horizontal rows across which there is a trend in properties.

Key areas in the Periodic Table

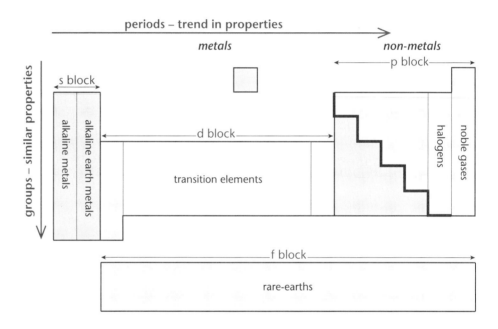

- The diagonal stepped line separates metals (to the left) from non-metals (to the right).
- Elements close to this line, such as silicon and germanium, are called **semi-metals** or **metalloids**. They show properties intermediate between those of a metal and a non-metal.
- The four blocks (s, p ,d and f) show the sub-shell being filled.

The trend in properties across a period is repeated across each period – this is called *periodicity*.

Periodicity

Periodicity is the periodic trend in properties, repeated across each period.

| e.g. | Period 2 | METAL | $\longrightarrow$ | NON-METAL |
| | Period 3 | METAL | $\longrightarrow$ | NON-METAL |

This periodicity of properties means that predictions can be made about the likely properties of an element and its compounds.

Group 4

C
Si
Ge
Sn
Pb

non-metal
to metal

Note, however, the metal/non-metal divide. On descending the Periodic Table, the changeover from metal to non-metal takes place further to the right. For example at the top of Group 4 carbon is a non-metal whereas, at the bottom of Group 4, tin and lead are metals. This means that subtle trends in properties take place down a group.

Trends in atomic radii and first ionisation energies

AQA	M1	OCR	
EDEXCEL	M1	SALTERS	
CCEA	M1	WJEC	M2

Ionisation energies measure the ease with which an atom of an element loses electrons (see page 26).

Across a period

Across a period, the atomic radius decreases. The trend in atomic radius across Period 2 is shown below:

number of protons increases

radius decreases

Atomic radius decreases across a period.

First ionisation energy increases across a period.

Across a period:

- the **nuclear charge increases**
- outer electrons are being added to the same shell
- the attraction between the nucleus and outer electrons increases
- the atomic radius **decreases**
- the first ionisation energy **increases**.

Down a group

Atomic radius increases down a group.

First ionisation energy decreases down a group.

Down a group:

- extra shells are added that are **further** from the nucleus
- there are more shells between the outer electrons and the nucleus leading to **greater shielding** of the nuclear charge
- the attraction between the nucleus and the outer electrons decreases
- the atomic radius **increases**
- the first ionisation energy **decreases**.

shells increase

shielding increases

atomic radius increases

Down a group, the nuclear charge also increases but this is more than compensated by the increase in atomic radius and shielding.

> **KEY POINT**
>
> Across a Period, nuclear charge is the key factor.
> Down a Group, atomic radius and shielding are the key factors (see also ionisation energies, page 26).

Trends in melting and boiling points

AQA	M1	OCR	M1
EDEXCEL	M1	SALTERS	M1
CCEA	M1	WJEC	M2

Trends in melting and boiling point provide information about structure: giant or simple molecular.

The graphs below show the variation in boiling points across Period 2 and Period 3 of the Periodic Table.

Period 2

Period 3

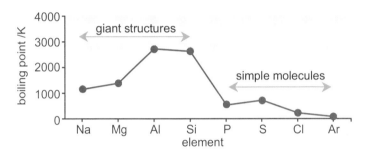

Across a period:

* the boiling point increases from Group 1 to Group 4
* there is a sharp decrease in boiling point between Group 4 and Group 5
* the boiling points are comparatively low from Group 5 to Group 8.

> **The sharp decrease in boiling point marks a change from giant structures to simple molecular structures.**
>
> **KEY POINT**

Note the periodicity in boiling point: the trend for Period 2 is repeated across Period 3. The trend in melting points is similar.

More details of the structures of these elements are shown below.

Notice the changes in the molecular formulae of P_4, S_8, Cl_2 and Ar. This shows up well in the graph of boiling points across Period 3.

With more atoms (and consequently electrons) in each molecule, the forces between molecules (van der Waals' forces) increase.

When a giant structure is melted or boiled:

* strong forces are broken
* a large input of energy is required
* the melting and boiling points are high.

When a simple molecular structure is melted or boiled:

* weak forces between molecules are broken
* a relatively small input of energy is required
* the melting and boiling points are low.

Comparison of the boiling points of the metals

Note the increase in boiling points of the metals Na → Al in Period 3.

* The number of delocalised electrons in the lattice increases.
* The charge on each cation increases.
* This results in increasing attractive forces within the metallic lattice.

The diagram below compares the attractive forces between atoms of sodium, magnesium and aluminium.

See also Metallic
bonding, pages 56, 70.

attraction increases: boiling point increases

The increasing number of delocalised electrons in the lattice also explains the increase in electrical conductivity from Na → Al.

Progress check

1 Using ideas about nuclear charge, attraction and shells, explain the trend in atomic radii across a period and down each group.
2 Why is the melting point of carbon much higher than that of nitrogen?

2 Carbon has a giant molecular structure; between the atoms there are strong covalent bonds which break on melting. Nitrogen has a simple molecular structure; between the molecules the molecules the weak van der Waals' forces that break on melting.

1 Period: the nuclear charge increases as electrons are being added to the same shell. The attraction between the nucleus and outer electrons increases ∴ the atomic radius decreases across a period. Group: extra shells are added that are further from the nucleus leading to an increased shielding of the outer electrons from the nucleus. Both these factors lead to less attraction between the nucleus and the outer electrons ∴ the atomic radius increases down a group.

2.7 The s-block elements: Group 1 and Group 2

After studying this section you should be able to:

LEARNING SUMMARY

- describe the characteristic properties of the s-block elements
- understand how flame tests can be used to identify s-block cations
- describe the reactions of the s-block elements with oxygen and with water
- recall some of the reactions of s-block oxides
- recall the relative solubilities of the Group 2 hydroxides and sulfates
- describe and explain the thermal stability of the s-block carbonates and nitrates.

General properties

AQA	M2	OCR	M1
EDEXCEL	M2	SALTERS	M1 M2
CCEA	M2	WJEC	M2

The elements in Group 1 and Group 2 have hydroxides that are alkaline and their common names reflect this.

- The Group 1 elements are the **Alkali Metals**.
- The Group 2 elements are the **Alkaline Earth Metals**.

Electron configuration

The elements in Group 1 and Group 2 have their highest energy electrons in an s sub-shell and these two groups are known as the *s-block elements*.

Each Group 1 element has:
- one electron more than the electron configuration of a noble gas
- an outer s sub-shell containing one electron.

Each Group 1 element reacts in a similar way as each atom has 1 electron in the outer shell.

Electron configuration of the s-block elements

Group 1		Group 2	
Li	[He] $2s^1$	Be	[He] $2s^2$
Na	[Ne] $3s^1$	Mg	[Ne] $3s^2$
K	[Ar] $4s^1$	Ca	[Ar] $4s^2$
Rb	[Kr] $5s^1$	Sr	[Kr] $5s^2$
Cs	[Xe] $6s^1$	Ba	[Xe] $6s^2$
Fr	[Rn] $7s^1$	Ra	[Rn] $7s^2$

The s-block elements
Group 1
Group 2

Each Group 2 element has:
- two electrons more than the electron configuration of a noble gas
- an outer s sub-shell containing **two** electrons.

Each Group 2 element reacts in a similar way as each atom has 2 electrons in the outer shell.

Physical properties

Group 1
- They are soft metals and can be cut with a knife.
- They have low melting and boiling points.
- They have low densities: Li, Na and K all float on water.
- They have colourless compounds.

Densities of s-block elements

Group 1 Density/g cm^{-3}		Group 2 Density/g cm^{-3}	
Li	0.53	Be	1.85
Na	0.97	Mg	1.74
K	0.86	Ca	1.54
		Sr	2.60
		Ba	3.51

Densities for comparison:
H_2O, 1.00 g cm^{-3}; Fe: 7.86 g cm^{-3}

Group 2

- They have reasonably high melting and boiling points.
- They have low densities although not as low as those in Group 1.
- They have colourless compounds.

Melting points show a general decrease down Group 1 and Group 2:

- the atoms increase in size
- the outer electrons are further away from the nucleus
- the attractive force of nuclei on the outer electrons decreases.

Melting points of s-block elements			
Group 1 Melting point/°C		Group 2 Melting point/°C	
Li	181		
Na	98	Mg	649
K	63	Ca	839
		Sr	769
		Ba	725

Melting points for comparison:
Fe: 1535°C; Cu 1083°C

Flame tests

EDEXCEL   M2   SALTERS
CCEA   M1, M2   WJEC   M2

When heated in a flame, compounds of Groups 1 and 2 show characteristic flame colours, allowing different cations to be identified.

cation	Li^+	Na^+	K^+	Ca^{2+}	Ba^{2+}	Sr^{2+}
flame colour	red	yellow	lilac	brick-red	green	red

The characteristic flame colours arise from electron transitions within the atom (see pages 28–29).

> Heat energy from the flame excites an electron to a higher energy level.
> The electron returns to a lower energy level.
> The excess energy is emitted as light energy.
> The frequency of the flame colour depends upon the difference between the energy levels involved.
>
> **KEY POINT**

Reactivity of the s-block elements

AQA   M2   OCR
EDEXCEL   M2   SALTERS
CCEA   M2   WJEC   M2

Li   Be
Na   Mg
K   Ca   reactivity
Rb   Sr   **increases**
Cs   Ba
Fr   Ra

Down Group 1 and Group 2

atomic radius increases

first ionisation energy decreases

See page 73 for detailed explanation.

The elements in the s-block are the most reactive metals and are strong reducing agents.

Group 1 elements are oxidised in reactions, each atom losing one electron from its outer s sub-shell to form a 1+ ion (+1 oxidation state):

$$M \longrightarrow M^+ + e^-$$

Group 2 elements are oxidised in reactions, each atom losing two electrons from its outer s sub-shell to form a 2+ ion (+2 oxidation state):

$$M \longrightarrow M^{2+} + 2e^-$$

Reactivity **increases** down each group reflecting the increasing ease of losing electrons. The ionisation energy of the metal is an important factor in this process.

> Within each group, the elements become **more** reactive as the group descends.
> First ionisation energy **decreases** down the group.
>
> **KEY POINT**

The Group 2 elements

AQA   M2   OCR
EDEXCEL   SALTERS
CCEA   M2   WJEC   M2

Reaction with oxygen

The Group 2 elements react vigorously with oxygen. Each element forms the expected ionic oxide with the general formula, $M^{2+}O^{2-}$:

e.g.
$$2Ca(s) + O_2(g) \longrightarrow 2CaO(s)$$

Reactions of Group 2 elements with oxygen and water are redox reactions.

Mg forms MgO with steam.

Action of water

Reactivity increases down the group reflecting the increasing ease with which electrons can be lost.

- Mg reacts very slowly with water, forming the **hydroxide** and hydrogen:

$$Mg(s) + 2H_2O(l) \longrightarrow Mg(OH)_2(aq) + H_2(g)$$

- With steam, reaction is much quicker forming the **oxide** and hydrogen:

$$Mg(s) + H_2O(g) \longrightarrow MgO(s) + H_2(g)$$

- Further down the group from calcium, each metal reacts vigorously with water:

$$Ca(s) + 2H_2O(l) \longrightarrow Ca(OH)_2(aq) + H_2(g)$$

Group 2 oxides and hydroxides

Reaction with water

The Group 2 oxides form alkaline solutions with water:

$Mg(OH)_2$ solubility
$Ca(OH)_2$ increases
$Sr(OH)_2$ alkalinity
$Ba(OH)_2$ increases

Thus solid barium hydroxide is reasonably soluble in water to form a strong alkaline solution with greater $OH^-(aq)$ concentration.

e.g. $$MgO(s) + H_2O(l) \longrightarrow Mg^{2+}(aq) + 2OH^-(aq)$$

- Magnesium hydroxide, $Mg(OH)_2(s)$ is only slightly soluble in water and so the resulting solution is only a dilute alkali.
- The solubility in water increases down the group and resulting solutions are more alkaline:

$$Ba(OH)_2(s) + aq \longrightarrow Ba^{2+}(aq) + 2OH^-(aq)$$

The alkalinity of the Group 2 hydroxides is exploited commercially. Some indigestion tablets contain $Mg(OH)_2$ to neutralise excess acid in the stomach. Farmers add $Ca(OH)_2$ as 'lime' to neutralise acid soils.

Reaction with acids

- The Group 2 oxides behave as bases and are neutralised by acids, such as $HCl(aq)$, forming salts and water:

e.g. $$MgO(s) + 2HCl(aq) \longrightarrow MgCl_2(aq) + H_2O(l)$$

In medicine, barium sulfate is used as a 'barium meal' which shows up imperfections in the gut when exposed to X-rays. Barium compounds are extremely poisonous in solution but the insolubility of $BaSO_4$ is such that no harm is caused to the patient by this treatment.

Solubility of Group 2 sulfates (AQA and Edexcel only)

Magnesium sulfate $MgSO_4$ is very soluble in water but the solubility decreases as the group is descended. The trend is so marked that barium sulfate $BaSO_4$ is virtually insoluble.

In the laboratory, the precipitation of $BaSO_4$ is used to test for the sulfate ion. A solution of a soluble barium salt (usually $BaCl_2(aq)$ or $Ba(NO_3)_2$) is added to a solution of a substance in dilute nitric acid. In the presence of aqueous sulfate ions, a dense white precipitate of barium sulfate is formed.

$$Ba^{2+}(aq) + SO_4^{2-}(aq) \longrightarrow BaSO_4(s)$$

$MgSO_4$
$CaSO_4$ solubility
$SrSO_4$ **decreases**
$BaSO_4$

The thermal stability of s-block compounds

| EDEXCEL | M2 | OCR | M1 |
| CCEA | M2 | SALTERS | M1 |

The s-block carbonates

Group 2 carbonates

The Group 2 carbonates are decomposed by heat:

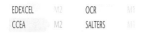

$MgCO_3$ ease of
$CaCO_3$ thermal
$SrCO_3$ decomposition
$BaCO_3$ **decreases**

e.g. $$MgCO_3(s) \longrightarrow MgO(s) + CO_2(g)$$

The Group 2 carbonates become more difficult to decompose as the group is descended.

Reason for trend (Edexcel and CCEA only)

The decomposition of the Group 2 carbonates is caused by the polarising effect of the Group 2 cation:

carbon–oxygen bond is weakened

It is increasingly more difficult to decompose the carbonate as the group is descended. The polarising power of the metal cation decreases as the charge : size ratio decreases (see also pages 58–59).

- As the size of the cation increases, the polarising effect becomes less.
- There is less weakening of the C–O bond.
- It becomes harder to decompose the carbonate.

thermal stability of carbonates increases

increased ionic size gives a less concentrated charge and less polarisation of the carbonate anion

> Polarisation occurs when a small cation distorts the electron shells of a larger anion.
>
> The polarising effect is greatest between a small densely charged cation and a large anion.

> Group 2 hydroxides show a similar trend – thermal stability increases down the group.

Group 1 carbonates (Edexcel only)

> The Group 1 carbonates are thermally stable (except Li_2CO_3)

The cations in the Group 1 carbonates are larger than those of the Group 2 carbonates and their polarising power is less. The Li^+ ion, however, is comparable in size with the Mg^{2+} ion and lithium carbonate is also decomposed by heat:

$$Li_2CO_3(s) \longrightarrow Li_2O(s) + CO_2(g)$$

The remaining Group 1 carbonates are **not** decomposed by heat.

The s-block nitrates (Edexcel only)

Group 2 nitrates

The Group 2 nitrates are decomposed by heat, releasing NO_2 and O_2 gases:

e.g. $\qquad Ca(NO_3)_2(s) \longrightarrow CaO(s) + 2NO_2(g) + \frac{1}{2}O_2(g)$

As with the Group 2 carbonates, the nitrates become more thermally stable as the group is descended. This again reflects the increasing size of the cations and the resultant decrease in polarising power.

Group 1 nitrates

> With heat, Li_2CO_3 and $LiNO_3$ behave similarly to the carbonates and nitrates of magnesium. This is known as a *diagonal relationship*. You can see why by comparing the positions of Li and Mg in the Periodic Table.

The Group 1 nitrates decompose with heat releasing only oxygen gas:

e.g. $\qquad NaNO_3(s) \longrightarrow NaNO_2(s) + \frac{1}{2}O_2(g)$

However, as with the carbonates above, the increasing polarising power of the smaller Li^+ ion makes lithium nitrate less thermally stable and lithium nitrate is decomposed by heat in a similar way to the Group 2 nitrates:

$$2LiNO_3(s) \longrightarrow Li_2O(s) + 2NO_2(g) + \frac{1}{2}O_2(g)$$

Progress check

1 Write down equations for the following reactions:
  (a) barium with water (b) calcium oxide with nitric acid.
2 Identify the oxidation number changes taking place during the thermal decomposition of sodium nitrate:
$$2NaNO_3(s) \longrightarrow 2NaNO_2(s) + O_2(g)$$

2 N, +5 ⟶ +3, O, −2 ⟶ 0.
(b) CaO(s) + 2HNO₃(aq) ⟶ Ca(NO₃)₂(aq) + H₂O(l)
1 (a) Ba(s) + 2H₂O(l) ⟶ Ba(OH)₂(aq) + H₂(g)

2.8 The Group 7 elements and their compounds

After studying this section you should be able to:

- describe the characteristic properties of the Group 7 elements
- recall the relative reactivity of the halogens as oxidising agents
- recall the characteristic tests for halide ions
- describe the characteristics of hydrogen halides
- describe the reactions of halides with concentrated sulfuric acid
- describe the use of thiosulfate titrations.

General properties

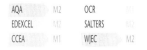

AQA	M2	OCR	
EDEXCEL		SALTERS	
CCEA	M1	WJEC	M2

The common name for the elements in Group 7 is the **halogens**.

Electron configuration

Each halogen has **seven** outer shell electrons; just one electron short of the electron configuration of a noble gas. The outer **p** sub-shell contains **five** electrons.

Electron configuration of the halogens

F [He] $2s^2 2p^5$
Cl [Ne] $3s^2 3p^5$
Br [Ar] $3d^{10} 4s^2 4p^5$
I [Kr] $4d^{10} 5s^2 5p^5$
At [Xe] $4f^{14} 5d^{10} 6p^2 6p^5$

Trend in physical states

The halogens exist as diatomic molecules, X_2. The boiling points of the halogens increase on descending the group.

- The physical states of the halogens at r.t.p. show the classic trend of gas $\longrightarrow$ liquid $\longrightarrow$ solid.
- On descending the group the number of electrons increases leading to an increase in van der Waals' forces between molecules.
- Therefore the boiling point increases.

F_2
Cl_2 boiling point **increases** down group
Br_2
I_2

Boiling points of the Halogens

	Boiling point/°C	State at r.t.p.
F_2	–188	gas
Cl_2	–35	gas
Br_2	59	liquid
I_2	184	solid
At_2	337	solid

Trend in electronegativity

AQA	M2	
CCEA	M1	

Electronegativity is a measure of the attraction of an atom for the pair of electrons in a covalent bond (see also page 62).

The hydrogen halides have polar molecules: $H^{\delta+}-X^{\delta-}$. The polarity decreases on descending the halogens and the order of polarity is:

most polar H–F > H–Cl > H–Br > H–I *least polar*

This trend in polarity results from the **decreasing** electronegativity of the halogen atom on descending the group.

- The atomic radius increases from F → Cl → I resulting in less nuclear attraction on the bonding electrons (this is despite the increase in nuclear charge!).
- There are more electron shells between the nucleus and the bonding electrons to shield the nuclear charge.

electronegativity of halogen **decreases**

polarity of H–X bond **decreases**

Electronegativity of the Halogens

	Pauling value
F	4.0
Cl	3.0
Br	2.8
I	2.5

The overall effect is that the smaller the halogen atom, the greater the nuclear attraction experienced by the bonding electrons. Thus the large electronegativity of fluorine results in a bond between hydrogen and fluorine that is more polar than between hydrogen and other halogens.

The relative reactivity of the halogens as oxidising agents

AQA	M2	OCR	
EDEXCEL	M2	SALTERS	
CCEA	M1	WJEC	M2

> **KEY POINT**
>
> The halogens are the most reactive non-metals and are strong oxidising agents.
>
> The halogens become **less** reactive as the group descends as their oxidising power decreases.

The halogens are reduced in reactions, each atom gaining one electron into a p sub-shell to form a –1 ion (–1 oxidation state):

e.g.
$$\tfrac{1}{2}F_2(g) \quad + \quad e^- \quad \longrightarrow \quad F^-(g)$$
$$[He]\ 2s^2 2p^5 \qquad\qquad\qquad [He]\ 2s^2 2p^6\ (or\ [Ne])$$

- An extra electron is captured by being attracted to the outer shell of an atom by the nuclear charge of an atom.

On descending the halogens:

F
Cl
Br   reactivity as
I   oxidising agent
At   **decreases**

- the atomic radii increase resulting in less nuclear attraction at the edge of the atom (despite the increase in nuclear charge).
- there are more electron shells between the nucleus and the edge of the atom to shield the nuclear charge.

 F

 Cl

Br

The overall effect is that most nuclear attraction is experienced at the edge of the small fluorine atoms. This attraction decreases down the group as the atoms get bigger.

Therefore the oxidising power of the halogens decreases down the group. Fluorine is the strongest oxidising agent and is able to attract an extra electron more strongly than other halogens.

Displacement reactions of the halogens

The decrease in reactivity as the group is descended can be demonstrated by displacement reactions of aqueous halides using Cl_2, Br_2 and I_2.

chlorine oxidises both Br^- and I^-:
$$Cl_2(aq) + 2Br^-(aq) \longrightarrow 2Cl^-(aq) + Br_2(aq)$$
$$Cl_2(aq) + 2I^-(aq) \longrightarrow 2Cl^-(aq) + I_2(aq)$$

bromine oxidises I^- only:
$$Br_2(aq) + 2I^-(aq) \longrightarrow 2Br^-(aq) + I_2(aq)$$

iodine does not oxidise either Cl^- or Br^-

The formation of halogens in displacement reactions is identified by colours, which are more distinctive in organic solvents:

halogen	water	hexane
Cl_2	pale green	pale green
Br_2	orange	orange
I_2	brown	purple

Industrial extraction of bromine

The main source of bromine is as bromide ions, Br^- in sea water. Bromine is extracted by oxidising sea water with chlorine. Because chlorine is a stronger oxidising agent, it displaces the bromide ions using the principles of the displacement reaction.

Reactions of halogens with metals and ions

Reaction of halogens with non-metals (CCEA only)

The halogens react readily with many non-metals.

Chlorine reacts explosively with hydrogen in the presence of sunlight.

$$H_2(g) + Cl_2(g) \longrightarrow 2HCl(g)$$

With phosphorus, trihalides are formed.

$$P_4(s) + 6Cl_2(g) \longrightarrow 4PCl_3(l)$$

Reaction of halogens with metals (Edexcel and CCEA only)

The halogens oxidise many metals forming ionic chlorides.

For example, sodium and magnesium are oxidised when heated with halogens.

$$2Na(s) + Cl_2(g) \longrightarrow 2NaCl(s)$$

$$Mg(s) + Cl_2(g) \longrightarrow MgCl_2(s)$$

Reaction of halogens with ions (Edexcel and CCEA only)

The halogens oxidise some ions to higher oxidation states.

For example, chlorine oxidises aqueous Fe(II) ions to Fe(III) ions.

$$2Fe^{2+}(g) + Cl_2(g) \longrightarrow 2Fe^{3+}(aq) + 2Cl^-(aq)$$

$$+2 \xrightarrow{\ +1\ } +3$$

$$0 \xrightarrow{\ -1\ } -1$$

Testing for halide ions

AQA	M2	OCR	M1
EDEXCEL	M2	SALTERS	M2
CCEA	M1	WJEC	M2

In sunlight, silver halides are reduced to silver. This reaction formed the basis of old photographic films.

Addition of aqueous silver ions (using $AgNO_3(aq)$) to a solution of halide ions in dilute nitric acid produces coloured precipitates that have different solubilities in aqueous ammonia.

fluoride: $Ag^+(aq) + F^-(aq) \not\longrightarrow$ no precipitate

chloride: $Ag^+(aq) + Cl^-(aq) \longrightarrow AgCl(s)$ white precipitate, soluble in dilute $NH_3(aq)$

bromide: $Ag^+(aq) + Br^-(aq) \longrightarrow AgBr(s)$ cream precipitate, soluble in conc. $NH_3(aq)$

iodide: $Ag^+(aq) + I^-(aq) \longrightarrow AgI(s)$ yellow precipitate, insoluble in conc. $NH_3(aq)$

Hydrogen halides

EDEXCEL	M2	CCEA	M1

Hydrogen halides react with ammonia gas to form ammonium salts:

$HCl(g) + NH_3(g) \rightarrow NH_4Cl(s)$

The hydrogen halides, HX, are colourless gases that are very soluble in water. They form strong acid solutions when dissolved in water. For example, hydrochloric acid forms when hydrogen chloride gas dissolves in water:

$$HCl(g) + aq \longrightarrow H^+(aq) + Cl^-(aq)$$

- As the group is descended, the H–X bond enthalpy decreases and the bond breaks more readily, resulting in an increase in acidity down the group.

The order of acidity is HI > HBr > HCl.

Reactions of halides with concentrated sulfuric acid

AQA	M2	CCEA	M1
EDEXCEL	M2		

HCl
HBr
HI

increased strength as reducing agents

HCl does **not** reduce H_2SO_4

Concentrated sulfuric acid is an oxidising agent. Halide salts react with concentrated sulfuric acid producing a range of products depending on the halide used. This difference is caused by the increasing reducing power of the hydrogen halides as the group is descended.

NaCl and H_2SO_4

Hydrogen chloride gas, HCl, is formed.

$$NaCl(s) + H_2SO_4(l) \longrightarrow NaHSO_4(s) + HCl(g)$$

The HCl formed is not a sufficiently strong reducing agent to reduce the sulfuric acid. No redox reaction takes place.

NaBr and H_2SO_4

Hydrogen bromide gas, HBr, is initially formed.

$$NaBr(s) + H_2SO_4(l) \longrightarrow NaHSO_4(s) + HBr(g)$$

Some of the hydrogen bromide reduces the sulfuric acid with the formation of sulfur dioxide and orange bromine fumes.

$$2HBr(g) + H_2SO_4(l) \longrightarrow SO_2(g) + Br_2(g) + 2H_2O(l)$$

$$+6 \xrightarrow{\ -2\ } +4 \qquad \qquad S \text{ reduced}$$

$$2 \times -1 \xrightarrow{\ +2\ } 0 \qquad \qquad Br \text{ oxidised}$$

HBr reduces H_2SO_4

$$H_2SO_4 \xrightarrow{\ HBr\ } Br_2 + SO_2$$

NaI and H_2SO_4

Hydrogen iodide gas, HI, is initially formed.

$$NaI(s) + H_2SO_4(l) \longrightarrow NaHSO_4(s) + HI(g)$$

Hydrogen iodide is a strong reducing agent and reduces the sulfuric acid to a mixture of reduced products including sulfur dioxide and hydrogen sulfide.

Reduction to SO_2 (ox no: +4)

$$2HI(g) + H_2SO_4(l) \longrightarrow SO_2(g) + I_2(s) + 2H_2O(l)$$

$$+6 \xrightarrow{\ -2\ } +4 \qquad \qquad S \text{ reduced}$$

$$2 \times -1 \xrightarrow{\ +2\ } 0 \qquad \qquad I \text{ oxidised}$$

HI reduces H_2SO_4 and SO_2

$$H_2SO_4 \xrightarrow{\ HI\ } I_2 + SO_2$$

$$SO_2 \xrightarrow{\ HI\ } I_2 + H_2S$$

Further reduction to H_2S (ox no: −2)

$$6HI(g) + SO_2(g) \longrightarrow H_2S(g) + 3I_2(s) + 2H_2O(l)$$

$$+4 \xrightarrow{\ -6\ } -2 \qquad \qquad S \text{ reduced}$$

$$6 \times -1 \xrightarrow{\ +6\ } 0 \qquad \qquad I \text{ oxidised}$$

Disproportionation

AQA   M2   OCR
EDEXCEL   SALTERS
CCEA   M1

> **KEY POINT**
>
> **Disproportionation** is a reaction in which the same element is both oxidised and reduced.

Disproportionation of chlorine in water

Chlorine reacts with water. This reaction is an example of *disproportionation* in which chlorine is both reduced (to chloride, Cl^-) and oxidised (to chlorate(I) ClO^-).

$$Cl_2(aq) + H_2O(l) \longrightarrow HClO(aq) + HCl(aq)$$

$$0 \xrightarrow{\ -1\ } -1 \qquad \qquad \textit{chlorine reduced}$$

$$0 \xrightarrow{\ +1\ } +1 \qquad \qquad \textit{chlorine oxidised}$$

A solution of chlorine in water is pale green, showing the presence of chlorine.

A small amount of chlorine is added to drinking water to kill bacteria that would make the water unsafe to drink. It has been claimed that chlorine treatment of water has done more to improve public health than any other treatment, preventing diseases such as cholera and typhoid. This is despite the toxicity of chlorine and possible risks from the formation of toxic chlorohydrocarbons by reaction with organic matter.

AQA ▸ M2   CCEA ▸ M1
EDEXCEL M2   OCR

In some areas fluoride ions are added to drinking water to reduce tooth decay.

Some people believe that fluoride ions pose other detrimental health risks and that the rights of the individual to use or reject fluoride have been compromised.

Disproportionation of chlorine in dilute aqueous alkalis

Dilute aqueous sodium hydroxide reacts with halogens when mixed at **room temperature**.

This reaction is another example of *disproportionation* in which chlorine is both reduced (to chloride, Cl^-) and oxidised (to chlorate(I) ClO^-).

$$Cl_2(aq) + 2NaOH(aq) \longrightarrow NaCl(aq) + NaClO(aq) + H_2O(l)$$

$$0 \xrightarrow{\quad -1 \quad} -1 \qquad \textit{chlorine reduced}$$

$$0 \xrightarrow{\quad +1 \quad} +1 \qquad \textit{chlorine oxidised}$$

The solution formed is the basis for common bleach.

Disproportionation of chlorine in concentrated aqueous alkalis (Edexcel and CCEA only)

Further disproportionation of the halogen takes place with **hot, concentrated** sodium hydroxide.

Iodine first undergoes disproportionation into iodide, I^- and iodate(I), IO^-.

$$I_2(aq) + 2NaOH(aq) \longrightarrow NaI(aq) + NaIO(aq) + H_2O(l)$$

Iodate(I) ions then undergo further disproportionation.

$$3NaIO(aq) \longrightarrow 2NaI(aq) + NaIO_3(aq)$$

$$2 \times +1 \xrightarrow{\quad -4 \quad} 2 \times -1 \qquad \textit{iodine reduced}$$

$$+1 \xrightarrow{\quad +4 \quad} +5 \qquad \textit{iodine oxidised}$$

Halogens from electrolysis

SALTERS

Halogens can be obtained by electrolysis of aqueous halide ions.

Halide ions are **oxidised**, discharging the halogen at the anode (positive electrode).

Electrolysis of brine (aqueous sodium chloride) is used for the industrial preparation of chlorine:

$$2Cl^-(aq) \longrightarrow Cl_2(g) + 2e^-$$

Iodine–thiosulfate titrations

EDEXCEL S2

Starch is a sensitive test for the presence of iodine. If it is added too early in this titration, the deep-blue colour perseveres and the end-point cannot be detected with accuracy.

Iodine-thiosulfate titrations can be used to determine iodine concentrations, either directly or by liberating iodine using a more powerful oxidising agent.

Iodine oxidises thiosulfate ions:

$$I_2(aq) + 2S_2O_3^{2-}(aq) \longrightarrow 2I^-(aq) + S_4O_6^{2-}(aq)$$

- Thiosulfate is added from the burette to the iodine until the deep-brown colour of iodine becomes a light-straw colour.
- Starch is now added, forming a deep-blue colour.
- Thiosulfate is added dropwise until the deep-blue colour becomes colourless.
- This is the end-point and shows that all the iodine has just been reacted.

The estimation of chlorine in a bleach

10.0 cm^3 of a sample of bleach was diluted to 250 cm^3 with water. 25.0 cm^3 of this solution was pipetted into a flask. An excess of aqueous potassium iodide was then added to liberate iodine:

> *equation 1* $Cl_2(aq) + 2I^-(aq) \longrightarrow I_2(aq) + 2Cl^-(aq)$

Titration of this solution required 23.2 cm^3 of $0.100 \text{ mol dm}^{-3}$ sodium thiosulfate.

> *equation 2* $2S_2O_3^{2-}(aq) + I_2(aq) \longrightarrow 2I^-(aq) + S_4O_6^{2-}(aq)$

Calculate the concentration of chlorine in the bleach.

From the titration results, amount (in moles) of $Na_2S_2O_3$ can be calculated:

> amount (in mol) of $Na_2S_2O_3 = c \times \dfrac{V}{1000} = 0.100 \times \dfrac{23.2}{1000} = 0.00232 \text{ mol}$

From the equations, the amount (in moles) of Cl_2 can be determined:

> From *equation 2*, 2 mol $S_2O_3^{2-}$ reacts with 1 mol I_2
>
> $\therefore$ 0.00232 mol $S_2O_3^{2-}$ reacts with $\dfrac{0.00232}{2}$ mol $I_2 = 0.00116$ mol I_2
>
> From *equation 1*, 1 mol Cl_2 forms 1 mol I_2
>
> $\therefore$ amount (in mol) of $Cl_2 = 0.00116$ mol

The concentration in mol dm^{-3} of the diluted bleach solution can now be calculated by scaling to 1000 cm^3:

> 25.0 cm^3 of diluted bleach contains 0.00116 mol Cl_2.
>
> 1 cm^3 of diluted bleach contains $\dfrac{0.00116}{25}$ mol Cl_2.
>
> 1 dm^3 of diluted bleach contains $\dfrac{0.00116}{25} \times 1000 = 0.0464$ mol Cl_2.

Finally, take into account the dilution with water from 10 cm^3 to 250 cm^3:

> The original bleach was diluted by a factor of $\dfrac{250}{10} = 25$
>
> $\therefore$ concentration of Cl_2 in the bleach is $25 \times 0.0464 = 1.16 \text{ mol dm}^{-3}$.

Compare acid-base titrations (page 40). The essential method is the same.

The concentration and volume of $Na_2S_2O_3$ are known.

Using the equations, determine the number of moles of Cl_2.

Work out the concentration of $Cl_2(aq)$, in mol dm^{-3}.

Don't forget to take into account any dilution.

Progress check

1 State and explain the trend in boiling points of the halogens fluorine to iodine.

2 How could you distinguish between NaCl, NaBr and NaI by a simple test?

3 Comment on the changes in oxidation number of chlorine in the following reaction: $Cl_2(aq) + H_2O(l) \longrightarrow HClO(aq) + H^+(aq) + Cl^-(aq)$

3 $Cl_2 \longrightarrow HClO$, Cl: $0 \longrightarrow +1$ (oxidation). $Cl_2 \longrightarrow Cl^-$, Cl: $0 \longrightarrow -1$ (reduction). Chlorine has been both oxidised and reduced (disproportionation).

2 Add $AgNO_3$(aq). NaCl gives a white precipitate, soluble in dilute ammonia. NaBr gives a cream precipitate, soluble in concentrated ammonia. NaI gives a yellow precipitate, insoluble in concentrated ammonia.

1 Boiling points increase down the group. From F_2 to I_2, the number of electrons increase leading to greater van der Waals' forces between molecules and higher boiling points.

2.9 Extraction of metals

After studying this section you should be able to:

- understand that carbon reduction is used to extract iron from its ore
- understand that electrolysis is used to extract aluminium from its ore
- understand that metal reduction is used to extract titanium from its ore
- understand the importance of economic factors and recycling

Reduction of metals

AQA ▶ M2

The main methods for extracting metals from their ores are:

- reduction of the ore with carbon
- reduction of the molten ore by electrolysis
- reduction of the ore with a more reactive metal.

Reduction of metal oxides with carbon

AQA ▶ M2

Production of iron

The main ore of iron, haematite, contains Fe_2O_3 and this is reduced by coke in the blast furnace. This is a continuous process in which high quality haematite, coke and limestone are fed in at the top of the furnace and hot air blown in near the bottom.

Coke contains a very high carbon content. Coke is cheap and provides an economic source of carbon and carbon dioxide as reducing agents.

The production of iron in the blast furnace is a **continuous** process. Raw materials are added at the top of the furnace and the products removed at the base. For further production, more raw materials are added and the process continues in a continuous flow, sometimes for months on end.

- Initially, hot coke reacts with air forming carbon monoxide:

$$2C(s) + O_2(g) \longrightarrow 2CO(g)$$

- The carbon monoxide reduces most of the iron oxide at around 1200°C:

$$Fe_2O_3(s) + 3CO(g) \longrightarrow 2Fe(l) + 3CO_2(g)$$

Both C and CO are reductants in the blast furnace.

- In hotter parts of the furnace, coke also reacts directly with iron oxide:

$$2Fe_2O_3(s) + 3C(s) \longrightarrow 4Fe(l) + 3CO_2(g)$$

This equation represents the overall reaction for the reduction.

Limestone removes SiO_2.

The limestone is used to remove acidic impurities such as silicon dioxide (sand). This produces a slag, largely of calcium silicate, $CaSiO_3$:

$$CaCO_3 + SiO_2 \longrightarrow CaSiO_3 + CO_2$$

• Molten iron collects at the bottom of the furnace and is run off as 'pig iron.'
• Pig iron is very impure and brittle, containing about 4% of carbon (as well as Mn, Si, P and S).

Production of other metals

Reduction with carbon is a common method for obtaining many other metals from their ores.

• Manganese

  Pyrolusite ore contains manganese(IV) oxide. The ore is heated with carbon to obtain manganese by reduction:

$$MnO_2 + 2C \longrightarrow Mn + 2CO$$
$$MnO_2 + 2CO \longrightarrow Mn + 2CO_2$$

• Copper

  Copper is usually reduced from its sulfide ores. It can also be obtained by high temperature reduction using carbon:

$$CuO + C \longrightarrow Cu + CO$$
$$CuO + CO \longrightarrow Cu + CO_2$$

Environmental pollution problems:

sulfide ores form sulfur dioxide;

reduction with carbon produces greenhouse gases.

Aluminium oxide is very stable and could only be reduced by carbon at very high temperatures. Electrolysis is used instead to extract aluminium (see page 88).

Limitations of carbon reduction

Metal ores often contain metal oxides or sulfides.

• Sulfide ores (such as galena, PbS, and sphalerite, ZnS) are first roasted in air to produce the metal oxide. The sulfur is oxidised as sulfur dioxide, which contributes to acid rain. The sulfur dioxide can be reacted to make sulfuric acid, from which a wide variety of useful products can be manufactured.

• Oxide ores (such as haematite, Fe_2O_3) are reduced to the metal using fossil fuels such as coke from coal. Carbon dioxide, emitted as the main gaseous product, is a greenhouse gas.

Ores of some metals (such as titanium and tungsten) react with carbon forming metal carbides, so this is not a practical method for extracting these metals.

$$TiO_2 + 3C \longrightarrow TiC + 2CO$$

Reduction of metal oxides by electrolysis of the molten ore

AQA   M2

For metals more reactive than zinc, reduction with carbon does not take place except at very high temperatures and these metals are usually extracted by electrolysis of the molten ore.

The extraction of aluminium

The melting point of aluminium oxide is 2045°C but this is decreased using molten cryolite. This reduces the energy requirements.

The main ore of aluminium, bauxite, contains Al_2O_3. Purified bauxite is dissolved in molten cryolite (Na_3AlF_6) at 970°C. This is a continuous process needing regular additions of aluminium oxide.

The electrodes are made of graphite and the cell reactions are:

at the cathode $Al^{3+} + 3e^- \longrightarrow Al$
at the anode $2O^{2-} \longrightarrow O_2 + 4e^-$

- Periodically the graphite anodes need replacing because, at the high temperatures used, the carbon anodes react with the liberated oxygen.
$$2C + O_2 \longrightarrow 2CO$$
$$\text{and } C + O_2 \longrightarrow CO_2$$
- The process consumes large amounts of electricity (as electricity is needed to melt the Al_2O_3 as well as to reduce it).
- The process is only economic where electricity is relatively inexpensive.

Reduction of metal halides with metals

AQA   M2

Some metals, such as titanium, become brittle if contaminated with traces of impurities such as carbon, oxygen or nitrogen.

Highly pure metals can be formed by reduction of a metal halide with a reactive metal.

Despite the high cost, pure titanium is manufactured by reduction of its molten chloride with a more reactive metal.

When pure, titanium is a light metal with a high strength and high resistance to corrosion.

When impure, it is brittle and of little use.

The production of titanium is a **batch** process. Raw materials are reacted together in a single 'batch' and the products are separated from the reaction mixture. For further production, the whole process needs to be repeated from scratch.

Production of titanium

The main ore of titanium, rutile, contains titanium(IV) oxide. Titanium is manufactured in a two-stage batch process.

- Rutile is first converted to titanium(IV) chloride using chlorine and coke at around 900°C:
$$TiO_2 + 2C + 2Cl_2 \longrightarrow TiCl_4 + 2CO$$
- The titanium(IV) chloride is purified from other chlorides (e.g. those of iron, silicon and chromium) by fractional distillation under argon or nitrogen.
- The chloride is then reduced by a more reactive metal such as sodium or magnesium:
$$TiCl_4 + 4Na \longrightarrow Ti + 4NaCl$$
$$TiCl_4 + 2Mg \longrightarrow Ti + 2MgCl_2$$
- An inert atmosphere of argon is used to prevent any contamination of the metal with oxygen or nitrogen.
- If sodium has been used, the sodium chloride by-product is washed out with dilute hydrochloric acid, leaving titanium as a granular powder.
- If magnesium has been used, the magnesium chloride by-product is removed by distillation at high temperature and low pressure.

Reduction of metal oxides with hydrogen

AQA   M2

The main ore of tungsten wolframite, is processed to precipitate tungsten(VI) oxide, WO_3. The tungsten(VI) oxide is reduced with hydrogen gas:

$$WO_3 + 3H_2 \longrightarrow W + 3H_2O$$

Hydrogen is more difficult to store than carbon and is highly flammable.

Economic factors and recycling

AQA   M2

Costs of extracting a metal

The method used to reduce a metal on an industrial scale is dependent upon several factors:

The cost of the reducing agent

A reducing agent that is naturally available such as carbon (from coal) is cheaper than one such as sodium, which has to be prepared first by a separate (and often costly) process.

The energy costs for the process

The lower the temperature of a process, the lower the energy requirement and the cheaper the process. Carbon reduction is a high temperature process with substantial energy costs.

Instead of using high temperatures to extract copper, scrap iron can be added to solutions containing copper ions. The more reactive iron displaces copper from solution.

This process also allows copper to be extracted from low grade ore, containing small proportions of copper.

The purity required of the metal

It is relatively expensive to produce a metal with very high purity. There needs to be a high demand for the metal if this extra expense is to be justified.

These factors need to be weighed against each other when considering the overall cost of an extraction process.

Recycling

Metals are a valuable resource and, instead of disposal, metals are often collected and recycled.

Recycling of iron

Recycling metals conserves metal ores and saves energy.

- Iron is recycled by collection of scrap iron which is melted and reused.
- The Earth's reserves of iron ore are conserved.
- The energy required to mine iron ore, transport it and smelt it is several times greater than the energy required to recycle scrap iron.

Recycling of aluminium

- Owing to its very high resistance to corrosion, used aluminium is as good as new.
- Recycling conserves the Earth's supply of aluminium.
- The electrolysis of aluminium oxide requires high temperatures and vast quantities of electricity.

The cost of re-using scrap aluminium is only one-twentieth of the cost of making the pure metal.

- It is also cheaper and easier to refine recycled aluminium than to mine bauxite and extract the metal by electrolysis.

2.10 Qualitative analysis

After studying this section you should be able to:

- *describe chemical tests for gases*
- *describe flame colours to identify metal ions*
- *describe tests for identifying and distinguishing between cations*

Using chemical tests to identify ions

CCEA   M2

Gas tests

gas	test
H_2	mild pop with lighted splint
O_2	relights glowing splint
Cl_2	turns moist indicator paper red, then bleached white
CO_2	turns lime water cloudy
SO_2	turns acidified dichromate paper from orange to green
HCl	forms dense white fumes with NH_3 fumes (from conc. NH_3(aq))
NH_3	forms dense white fumes with HCl fumes (from conc. HCl(aq))

Flame colours

Many potassium compounds contain traces of sodium compounds. The intense yellow Na^+ colour conceals the lilac K^+ colour.

Observing through blue glass filters out sodium's yellow colour, allowing potassium's lilac colour to be seen.

Flame colours can be used to identify some metal cations.

cation	Li^+	Na^+	K^+	Ca^{2+}	Ba^{2+}	Cu^{2+}
flame colour	red	yellow	lilac	brick-red	green	blue / green

Anion tests

SO_4^{2-}(aq) test
- Acidify solution with dilute nitric acid.
- Add aqueous barium chloride, $BaCl_2$(aq).
- SO_4^{2-} produces a white precipitate of barium sulfate, $BaSO_4$(s).

Distinguishing between CO_3^{2-}(aq) and HCO_3^-(aq)
- Add aqueous magnesium sulfate, $MgSO_4$(aq).
- CO_3^{2-} produces a white precipitate of $MgCO_3$(s).
- HCO_3^- produces no precipitate.

Tests for aqueous halide ions

See page 82 for more details of the reactions.

- Acidify with dilute nitric acid.
- Add aqueous silver nitrate, $AgNO_3$(aq).

halide	precipitate	solubility in NH_3(aq)
Cl^-	white precipitate of AgCl(s)	soluble in dilute NH_3(aq)
Br^-	cream precipitate of AgBr(s)	soluble in conc. NH_3(aq)
I^-	yellow precipitate of AgI(s)	insoluble in conc. NH_3(aq)

Cation tests

$Ba^{2+}(aq)$ test

- Acidify with dilute nitric acid.
- Add aqueous potassium chromate, $K_2CrO_4(aq)$.
- Ba^{2+} produces a deep-yellow precipitate of barium chromate, $BaCrO_4(s)$.

$Cu^{2+}(aq)$ test

- Add an excess of aqueous ammonia, $NH_3(aq)$.
- Cu^{2+} produces a deep-blue solution.

$NH_4^+(aq)$ test

- Heat with aqueous sodium hydroxide, $NaOH(aq)$.
- NH_4^+ produces alkaline $NH_3(g)$, which turns damp indicator paper blue.

Distinguishing between $Fe^{2+}(aq)$ and $Fe^{3+}(aq)$

- Add aqueous sodium hydroxide, $NaOH(aq)$.
- Fe^{2+} produces a **pale-green** precipitate of iron(II) hydroxide, $Fe(OH)_2(s)$.
- Fe^{3+} produces an **orange-brown** precipitate of iron(III) hydroxide, $Fe(OH)_3(s)$.

Distinguishing between $Al^{3+}(aq)$, $Mg^{2+}(aq)$ and $Zn^{2+}(aq)$

Notice that results from both NaOH(aq) and NH_3(aq) are needed to distinguish between the three ions.

- Add a few drops of aqueous $NaOH(aq)$ and $NH_3(aq)$ separately.
- All three ions form white precipitates.
- Ions are distinguished by solubility in excess $NaOH(aq)$ and $NH_3(aq)$.

	NaOH(aq)		NH₃(aq)	
	precipitate with few drops	*solubility in excess*	*precipitate with few drops*	*solubility in excess*
$Al^{3+}(aq)$	white	dissolves	white	insoluble
$Mg^{2+}(aq)$	white	**insoluble**	white	insoluble
$Zn^{2+}(aq)$	white	dissolves	white	**dissolves**

$Fe^{3+}(aq)$ test

- Add aqueous potassium thiocyanate, $KSCN(aq)$.
- Fe^{3+} produces a deep-blood-red solution.

Sample question and model answer

Chemical bonding helps to explain different properties of materials.

The phrase 'electrostatic attraction between ions' is essential when describing an ionic bond.

(a) Using suitable diagrams, explain what is meant by *ionic*, *covalent* and *metallic* bonding.

An ionic bond is the electrostatic attraction between ions. ✓
An example is sodium chloride:

Na⁺ ion   Cl⁻ ion

✓ (dot and cross)
✓ (charges on ions)

DO use diagrams in your answers. Here, dot and cross diagrams are essential for showing ionic and covalent compounds.

A covalent bond is a **shared** ✓ **pair** of electrons. ✓
An example is hydrogen chloride.

✓ (dot and cross)

The word 'pair' is essential when describing a covalent bond. Omit it and you risk losing the mark.

Metallic bonding occurs between positive centres/ions ✓
surrounded by mobile or delocalised electrons: ✓

There are plenty of examples. Keep them simple and make sure that you choose correctly. Don't use NaCl as a covalent example – it is ionic!

Many candidates do actually make this mistake.

Metallic bonding is the attraction between positive ions and mobile electrons. ✓

(b) Three substances have ionic, covalent and metallic bonding respectively. Compare and explain the electrical conductivity of the three materials.

The ionic compound does not conduct electricity when solid ✓ because the ions are fixed ✓ in a lattice.
The ionic compound does conduct electricity when aqueous or molten ✓ because the ions are mobile. ✓
The covalent compound does not conduct ✓ at all because there are no free charge carriers ✓ (electrons or ions).
The metallic compound does conduct electricity ✓ because the delocalised electrons are able to move ✓ across a potential difference.

Special care needed throughout with language. A-grade students will score all of these marks.

Ionic bonding: positive ions and negative ions – **both** carry electricity.

Metallic bonding: positive ions and delocalised electrons – **only** the electrons move and carry electricity.

17 marking points ⟶ maximum of [15]

Practice examination questions

1 (a) A water molecule, H_2O, is bonded by covalent bonds. What is meant by the term *covalent bond*? [2]

(b) Use the formation of the H_3O^+ ion from water to explain what is meant by a *dative covalent bond*. [2]

(c) State the bond angle in a water molecule and predict, with an explanation, the bond angle in an H_3O^+ ion. [4]

(d) Name the major force of attraction that exists between molecules in water and explain how this type of force arises. [3]

[Total: 11]

2 Describe how *electron pair repulsion* explains the shapes of simple molecules. Show, using diagrams, how electron pair repulsion determines the molecular shapes and bond angles in molecules of (a) SiH_4; (b) PH_3; (c) BCl_3. [13]

[Total: 13]

3 When liquid bromine, Br_2, is heated gently, it forms an orange–brown vapour. When a crystal of potassium chloride is heated to the same temperature, it does not change state.

(a) Name the type of bonding or force that occurs:

(i) between bromine atoms in a molecule of bromine

(ii) between bromine molecules in liquid bromine. [2]

(b) Explain why liquid bromine turns into a vapour when heated gently. [1]

(c) Explain, in terms of its bonding, why the crystal of potassium chloride does not melt or vaporise when heated gently. [2]

(d) Describe what happens to the particles in potassium chloride when the solid is heated above room temperature but below its melting point. [1]

(e) Suggest why much more energy is required to vaporise potassium chloride than to melt it. [2]

[Total: 8]

4 The boiling points of (a) water, (b) hydrogen chloride and (c) krypton are shown below.

liquid	H_2O	HCl	Kr
boiling point /°C	100	−85	−152

Suggest reasons for the different boiling points by considering the nature and strength of the intermolecular forces in each case.

[Total: 12]

Practice examination questions

5 In the Periodic Table, describe and explain the trend in the atomic radii of the elements in Period 3 and in Group 2. [Total: 8]

6 Barium and magnesium are elements in Group 2 of the Periodic Table.
(a) Complete and balance the following equations:
   (i) $Ba(s) + H_2O(l) \longrightarrow$
   (ii) $Mg(s) + O_2(g) \longrightarrow$
   (iii) $Mg(s) + HCl(aq) \longrightarrow$ [3]

(b) (i) Suggest why the reactivity of the Group 2 elements increases on descending the group.
   (ii) Name one reaction of Group 2 elements that illustrates this trend of increasing reactivity. [4]

(c) (i) What is the property of magnesium oxide that makes it suitable for its use as a lining in some furnaces?
   (ii) Name **two** other Group 2 compounds, and state a use for each of them. [3]
[Total: 10]

7 This question is about the Group 7 elements: chlorine, bromine and iodine.
(a) Describe and explain the trend in oxidising ability shown by the Group 7 elements. [5]

(b) Chlorine reacts with hot concentrated sodium hydroxide as in the equation below.
$$3Cl_2(g) + 6NaOH(aq) \longrightarrow 5NaCl(aq) + NaClO_3(aq) + 3H_2O(l)$$
   (i)   Use changes in oxidation numbers to show that this is a redox reaction.
   (ii) Calculate the maximum mass of $NaClO_3$ that could be prepared from the reaction of 65 dm³ of chlorine with hot concentrated sodium hydroxide at room temperature and pressure.
   (iii) Chlorine reacts with **dilute** aqueous sodium hydroxide at room temperature. Write an equation for this reaction and state what the resulting solution could be used for. [9]
[Total: 14]

8 (a) (i) What does the term *electronegativity* mean?
       (ii) What is the trend in electronegativity of the halogens fluorine to iodine? Explain why this trend happens. [5]

(b) State and explain the trend in volatility of the halogens fluorine to iodine. [3]

(c) Aqueous bromine was added separately to aqueous solutions of potassium chloride and potassium iodide. Describe what would be observed and write equation(s) for any reaction(s) that take place. [4]

(d) State and explain the trend in reactivity shown by the experiments in part (c). [2]
[Total: 14]

Chapter 3

Energetics, rates and equilibrium

The following topics are covered in this chapter:

- Enthalpy changes
- Determination of enthalpy changes
- Bond enthalpy

- Reaction rates
- Catalysis
- Chemical equilibrium

3.1 Enthalpy changes

After studying this section you should be able to:

- understand that reactions can be exothermic or endothermic
- construct a simple enthalpy profile diagram for a reaction
- explain and use the terms: standard conditions, enthalpy changes of reaction, formation and combustion

LEARNING SUMMARY

Energy out, energy in

AQA	M2	OCR	M2
EDEXCEL		SALTERS	M1
CCEA	M2	WJEC	M1

Enthalpy, *H*, is the **heat energy** that is stored in a chemical system.

Enthalpy cannot be measured experimentally. However, an **enthalpy change** can be measured from the temperature change in a chemical reaction.

> **KEY POINT**
>
> An **enthalpy change** ΔH is the heat energy exchange with the surroundings at constant pressure.

The conservation of energy is an important principle in science. This is often summarised by the first law of thermodynamics.

The first law of thermodynamics underpins all of this section.

> **KEY POINT**
>
> **The first law of thermodynamics** states that energy may be exchanged between a chemical system and the surroundings but the *total* energy remains constant.

Exothermic reactions

During an exothermic reaction, heat energy is **released** to the surroundings.

Any energy **loss** from the chemicals is balanced by the same energy **gain** to the surroundings, which rise in temperature.

exothermic

chemicals lose energy:
ΔH **–ve**

surroundings gain energy and rise in temperature:
ΔT **+ve**

Chemists refer to surroundings as anything other than the reacting chemicals.

Surroundings often just means the water in which chemicals are dissolved.

Energy pathway diagram (reaction profile)

*Energy gain by the surroundings (identified by a temperature **rise**: ΔT +ve)*

> **KEY POINT**
>
> In an exothermic reaction, ΔH is negative:
> - heat is given out (**to** the surroundings)
> - the reacting chemicals lose energy.

Endothermic reactions

During an endothermic reaction, heat energy is **taken in** from the surroundings.

Any energy **gain** to the chemicals is provided by the same energy **loss** from the surroundings, which fall in temperature.

Energy pathway diagram (reaction profile)

endothermic

chemicals gain energy:
ΔH +ve

surroundings lose energy and fall in temperature:
ΔT −ve

*Energy loss from the surroundings
(identified by a temperature **fall**: ΔT −ve)*

> **KEY POINT**
> In an endothermic reaction, ΔH is positive:
> - heat is taken in (**from** the surroundings)
> - chemicals gain energy.

Standard enthalpy changes

AQA	M2	OCR		
EDEXCEL	M1	SALTERS		M1
CCEA	M2	WJEC		M1

Enthalpy changes have been measured for many reactions. Many are recorded in data books as *standard enthalpy changes* and these are discussed below.

Standard conditions

$\Delta H^{\ominus}$ refers to an enthalpy (H) change (Δ) under standard conditions ($^{\ominus}$).

Standard pressure is 100 kPa or 1 bar.

The former standard pressure of 101 kPa or 1 atmosphere is still quoted in many books.

> **KEY POINT**
> *Standard conditions* are:
> - a pressure of 100 kPa
> - a stated temperature: 298 K (25°C) is usually used
> - a concentration of 1 mol dm⁻³ (*for aqueous solutions*).
>
> A *standard state* is the physical state of a substance under standard conditions.

The standard state of water at 298 K and 100 kPa is a liquid.

Standard enthalpy change of reaction

Always use a chemical equation or an unambiguous definition with a stated enthalpy change.

> **KEY POINT**
> The *standard enthalpy change of reaction* $\Delta H^{\ominus}_r$ is the enthalpy change that accompanies a reaction in the molar quantities that are expressed in a chemical equation under standard conditions, all reactants and products being in their standard states.

The enthalpy change of reaction $\Delta H^{\ominus}_r$ depends upon the quantities shown in a chemical equation. $\Delta H^{\ominus}_r$ should always be quoted with an equation.

$\Delta H^{\ominus}_r$ only has a meaning with an equation.

$\Delta H^{\ominus}_r$ has units of kJ mol⁻¹.

mol⁻¹ means 'for the amount (in moles) shown in the equation'.

For the reaction:

$$H_2(g) \quad + \quad \tfrac{1}{2} O_2(g) \longrightarrow H_2O(l) \qquad \Delta H^{\ominus}_r = -286 \text{ kJ mol}^{-1}$$
$$1 \text{ mol} \qquad\qquad \tfrac{1}{2} \text{ mol} \qquad\qquad 1 \text{ mol}$$

but with twice the quantities, there is twice the enthalpy change:

$$2H_2(g) \quad + \quad O_2(g) \longrightarrow 2H_2O(l) \qquad \Delta H^{\ominus}_r = -572 \text{ kJ mol}^{-1}$$
$$2 \text{ mol} \qquad\qquad 1 \text{ mol} \qquad\qquad 2 \text{ mol}$$

Standard enthalpy change of combustion

The *standard enthalpy change of combustion* $\Delta H^{\ominus}_c$ is the enthalpy change that takes place when one mole of a substance reacts completely with oxygen under standard conditions, all reactants and products being in their standard states.

This means that complete combustion of 1 mole of $C_2H_4(g)$ releases 1411 kJ of heat energy to the surroundings at 298 K and 100 kPa.

e.g. $C_2H_4(g) + 3O_2(g) \longrightarrow 2CO_2(g) + 2H_2O(l)$ $\Delta H^{\ominus}_c = -1411 \text{ kJ mol}^{-1}$

Standard enthalpy change of formation

The *standard enthalpy change of formation* $\Delta H^{\ominus}_f$ is the enthalpy change that takes place when one mole of a compound in its standard state is formed from its constituent elements in their standard states under standard conditions.

This means that the formation of 1 mole of $H_2O(l)$ from 1 mole of $H_2(g)$ and ½ mole of $O_2(g)$ releases 286 kJ of heat energy to the surroundings at 298 K and 100 kPa.

e.g. $H_2(g) + \frac{1}{2} O_2(g) \longrightarrow H_2O(l)$ $\Delta H^{\ominus}_f = -286 \text{ kJ mol}^{-1}$

For an element, the standard enthalpy change of formation is defined as zero.

The formation of $H_2(g)$ from $H_2(g)$ does not involve a chemical change so there is no enthalpy change.

Standard enthalpy change of neutralisation (Edexcel and CCEA only)

Notice that
$\Delta H^{\ominus}_f (O_2) = 0 \text{ kJ mol}^{-1}$
$\Delta H^{\ominus}_r$ for any element must be zero (formation of the element from the element – no change).

The *standard enthalpy change of neutralisation* $\Delta H^{\ominus}_{neut}$ is the energy change that accompanies the neutralisation of an acid by a base to form 1 mole of $H_2O(l)$, under standard conditions.

$$HCl(aq) + NaOH(aq) \longrightarrow NaCl(aq) + H_2O(l) \quad \Delta H^{\ominus}_{neut} = -57.9 \text{ kJ mol}^{-1}$$
acid   base $\longrightarrow$ 1 mol

- In aqueous solution, strong acids and bases are completely dissociated and $\Delta H^{\ominus}_{neut}$ is approximately equal to the value above. This neutralisation process corresponds to the reaction: $H^+(aq) + OH^-(aq) \longrightarrow H_2O(l)$.
- For weak acids, this enthalpy change is less exothermic because some input of energy is required to ionise the acid (see also page 45).

Progress check

1   Draw energy profile diagrams for the following reactions:
    (a) $N_2O_4(g) \longrightarrow 2NO_2(g)$ $\Delta H = +58 \text{ kJ mol}^{-1}$
    (b) $N_2(g) + 3H_2(g) \longrightarrow 2NH_3(g)$ $\Delta H = -92 \text{ kJ mol}^{-1}$

2   Write an equation to represent the enthalpy change for:
    (a) $\Delta H^{\ominus}_c$ (CH$_4$);   (b) $\Delta H^{\ominus}_f$ (NO$_2$)

2 (a) $CH_4(g) + 2O_2(g) \longrightarrow CO_2(g) + 2H_2O(l)$
  (b) $\frac{1}{2}N_2(g) + O_2(g) \longrightarrow NO_2(g)$

3.2 Determination of enthalpy changes

After studying this section you should be able to:

- *calculate enthalpy changes from direct experimental results, using the relationship: energy change = $mc\Delta T$*
- *use Hess' Law to construct enthalpy cycles*
- *determine enthalpy changes indirectly, using enthalpy cycles and enthalpy changes of formation and combustion*

LEARNING SUMMARY

Direct determination of enthalpy changes

AQA	M2	OCR	M2
EDEXCEL	M1	SALTERS	M1
CCEA	M2	WJEC	M1

Calculating enthalpy changes

The **heat energy change, Q,** in the surroundings can be calculated using the relationship below.

$$Q = mc\Delta T \text{ Joules}$$

- *m* is the **mass** of the surroundings that experience the temperature change
- *c* is the **specific heat capacity** of the surroundings
- ΔT is the **temperature change** (final temperature – initial temperature)

The **specific heat capacity** of a substance is the energy required to raise the temperature of 1 g of a substance by 1°C.

Example

Addition of an excess of magnesium to 100 cm³ of 2.00 mol dm⁻³ $CuSO_4(aq)$ raised the temperature from 20.0°C to 65.0°C. Find the enthalpy change for the reaction:

$$Mg(s) + CuSO_4(aq) \longrightarrow MgSO_4(aq) + Cu(s)$$

specific heat capacity of solution $c = 4.18$ J g⁻¹ K⁻¹
density of solution = 1.00 g cm⁻³.

Assume that solids, such as Mg(s) make little difference to the energy change.

Find the heat energy change

100 cm³ of solution has a mass of 100 g;

Temperature change, ΔT	= (65.0–20.0)°C
	= +45.0°C
Heat energy **gain** to surroundings, $Q = mc\Delta T$	= 100 × 4.18 × 45.0 J
	= **+18810 J**
∴ heat energy **loss** from the reacting chemicals	= **–18810 J**

Any energy **gain** by the surroundings must have come from the same energy **loss** in the chemical reaction.

Find out the amount (in moles) that reacted

Amount (in mol) of $CuSO_4$ that reacted = $2.00 \times \dfrac{100}{1000}$ mol = 0.200 mol

Scale the quantities to those in the equation

$$Mg(s) + CuSO_4(aq) \longrightarrow MgSO_4(aq) + Cu(s)$$
1 mol   1 mol   1 mol   1 mol

For 0.200 mol (1/5th mol) of $CuSO_4$, $\Delta H = -18810$ J
For 1 mol of $CuSO_4$, $\Delta H = 5 \times -18810 = -94050$ J
                               $\Delta H = -94.1$ kJ mol⁻¹ (to 3 sig. figs)

All values in this example are to 3 significant figures. This indicates the expected accuracy of the answer which should also be expressed to 3 significant figures.

∴ enthalpy change of reaction is given by:

$$Mg(s) + CuSO_4(aq) \longrightarrow MgSO_4(aq) + Cu(s) \quad \Delta H = -94.1 \text{ kJ mol}^{-1}$$

In this case,
$\Delta H = -94.1$ kJ mol⁻¹.

Indirect determination of enthalpy changes

AQA	M2	OCR	M2	
EDEXCEL		SALTERS		
CCEA	M2	WJEC	M1	

Hess' Law

Hess' Law is an extension of the First Law of Thermodynamics.

Many reactions have enthalpy changes that cannot be found directly from a single experiment. Hess' Law provides a method for the *indirect* determination of enthalpy changes.

> **KEY POINT**
>
> *Hess' Law* states that, if a reaction can take place by more than one route and the initial and final conditions are the same, the total enthalpy change is the same for each route.

The diagram below shows two routes for converting reactants into products.

Following the arrows from reactants to products,

Indirect determination of an enthalpy change uses an energy cycle based on Hess' Law.

This method is used when the reaction is very difficult to carry out and related reactions can be measured more easily.

Following the arrows from reactants to products,

Route 1: **A + B + C**
Route 2: **D**

By Hess' Law, the total enthalpy change is the same for each route.

∴ **A + B + C = D**

If three of these enthalpy changes are known, the fourth can always be calculated.

Enthalpy changes of combustion are required for all reactants and products.

Using $\Delta H^{\ominus}_c$ values to determine an enthalpy change indirectly

Example:

Find the enthalpy change for the reaction:

$$C(s) \ + \ 2H_2(g) \longrightarrow CH_4(g)$$

substance	C(s)	H₂(g)	CH₄(g)
$\Delta H^{\ominus}_c$ / kJ mol⁻¹	–394	–286	–890

Use the $\Delta H^{\ominus}_c$ data as a 'link' to construct an energy cycle.

Always show your working. In exams, you are rewarded for a good method.

If you make one small slip in a calculation, you may only lose 1 mark provided that you show clear working.

- An energy cycle is constructed by linking the reactants and products to their **combustion products**.
- Note the direction of the arrows from the reactants and products of the reaction to the common combustion products.

The common combustion products here are $CO_2(g)$ and $H_2O(l)$. You can include these in your cycle but they are not required and 'combustion products' have been used here.

Calculate the unknown enthalpy change

By Hess' Law,

Route 1: $\Delta H + [(-890)]$

Route 2: $[(-394) + (2 \times -286)]$

$\therefore \underbrace{\Delta H + [(-890)]}_{\text{Route 1}} = \underbrace{[(-394) + (2 \times -286)]}_{\text{Route 2}}$

$\Delta H = [(-394) + (2 \times -286)] - [(-890)]$

$\Delta H = \mathbf{-76\ kJ\ mol^{-1}}$

$\therefore C(s) + 2H_2(g) \longrightarrow CH_4(g) \qquad \Delta H^{\ominus} = -76\ kJ\ mol^{-1}$

KEY POINT

For enthalpy changes of combustion data only,

$\Delta H = \Sigma \Delta H^{\ominus}_{c}\ (\text{reactants}) - \Sigma \Delta H^{\ominus}_{c}\ (\text{products})$

> Enthalpy changes of formation are required for all reactants and products that are compounds.
>
> For elements, $\Delta H^{\ominus}_{f} = 0$ (formation of the element from the element – no change).

Using $\Delta H^{\ominus}_{f}$ values to determine an enthalpy change indirectly

Example:

Find the enthalpy change for the reaction:

$$C_2H_6(g) + 3\tfrac{1}{2}O_2(g) \longrightarrow 2CO_2(g) + 3H_2O(l)$$

substance	$C_2H_6(g)$	$CO_2(g)$	$H_2O(l)$
$\Delta H^{\ominus}_{f}$/ kJ mol^{-1}	–85	–394	–286

Use the $\Delta H^{\ominus}_{f}$ data as a 'link' to construct an energy cycle

- An energy cycle is constructed linking the reactants and products with their constituent elements.
- Note the direction of the arrows from the constituent elements to the reactants and products of the reaction.

> Notice that
> $\Delta H^{\ominus}(O_2) = 0\ kJ\ mol^{-1}$
> This can be omitted from the cycle.

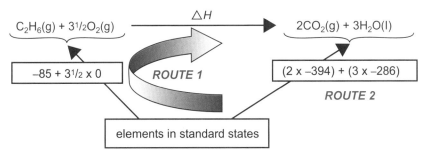

Calculate the unknown enthalpy change

By Hess' Law,

Route 1: $[(-85) + (3\tfrac{1}{2} \times 0)] + \Delta H$

Route 2: $[(2 \times -394) + (3 \times -286)]$

$\therefore \underbrace{[(-85) + 0] + \Delta H}_{\text{Route 1}} = \underbrace{[(2 \times -394) + (3 \times -286)]}_{\text{Route 2}}$

$\therefore \Delta H = [(2 \times -394) + (3 \times -286)] - [(-85) + 0]$

$\therefore \Delta H = \mathbf{-1561}\ kJ\ mol^{-1}$

$C_2H_6(g) + 3\tfrac{1}{2}O_2(g) \longrightarrow 2CO_2(g) + 3H_2O(l) \qquad \Delta H^{\ominus} = -1561\ kJ\ mol^{-1}$

KEY POINT

For enthalpy changes of formation data only,

$\Delta H = \Sigma \Delta H^{\ominus}_{f}\ (\text{products}) - \Sigma \Delta H^{\ominus}_{f}\ (\text{reactants})$

Progress check

For questions 1 and 2 assume that:

specific heat capacity of solution
$c = 4.18$ J g^{-1} K^{-1};

density of solution
$= 1.00$ g cm^{-3}.

1 Addition of zinc powder to 55.0 cm³ of aqueous copper(II) sulfate at 22.8 °C raised the temperature to 32.3 °C. 0.3175 g of copper were obtained.
(a) Calculate the energy released in this reaction.
(b) Write an equation, including state symbols, for this reaction.
(c) Calculate the enthalpy change for this reaction per mole of copper formed.

2 Combustion of 1.60 g of ethanol, C_2H_5OH, raised the temperature of 150 g of water from 22.0 °C to 71.0 °C. Find the enthalpy change of combustion of ethanol.

3 Use the $\Delta H^{\ominus}{}_c$ data below to calculate enthalpy changes for:
(a) $3C(s) + 4H_2(g) \longrightarrow C_3H_8(g)$
(b) $C(s) + 2H_2(g) + \frac{1}{2}O_2(g) \longrightarrow CH_3OH(l)$

substance	$\Delta H^{\ominus}{}_c$ / kJ mol^{-1}
C(s)	−394
$H_2(g)$	−286
$C_3H_8(g)$	−2219
$CH_3OH(l)$	−726

4 Use the $\Delta H^{\ominus}{}_f$ data below to calculate enthalpy changes for:
(a) $C_2H_4(g) + H_2(g) \longrightarrow C_2H_6(g)$
(b) $SO_2(g) + 2H_2S(g) \longrightarrow 2H_2O(l) + 3S(s)$

compound	$\Delta H^{\ominus}{}_f$ / kJ mol^{-1}
$C_2H_4(g)$	+52
$C_2H_6(g)$	−85
$SO_2(g)$	−297
$H_2S(g)$	−21
$H_2O(l)$	−286

4 (a) −137 kJ mol⁻¹; (b) −233 kJ mol⁻¹.
3 (a) −107 kJ mol⁻¹; (b) −240 kJ mol⁻¹.
2 −883 kJ mol⁻¹.
1 (a) 2.18 kJ; (b) Zn(s) + CuSO₄(aq) ⟶ Cu(s) + ZnSO₄(aq); (c) −437 kJ mol⁻¹.

3.3 Bond enthalpy

After studying this section you should be able to:

- *understand and use the term bond enthalpy*
- *explain chemical reactions in terms of enthalpy changes associated with the breaking and making of chemical bonds*
- *determine enthalpy changes indirectly, using average bond enthalpies*

Bond enthalpy

AQA	M2	OCR	M2
EDEXCEL	M1	SALTERS	M1
CCEA	M2	WJEC	M1

AQA (A2)	M5

Bond enthalpies are **positive** and refer to **bond breaking** – this process requires energy.

Enthalpy is stored within chemical bonds and **bond enthalpy** indicates the strength of a chemical bond in a gaseous molecule. For simple molecules, such as $H_2(g)$ and $HCl(g)$, bond enthalpy applies to the following processes:

$$H-H(g) \longrightarrow 2H(g) \qquad \Delta H = +436 \text{ kJ mol}^{-1}$$
$$H-Cl(g) \longrightarrow H(g) + Cl(g) \quad \Delta H = +432 \text{ kJ mol}^{-1}$$

> **KEY POINT**
>
> *Bond enthalpy* is the enthalpy change required to **break** and separate **1 mole of bonds** in the molecules of a gaseous element or compound so that the resulting gaseous species exert no forces upon each other.

See also Activation Energy: page 105.

Average bond enthalpies

Bond enthalpies such as those above (H–H and H–Cl) apply to specific compounds. Only H_2 can have H–H bonds and the H–H bond enthalpy shown above has a definite value. However, some bonds can have different strengths in different environments.

Not all C–H bonds are created equal.

The Cl atom in CH_3Cl affects the environment of the C–H bonds.

C–H bonds have different strengths and different bond enthalpies

Data books provide an indication of the likely bond enthalpy of a particular bond by listing **average** or **mean bond enthalpies**.

An *average* bond enthalpy indicates the strength of a *typical* bond.
The average bond enthalpy for the C–H bond is +413 kJ mol⁻¹.

Bond making and bond breaking

Bond breaking requires energy: ENDOTHERMIC .
Bond making releases energy: EXOTHERMIC.

Chemical reactions involve bond breaking followed by bond making.

- Energy is first needed to break bonds in the reactants.
  Bond breaking is an endothermic process and **requires** energy.
- Energy is then released as new bonds are formed in the products.
  Bond making is an exothermic process and **releases** energy.

Bond enthalpy is an endothermic change (ΔH +ve) for bonds being broken.

When bonds are made, the enthalpy change will be the same magnitude but the opposite sign (ΔH –ve).

Using bond enthalpies to determine enthalpy changes

The enthalpy change for a reaction involving simple gaseous molecules can be determined using average bond enthalpies in an **energy cycle**:

- Enthalpy required to break bonds = Σ(bond enthalpies in reactants)
- Enthalpy released to make bonds = –Σ(bond enthalpies in products)

Notice that the relative strengths of the bonds in the reactants and the bonds in the products decide whether a reaction is exothermic or endothermic.

$\Delta H = \Sigma(\text{bond enthalpies in reactants}) - \Sigma(\text{bond enthalpies in products})$.

For the reaction: $CH_4(g) + 2O_2(g) \longrightarrow CO_2(g) + 2H_2O(g)$,

$$
\underset{\text{H}}{\overset{\text{H}}{\text{H}-\text{C}-\text{H}}} + \begin{matrix} \text{O}=\text{O} \\ \text{O}=\text{O} \end{matrix} \longrightarrow \text{O}=\text{C}=\text{O} + \begin{matrix} \text{H} \diagdown {}_\text{O}\diagup \text{H} \\ \text{H} \diagdown {}_\text{O}\diagup \text{H} \end{matrix}
$$

	4 (C–H) + 2 (O=O)	2 (C=O) + 4 (O–H)
ΔH/kJ mol^{-1}	$(4 \times 413) + (2 \times 497)$	$(2 \times 805) + (4 \times 463)$

Bonds broken: (endothermic)   Bonds made: (exothermic)

$\Delta H = \Sigma(\text{bond enthalpies in reactants}) - \Sigma(\text{bond enthalpies in products})$

$\therefore \ \Delta H = [\ (4 \times 413) + (2 \times 497)\ ] - [\ (2 \times 805) + (4 \times 463)\ ]$ kJ mol^{-1}

$\quad\quad = -816$ **kJ mol^{-1}**

Note that the calculated value is only approximate – the actual bond enthalpies involved may differ from the average values.

- Bonds with small bond enthalpies will break first.
- Low bond enthalpies indicate that a reaction will take place quickly.

Entropy

SALTERS

Entropy is a measure of the number of ways that molecules and their associated energy quanta can be arranged.

In chemical reactions, energy changes depend not only on enthalpy but also on temperature and entropy.

Entropy, S, is a measure of the number of ways that molecules can be arranged.

Entropy increases whenever particles become more disordered, e.g.

- when a gas spreads spontaneously through a room
- with increasing temperature
- during changes in state: solid $\longrightarrow$ liquid $\longrightarrow$ gas
- when a solid lattice dissolves

$$NaCl(aq) + aq \longrightarrow Na^+(aq) + Cl^-(aq)$$

At 0 K, perfect crystals have zero entropy.

- in a reaction in which there is an increase in the number of gaseous molecules (i.e. when a gas is evolved).

$$NaHCO_3(s) + HCl(aq) \longrightarrow NaCl(aq) + H_2O(l) + CO_2(g)$$
$$MgCO_3(s) \longrightarrow MgO(s) + CO_2(g).$$

Progress check

1 Use the data in the table to find the enthalpy change for each reaction.
   (a) $C_2H_4(g) + 3O_2(g) \longrightarrow 2CO_2(g) + 2H_2O(g)$
   (b) $N_2(g) + 3H_2(g) \longrightarrow 2NH_3(g)$

bond	C–H	C=O	O=O	O–H	C=C	C–C	N≡N	H–H	N–H
bond average enthalpy /kJmol^{-1}	+413	+805	+497	+463	+612	+347	+945	+436	+391

1 (a) -1317 kJ mol^{-1}; (b) -93 kJ mol^{-1}.

3.4 Reaction rates

What is a reaction rate?

AQA	M2	OCR	M2
EDEXCEL	M2	SALTERS	M2
CCEA	M2	WJEC	M1

The rate of a chemical reaction is a measure of how quickly a reaction takes place.

> **KEY POINT**
>
> The rate of a reaction is usually measured as the rate of change of **concentration** of a stated species in a reaction.
>
> The units of rate are mol dm^{-3} s^{-1}.

Measuring reaction rates

Although concentration can be measured, other quantities proportional to concentration, such as gas volumes, may be easier to monitor. Rates of different reactions show a very wide variation and it may be more convenient to measure a reaction rate per minute or per hour rather than per second.

What affects a reaction rate?

The reasons why reaction rates are affected by these factors is discussed in the rest of this chapter.

The rate of a reaction may be affected by the following factors:

- the concentration of the reactants
- the surface area of solid reactants
- a temperature change
- the presence of a catalyst.

In some reactions, such as photosynthesis, the rate is affected by the presence and intensity of radiation.

How does reaction rate change during a reaction?

A is being used up
– its concentration decreases.

B is being formed
– its concentration increases.

For a reaction: **A** $\longrightarrow$ **B**, the reaction rate = $\dfrac{\text{change of concentration}}{\text{time}}$

The reaction rate can be measured as:

- the rate of **decrease** in concentration of **A**
- the rate of **increase** in concentration of **B**.

The graphs below show how the concentrations of **A** and **B** change during the course of the reaction: **A** ⟶ **B**.

A is used up and its concentration falls. **B** is formed and its concentration increases.

In the graph, the gradient indicates the reaction rate at any time.

- The steeper the gradient, the faster the rate of the reaction.
- The reaction is fastest at the start when the concentration of **A** is greatest.
- As the reaction proceeds, the rate slows down because the concentration of **A** decreases.
- When the reaction is complete, the graph levels off and the gradient becomes zero.

Activation energy

AQA	M2	OCR	M2
EDEXCEL	M2	SALTERS	M2
CCEA	M2	WJEC	M1

The activation energy of a reaction is the minimum energy required for the reaction to occur.

Only those collisions of sufficient energy to overcome the activation energy lead to a reaction.

In a gas or solution, particles are in constant motion and they collide with each other, with any solid species and with the walls of their container. When particles collide, a reaction can only take place if the energy of the collision exceeds the activation energy of the reaction.

The *activation energy* of a reaction is the energy required to start a reaction by breaking bonds (see page 102).

- Activation energy is often supplied by a spark or by heating the reactants.
- Reactions with a small activation energy often take place very readily.
- A large activation energy may 'protect' the reactants from taking part in the reaction (at room temperature).

The Boltzmann distribution

AQA	M2	OCR	M2
EDEXCEL	M2	SALTERS	M2
CCEA	M2	WJEC	M1

The Boltzmann, or Maxwell-Boltzmann, distribution shows the distribution of molecular energies in a gas at constant temperature.

The Boltzmann distribution

Characteristics of the Boltzmann distribution

* Most gas molecules have energies within a comparatively narrow range.
* The curve will only meet the energy axis at infinity energy. No molecules have zero energy.
* The area under the distribution curve gives the total number of gas molecules.
* Only those molecules with more energy than the activation energy of the reaction are able to react.

The effect of a concentration change on reaction rate

If the concentration of a reactant in solution or in a gas mixture is increased,

* there are more particles present per volume
* more collisions take place each second
* more collisions exceed the activation energy every second
* therefore the rate of reaction increases.

The diagram below shows distribution curves for two concentrations, C_1 and C_2. where concentration, C_2 > concentration, C_1.

* Providing the temperature is the **same**, distribution curves for different concentrations have the **same** shape.

> A reaction can only take place if the activation energy is exceeded.

> Note that the proportion of the total number of molecules exceeding the activation energy is the same. The rate increases because there are more molecules per volume and more molecules must now exceed the activation energy.

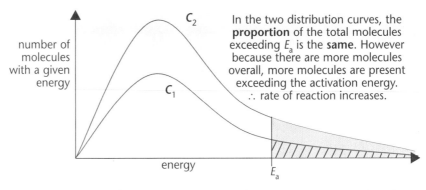

In the two distribution curves, the **proportion** of the total molecules exceeding E_a is the **same**. However because there are more molecules overall, more molecules are present exceeding the activation energy.
∴ rate of reaction increases.

For reactions involving gases, increasing the pressure also increases the concentration of any gas. This results in an increased reaction rate.

The effect of a change in surface area on reaction rate

For a reaction involving a solid, the reaction takes place at a faster rate when the solid is in a powdered form rather than as lumps.
Calcium carbonate reacts with hydrochloric acid producing carbon dioxide gas:

$$CaCO_3(s) + 2HCl(aq) \longrightarrow CaCl_2(aq) + H_2O(l) + CO_2(g)$$

By measuring the volume of carbon dioxide gas evolved with time, the rate of this reaction can be monitored. The graph below compares the reaction rates when using an excess of calcium carbonate as powder and as lumps. The same volume of the same concentration of hydrochloric acid has been reacted in both experiments.

> The gradient provides an indication of reaction rate.
>
> The steeper the curve, the faster the reaction.

The final volume is the same in both experiments since the same amount (in mol) of hydrochloric acid has been reacted with an excess of calcium carbonate.

The gradient of the graph is much steeper with powdered carbonate because the surface area available for the reaction with hydrochloric acid is much greater than with lumps, allowing more collisions per second.

The effect of a temperature change on reaction rate

The average kinetic energy of the particles is proportional to temperature. As temperature increases, so does the kinetic energy of gas molecules.

The diagram below shows distribution curves for a sample of gas at two temperatures, T_1 and T_2, where temperature, T_2 > temperature, T_1

Increasing the temperature **moves** the distribution curve to the right.

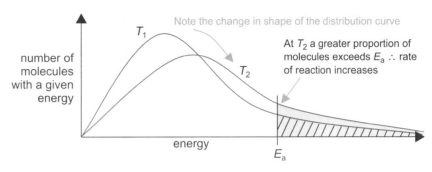

The Boltzmann distribution curve is displaced to the right with the peak lower. The average energy is now increased.

The total area under each curve is a measure of the total number of molecules present, and this is the same for each curve.

Increasing the temperature **does not change** the activation energy or the total number of molecules – the **shape of the curve changes**.

A **greater proportion** of molecules **exceeds the activation energy at higher temperature**.

An increase in temperature increases the rate of a reaction because:

- the molecules move faster and have more kinetic energy
- there are more collisions each second
- the increased kinetic energy produces more energetic collisions
- **a greater proportion of molecules exceed the activation energy.**

With even small increases in temperature, the shift in the distribution curve greatly increases the number of molecules exceeding the activation energy. This means that a small temperature rise can lead to a large increase in rate. Many reactions double their rate for each 10°C increase in temperature.

Progress check

1 Explain the following, in terms of collision theory and distribution graphs.
   (a) Reactions take place quicker when the reactants are more concentrated.
   (b) Reactions take place quicker at higher temperatures.

1 (a) Because there are more molecules per volume, more collisions take place each second. More collisions exceed the activation energy every second, increasing the reaction rate.
(b) The molecules move faster and have more kinetic energy. There are more collisions each second and the increased kinetic energy produces more energetic collisions. A greater proportion of molecules exceeds the activation energy.

3.5 Catalysis

After studying this section you should be able to:

- *understand what is meant by a catalyst*
- *explain how a catalyst changes the activation energy of a reaction*
- *explain what is meant by homogeneous and heterogeneous catalysis*

LEARNING SUMMARY

How do catalysts work?

AQA	M2	OCR	M2
EDEXCEL	M2	SALTERS	M1, M3
CCEA	M2	WJEC	M1

What is a catalyst?

A catalyst changes the rate of a chemical reaction, but is unchanged at the end of the reaction. Most catalysts speed up reaction rates although there are some (called inhibitors) that slow down reactions.

These key points are often tested in exams.

> **KEY POINT**
>
> A catalyst speeds up the rate of a reaction by providing an **alternative route** for the reaction with a **lower activation energy**.
> Activation energy with catalyst, E_c < Activation energy without catalyst, E_a

The effects of a catalyst on reaction route and activation energy are shown in the diagrams below.

Energy profile diagram

Notice the way the catalyst reduces the energy barrier to the reaction.

The catalyst provides an alternative route in which $E_c < E_a$

*The catalyst does **not** change the distribution curve.*

Notice how more molecules have an energy exceeding the new, lower activation energy.

Boltzmann distribution curve

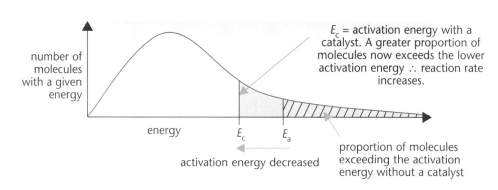

E_c = activation energy with a catalyst. A greater proportion of molecules now exceeds the lower activation energy ∴ reaction rate increases.

Heterogeneous and homogeneous catalysis

CCEA	M2	SALTERS	M1, M2
OCR	M2	WJEC	M1

A **hetero**geneous catalyst is in a **different** phase from the reactants. Many reactions between **gases** are catalysed using a **solid** catalyst.

A **homo**geneous catalyst is in the **same** phase as the reactants. Many reactions between **gases** in the atmosphere are catalysed using a **gaseous** catalyst. Homogeneous catalysis also takes place in many **aqueous** reactions in which both the reactants and the catalyst are dissolved in water.

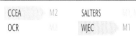

hetero- = different
homo- = same

Heterogeneous catalysis

Many heterogeneous catalysts are transition metals or their compounds.

The examples below are gaseous reactions catalysed by a solid catalyst.

Haber Process for ammonia (see also pages 113–115)

$$N_2(g) + 3H_2(g) \rightleftharpoons 2NH_3(g)$$

• The **gaseous** equilibrium process is catalysed by a **solid** iron catalyst, Fe(s).

surface of iron catalyst
Reactant molecules diffuse towards the catalyst surface

Reactant molecules adsorbed at active sites on the catalyst surface
Bonds weaken allowing reaction to take place

Following reaction, the product molecules diffuse away from the catalyst surface

Polluting gases temporarily bond to the catalyst's surface and it is here that reactions are able to take place. The products then leave the surface, allowing processing of more polluting gases by the catalyst.

Contact process for sulfur trioxide (and subsequently sulfuric acid)

$$2SO_2(g) + O_2(g) \rightleftharpoons 2SO_3(g)$$

• The **gaseous** equilibrium process is catalysed by **solid** vanadium(V) oxide, V_2O_5(s).

Catalytic converters

Many polluting gases, such as carbon monoxide, CO, nitrogen monoxide, NO and unburnt hydrocarbons are produced in a car engine.

These gases are formed by:

The catalytic converter has a surface area equivalent to two football pitches! The reacting gases now have a much greater chance of being adsorbed onto the catalyst so that a reaction can take place.

• incomplete combustion of hydrocarbons forming CO
• reaction between N_2 and O_2 in a hot car engine forming NO.

These polluting **gases** are made to react using the catalytic converter in a car's exhaust system. The **solid** Rh/Pt/Pd catalyst is supported on a ceramic honeycomb which gives a large surface area for catalysis, a feature that greatly improves the effectiveness of many solid heterogeneous catalysts.

The catalytic converter removes much of this pollution in a series of reactions, e.g.

$$2CO(g) + 2NO(g) \longrightarrow 2CO_2(g) + N_2(g)$$

Homogeneous catalysis (not CCEA)

A homogeneous catalyst does react but it is then regenerated, to be used again in a chain reaction – the middleman of chemistry.

During homogeneous catalysis, the catalyst **does** react forming an *intermediate state* with **lower activation energy**. Subsequently, the catalyst is regenerated, allowing further reaction to take place. Overall the catalyst is not used up, but it is actively involved in the process.

The breakdown of atmospheric ozone, catalysed by Cl· radicals

The Cl· radicals are recycled in a chain reaction. A single Cl radical can break down thousands of O_3 molecules.

Chlorine radicals, Cl· are formed in the stratosphere by the action of ultraviolet radiation on chlorofluorocarbons, CFCs. The **gaseous** Cl· radicals catalyse the breakdown of **gaseous** ozone – the catalysis is homogeneous.

For further details of the role of Cl· radicals in ozone depletion see page 149.

chain reaction: $O_3(g) + Cl·(g) \longrightarrow O_2(g) + ClO·(g)$
              $O(g) + ClO·(g) \longrightarrow O_2(g) + Cl·(g)$
overall: $O_3(g) + O(g) \longrightarrow 2O_2(g)$

See also radical substitution of alkanes, page 128.

The O(g) radicals are formed naturally in the stratosphere by the action of ultraviolet radiation on O_2(g).

3.6 Chemical equilibrium

Reversible reactions and dynamic equilibrium

AQA	M2	OCR	M2
EDEXCEL	M2	SALTERS	M2
CCEA	M2	WJEC	M1

In a chemical equation,

reactants are on the left-hand side;

products are on the right-hand side.

Reversible reactions

Many chemical reactions take place until the reactants are completely used up and such reactions *'go to completion'*.

An example of a reaction that goes to completion is the oxidation of magnesium:

$$2Mg(s) + O_2(g) \longrightarrow 2MgO(s)$$

- The sign '$\longrightarrow$' indicates that the reaction takes place from left to right: this is called the *forward direction*.

Many reactions are **reversible**: they can take place in either direction.

An example of a reaction that is reversible is the formation of ammonia $NH_3(g)$ from nitrogen $N_2(g)$ and hydrogen $H_2(g)$.

$$N_2(g) + 3H_2(g) \rightleftharpoons 2NH_3(g)$$

- The sign '$\rightleftharpoons$' indicates that the reaction is reversible.

The reversible reaction:
$N_2(g) + 3H_2(g) \rightleftharpoons 2NH_3(g)$
is used to illustrate dynamic equilibrium throughout this chapter.

This is a homogeneous equilibrium – one in which all reactants and products have the same phase. In this system, all are gases.

Approaching equilibrium

- If nitrogen and hydrogen are mixed together, the reaction can proceed only in the forward direction because no products are yet present:

$$N_2(g) + 3H_2(g) \longrightarrow$$

- As the reaction proceeds, ammonia is formed. The ammonia starts to react in the *reverse direction*, indicated by '$\longleftarrow$', forming nitrogen and hydrogen. At first, there is so little ammonia present that the reverse process takes place extremely slowly:

$$N_2(g) + 3H_2(g) \overset{\longrightarrow}{\leftharpoondown} 2NH_3(g)$$

- As more ammonia is formed the reverse process takes place faster. Nitrogen and hydrogen are being used up and the forward reaction slows down. Eventually, the reverse process takes place at the same rate as the forward reaction and **equilibrium** is attained:

$$N_2(g) + 3H_2(g) \rightleftharpoons 2NH_3(g)$$

Dynamic equilibrium

At equilibrium, there is a balance: reactants and products are both present and the reaction *appears* to have stopped.

- Although there is no apparent change, both forward and reverse processes continue to take place – the equilibrium is *dynamic*.
- The forward reaction proceeds at the same rate as the reverse reaction.

A reversible reaction can be approached from either direction and the terms *reactants* and *products* need to be used with caution.

- The concentrations of reactants and products are constant.

The equilibrium position:

- can be reached from either forward or reverse directions
- can only be achieved in a *closed system* – one in which no materials are being added or removed.

Equilibrium may be approached from either direction

$$A + B \rightleftharpoons C + D$$

REACTANTS   PRODUCTS

Only in a closed system

Factors affecting equilibrium

AQA	M2	OCR	M2
EDEXCEL	M2	SALTERS	M2
CCEA	M2	WJEC	M1

By opening up a closed equilibrium system, conditions can be changed after which the system can be allowed to reach equilibrium again.

The equilibrium position of the system may be altered by the following changes:

- Changing the concentration of a reactant or product.
- Changing the pressure of a gaseous equilibrium.
- Changing the temperature.

The likely effect on the equilibrium position can be predicted using Le Chatelier's Principle.

> **KEY POINT**
>
> *Le Chatelier's Principle* states that if a system in dynamic equilibrium is subjected to a change, processes will occur to minimise this change.

The effect of concentration changes

A change in the concentration of a reactant will alter the rate of the forward direction. A change in the concentration of a product will alter the rate of the reverse direction. Either change results in a shift in the equilibrium position.

$$N_2(g) + 3H_2(g) \rightleftharpoons 2NH_3(g)$$

If the concentration of a reactant is increased, or the concentration of the product is decreased, e.g. by removing some of it, the equilibrium is displaced to the right and more product is obtained.

Change: **Increase** concentration of a reactant: $N_2(g)$ or $H_2(g)$

Change opposed: The equilibrium system will **decrease** the concentration of the reactant by removing it. This is achieved by a shift in equilibrium to the right, forming more $NH_3(g)$.

The effects of changes in concentration for this equilibrium are shown below. These effects will apply to any system in equilibrium.

> **KEY POINT**
>
> increase concentration of reactant OR reduce concentration of product
>
> $$N_2(g) + 3H_2(g) \rightleftharpoons 2NH_3(g)$$
>
> decrease concentration of reactant OR increase concentration of product

The effect of pressure changes

A change in total pressure may alter the equilibrium position of a system involving gases. **The direction favoured depends upon the total number of gas molecules on each side of the equilibrium.**

$$N_2(g) + 3H_2(g) \rightleftharpoons 2NH_3(g)$$

Change: **Increase** the total pressure.

Change opposed: The equilibrium system will **decrease** the pressure by reducing the total number of moles of gas molecules. This is achieved by a shift in equilibrium to the right:

$$N_2(g) \quad + \quad 3H_2(g) \quad \rightleftharpoons \quad 2NH_3(g)$$

1 mol   3 mol   2 mol

            4 mol   2 mol

> *Movement of gas molecules causes pressure. The more gas molecules in a volume, the greater the pressure.*

> *If there are more moles of gaseous reactant than there are moles of gaseous product, an increase in total pressure will displace the reaction to the right.*

KEY POINT

| pressure increase |
| more gas molecules $N_2(g) + 3H_2(g) \rightleftharpoons 2NH_3(g)$ fewer gas molecules |
| pressure decrease |

Increasing the pressure also increases the concentration of any gas present and this **speeds** up the reaction.

The equilibrium position is also changed if:

- at least one of the equilibrium species is a gas
- there are different numbers of gaseous moles of reactants and of products.

The effect of temperature changes

A change in temperature alters the rates of the forward and reverse reactions by different amounts, resulting in a shift in the equilibrium position. **The direction favoured depends upon the sign of the enthalpy change.**

$$N_2(g) + 3H_2(g) \rightleftharpoons 2NH_3(g) \qquad \Delta H^\ominus = -92 \text{ kJ mol}^{-1}$$

Change: **Decrease** the temperature.

Change opposed: The equilibrium system will **increase** the temperature by releasing more heat energy. This is achieved by a shift in equilibrium in the exothermic direction, to the right.

> *An exothermic process is favoured by low temperatures.*
> *An endothermic process is favoured by high temperatures.*

> *An exothermic reaction in one direction is endothermic in the opposite direction.*
> *The value of the enthalpy changes is the same – the sign is different.*

KEY POINT

exothermic process favoured by low temperatures

$$\Delta H^\ominus = +92 \text{ kJ mol}^{-1} \quad N_2(g) + 3H_2(g) \rightleftharpoons 2NH_3(g) \qquad \Delta H^\ominus = -92 \text{ kJ mol}^{-1}$$

endothermic process favoured by high temperatures

Increasing the temperature *speeds* up the reaction and less time is needed to reach equilibrium. However, the equilibrium position may change to produce a lower equilibrium yield of products (see also pages 113–115).

The effect of a catalyst

There is **no change** in the equilibrium position. However, the catalyst **speeds** up both forward and reverse reactions and equilibrium is reached quicker.

The hydrogencarbonate / carbonate equilibrium

WJEC   M1

Carbon dioxide is an acidic gas that is slightly soluble in water. Dissolved carbon dioxide forms an acidic solution, containing H^+ ions and hydrogencarbonate ions, HCO_3^-.

$$CO_2(aq) + H_2O(l) \rightleftharpoons H^+(aq) + HCO_3^-(aq)$$

Any increase in CO_2 would move this equilibrium to the right, increasing the acidity of the water by increasing the H^+ concentration.

In sea-water, a second natural equilibrium exists between hydrogencarbonate ions, HCO_3^-, and carbonate ions, CO_3^{2-}.

$$HCO_3^-(aq) \rightleftharpoons H^+(aq) + CO_3^{2-}(aq)$$

If the H^+ concentration of sea-water increases, this equilibrium will move to the left to reduce the change.

This self-regulating process maintains the acidity of sea-water at a near constant pH.

Equilibria, rates and industrial processes

AQA   M2   OCR
CCEA   M2

For many industrial processes, the overall yield of a chemical product can vary considerably, depending on the conditions used.

In deciding the conditions to use, industrial chemists need to consider:

All courses study one or more industrial processes.

You should learn outline details of any process on your course.

The principles, however, are common for any similar situation.

- the availability of the starting materials required for a process
- the equilibrium conditions required to ensure a good yield
- the rate of reaction, which should be fast but manageable
- the cost, which should be as low as possible taking into account energy, cost of materials and of the chemical plant
- the safety of workers from hazardous chemicals, pressure, temperature, etc.
- any effects on the environment from waste discharges, toxic fumes, etc.

Industrial chemists compare each of these factors to arrive at compromise conditions. Although the equilibrium yield might be **optimised** by using a very high temperature and pressure, this may prove **impracticable** for reasons of cost and safety.

The importance of compromise in industrial processes

AQA   M2   OCR
CCEA   M2

Four industrial processes are discussed below to illustrate the importance of compromise.

In each of the four processes:

- the right-hand side, with the desired product, has **fewer moles of gas** than the left-hand side (reactants)
- the forward reaction, forming the desired product, is **exothermic**.

The raw materials used should be readily available.

$N_2(g)$ is obtained from the air.

$H_2(g)$ is obtained, together with $CO(g)$, by reacting together natural gas, $CH_4(g)$, and steam, $H_2O(g)$.

$C_2H_4(g)$ is obtained from the cracking of crude oil fractions.

The Haber process for the production of ammonia

$$N_2(g) + 3H_2(g) \rightleftharpoons 2NH_3(g) \qquad \Delta H^\ominus = -92 \text{ kJ mol}^{-1}$$

The contact process during the production of sulfuric acid

$$2SO_2(g) + O_2(g) \rightleftharpoons 2SO_3(g) \qquad \Delta H^\ominus = -197 \text{ kJ mol}^{-1}$$

The industrial production of ethanol by hydration of ethene

$$C_2H_4(g) \ + \ H_2O(g) \ \rightleftharpoons \ C_2H_5OH(g) \qquad \Delta H^\ominus = -46 \text{ kJ mol}^{-1}$$

The industrial production of methanol from carbon monoxide with hydrogen

$$CO(g) \ + \ 2H_2(g) \ \rightleftharpoons \ CH_3OH(g) \qquad \Delta H^\ominus = -91 \text{ kJ mol}^{-1}$$

The optimum equilibrium conditions

Optimising the process

These are the ideal conditions to give a maximum equilibrium yield.

Consider le Chatelier's Principle:

- The forward reaction produces fewer moles of gas, favoured by a **high** pressure.
- The forward reaction producing the desired product is exothermic, favoured by a **low** temperature.

The optimum equilibrium conditions for maximum equilibrium yield are:

- **high pressure and low temperature.**

The need for compromise

Compromising

Optimum conditions, reaction rate, feasibility, reality, safety and economics must be considered.

Having arrived at optimum equilibrium conditions, each must be considered to arrive at compromise conditions.

optimum condition	advantages	disadvantages
HIGH PRESSURE	equilibrium yield of desired product is high	energy costs are high – it is expensive to compress gases
	the concentration of gases is high, increasing the rate	there are considerable safety implications of using very high pressures – vessel walls need to be very thick to withstand high pressures, and weaknesses cause danger to workers and potential leakage of chemicals into the environment
LOW TEMPERATURE	equilibrium yield of product is high	the reaction takes place very slowly as few molecules possess the activation energy of the reaction

The use of a catalyst

A catalyst speeds up the rates of both the forward and backward reactions and hence less time is needed for the reaction to reach equilibrium. The increase in reaction rate allows lower temperatures to be used for a realistic reaction rate. The use of lower temperatures has the added bonus of supporting the optimum conditions for this process, increasing the equilibrium yield of the desired product.

Compromise conditions

Exam papers always test the wider aspects of chemistry to society.

Exam answers need to be sensible and related to relevant issues arising from the process itself. It is pointless to use terms such as 'dangerous', 'toxic', etc. unless substantiated with the reasons why the hazard is present.

The compromise conditions used in each process are such that:

- the temperature is sufficiently high to allow the reaction to occur at a realistic rate, but not too high to give a minimal equilibrium yield of the desired product
- a high pressure is used but not too high as to be impracticable
- the reaction rate is increased by using a catalyst. This enables the process to take place at lower temperatures, saving on energy costs.

The compromise conditions used for a process may only yield a small percentage of the desired product. Unreacted reactants are recycled.

Use of catalysts in industry

EDEXCEL   M2
OCR   M2
WJEC   M2

Catalysis is of great economic importance, for example:

iron in the Haber process;

vanadium(V) oxide, V_2O_5, in the contact process;

Ziegler–Natta catalyst in poly(ethene) production;

platinum / palladium / rhodium in catalytic converters.

Industrial preparations often use catalysts, allowing reactions to occur that would not otherwise take place.

Catalysis can reduce production costs and also lead to benefits for the environment.

- Catalysts affect the conditions that are needed, often requiring lower temperatures.
- With lower temperatures, energy demand is reduced. If energy is generated by the combustion of fossil fuels, there will be fewer CO_2 emissions.
- Catalysts may allow different reactions to be used, with better atom economy and reduced waste. (See also page 124.)
- Catalysts are often enzymes, generating very specific products, and operating effectively close to room temperature and pressure. This increases percentage yields and saves on energy. Conventional catalysts are often toxic and their disposal may present environmental problems. Enzymes are usually non-toxic, making for a 'greener' process.

Progress check

1 What will be the result of an increase in pressure on the following equilibria?
(a) $N_2O_4(g) \rightleftharpoons 2NO_2(g)$
(b) $CO(g) + 2H_2(g) \rightleftharpoons CH_3OH(g)$
(c) $H_2(g) + Br_2(g) \rightleftharpoons 2HBr(g)$

2 What will be the result of an increase in temperature on the following equilibria?
(a) $N_2(g) + O_2(g) \rightleftharpoons 2NO(g)$ $\Delta H^\ominus = +180$ kJ mol^{-1}
(b) $H_2(g) + Br_2(g) \rightleftharpoons 2HBr(g)$ $\Delta H^\ominus = -9.6$ kJ mol^{-1}
(c) $2SO_2(g) + O_2(g) \rightleftharpoons 2SO_3(g)$ $\Delta H^\ominus = -197$ kJ mol^{-1}

3 What are the optimum conditions of temperature **and** pressure for a high yield of the products in the equilibria below?
(a) $CO(g) + 2H_2(g) \rightleftharpoons CH_3OH(g)$ $\Delta H^\ominus = -92$ kJ mol^{-1}
(b) $PCl_5(g) \rightleftharpoons PCl_3(g) + Cl_2(g)$ $\Delta H^\ominus = +124$ kJ mol^{-1}

3 (a) low temperature and high pressure;
(b) high temperature and low pressure.
2 (a) moves to right; (b) moves to left; (c) moves to left.
1 (a) moves to left; (b) moves to right; (c) no change.

Sample question and model answer

1

The diagram below shows the Boltzmann distribution at a temperature, T_1, for a mixture of gases that react with each other. The activation energy for the reaction is labelled **X**.

Notice that the distribution curve starts at the origin but it never quite reaches the x-axis, even at high energies. Remember also that the area under the curve is equal to the total number of molecules.

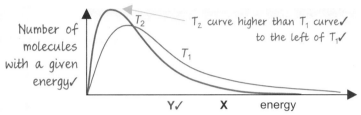

(a) (i) Explain the meaning of the term *activation energy*.

This is the minimum ✓ energy required for a reaction to take place. ✓

(ii) Label the y-axis.

(iii) Draw a second curve on the diagram above for the same mixture at a lower temperature. Label the second curve T_2.

At different temperatures,
the curve changes
the activation energy stays the same.

In the presence of a catalyst,
the activation energy changes
the curve stays the same.

(iv) Explain how a lower temperature affects the rate of this reaction.

Reduced temperature slows down the reaction rate. ✓
The overall kinetic energy of the molecules is reduced ✓ and fewer molecules now exceed the activation energy. ✓   [8]

(b) A catalyst is added to the mixture of gases.

(i) On the diagram above, label a possible activation energy **Y** for the catalysed reaction.

(ii) Explain, in terms of activation energy and the Boltzmann distribution, how a catalyst affects the rate of this reaction.

The catalyst speeds up the reaction rate. ✓ A catalyst allows the reaction to proceed via a different route ✓ with a lower activation energy. This means that a larger proportion of molecules now exceeds the activation energy. ✓   [4]

Always pay attention to the words that examiners use.

In parts (i) and (ii), the examiner has used 'Explain'. You need to justify your answer.

In part (iii), the examiner has used 'State'. You only need to 'state' your answer. You waste time here if you 'explain'.

(c) The equation for the industrial manufacture of ethanol, C_2H_5OH, from ethene, C_2H_4, is shown below.
$$C_2H_4(g) + H_2O(g) \rightleftharpoons C_2H_5OH(g) \qquad \Delta H = -46 \text{ kJ mol}^{-1}$$
The industrial conditions used are a high temperature and high pressure.

(i) Explain why the reaction is carried out at a high pressure.
There is a greater equilibrium yield. ✓ The increase in pressure is relieved by reducing the number of gas molecules. ✓ The equilibrium shifts in favour to the right because there are fewer gas molecules to the right. ✓

(ii) Explain why pressures higher than 1000 atmospheres are not used.
High pressures are very costly to generate in terms of energy. ✓

(iii) State **one** advantage and **one** disadvantage of carrying out the reaction at high temperature.
An advantage is that there is a greater reaction rate. ✓
A disadvantage is that the equilibrium yield is reduced. ✓   [6]
[Total: 18]

Practice examination questions

1 (a) Write a chemical equation, including state symbols, for the reaction that is used to define the enthalpy change of formation of one mole of sodium carbonate, $Na_2CO_3(s)$. [?]

(b) State the standard conditions used for a standard enthalpy change of formation, $\Delta H^{\ominus}_f$. [1]

(c) Use the standard enthalpy changes of formation given below to calculate a value for the standard enthalpy change for the following reaction:

$$Na_2CO_3.10H_2O(s) \longrightarrow Na_2CO_3(s) + 10H_2O(l)$$

compound	$Na_2CO_3.10H_2O(s)$	$Na_2CO_3(s)$	$H_2O(l)$
$\Delta H^{\ominus}_f$ / kJ mol^{-1}	−4081	−1131	−286

[3]

[Total: 6]

2 Propan-1-ol, $CH_3CH_2CH_2OH$, reacts with oxygen in a combustion reaction.

(a) (i) Define the term *standard enthalpy change of combustion*.
(ii) State the temperature that is conventionally chosen for standard enthalpy changes. [4]

(b) (i) Write a balanced equation for the combustion of propan-1-ol.
(ii) Calculate the standard enthalpy change of combustion of propan-1-ol using the following data.

compound	$\Delta H^{\ominus}_f$ /kJ mol^{-1}
$CH_3CH_2CH_2OH(l)$	−303
$CO_2(g)$	−394
$H_2O(l)$	−286

[4]

[Total: 8]

3 (a) Define the term *standard enthalpy change of formation* ($\Delta H^{\ominus}_f$). [3]

(b) Give the equation for the change that represents the standard enthalpy change of formation of methane. [2]

(c) Use the following data to calculate a value for the standard enthalpy change of formation of methane.

compound	$\Delta H^{\ominus}_c$ /kJ mol^{-1}
C(s)	−394
$H_2(g)$	−242
$CH_4(g)$	−802

[3]

(d) Use the data below to calculate average bond enthalpy values for the C–H and the C–C bonds.

$$CH_4(g) \longrightarrow C(g) + 4H(g) \qquad \Delta H = 1648 \text{ kJ mol}^{-1}$$
$$C_2H_6(g) \longrightarrow 2C(g) + 6H(g) \qquad \Delta H = 2820 \text{ kJ mol}^{-1}$$

[3]

[Total: 11]

Practice examination questions

4 (a) The diagram below shows a Boltzmann distribution for the mixture of gases at a temperature, T_1.

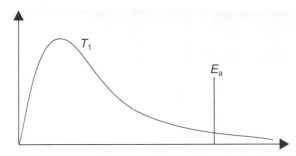

   (i) Label the axes and add a second curve for the Boltzmann distribution of this mixture at a higher temperature, T_2.

   (ii) On the diagram, add a label: 'E_a catalysed' for a possible activation energy of the reaction when catalysed. [5]

   (b) The elimination of steam from ethanol is an endothermic reaction:

   $$C_2H_5OH(g) \longrightarrow C_2H_4(g) + H_2O(g) \qquad \Delta H = +46 \text{ kJ mol}^{-1}$$

   (i) Sketch the reaction profile diagram for this reaction. On your diagram, label clearly the enthalpy change for the reaction, ΔH, and the activation energy, E_a.

   (ii) Add to your diagram a labelled reaction profile diagram for this reaction when catalysed.

   (iii) Explain why a catalyst increases the rate of a chemical reaction.
   [7]
   [Total: 12]

5 When hydrogen and carbon dioxide react, a dynamic homogenous equilibrium is set up as shown below:

$$H_2(g) + CO_2(g) \rightleftharpoons H_2O(g) + CO(g)$$

   (a) (i) Why is this reaction a *homogeneous equilibrium*?

   (ii) State **three** features of a *dynamic equilibrium*. [5]

   (b) For this equilibrium, explain the effect of an increase in pressure on:

   (i) the equilibrium position

   (ii) the reaction rate. [4]

   (c) The equilibrium yield of steam and carbon monoxide increases as the temperature is increased. Determine whether the forward reaction in the equilibrium above is exothermic or endothermic. Explain your answer. [2]
   [Total 11]

Chapter 4

Organic chemistry, analysis and the environment

The following topics are covered in this chapter:

- Basic concepts
- Hydrocarbons from oil
- Alkanes
- Alkenes

- Alcohols
- Halogenoalkanes (Haloalkanes)
- Analysis
- Chemistry in the environment

4.1 Basic concepts

After studying this section you should be able to:

- understand the different types of formula used for organic compounds
- recognise types of hydrocarbon
- recognise common functional groups
- understand what is meant by structural isomerism
- apply rules for naming simple organic compounds
- understand the difference between homolytic and heterolytic fission
- calculate percentage yields and atom economies

LEARNING SUMMARY

Types of formula

AQA	M1	OCR	M2
EDEXCEL		SALTERS	
CCEA	M2	WJEC	M2

In organic chemistry, there are many ways of representing a formula.

For the compound butane, with 4 carbon atoms and 10 hydrogen atoms:

the *empirical* formula is:	C_2H_5	The simplest, whole-number ratio of elements in a compound.
the *molecular* formula is:	C_4H_{10}	The *actual* number of atoms of each element in a molecule.
the *structural* formula is:	$CH_3CH_2CH_2CH_3$	The minimal detail for an unambiguous structure.
the *displayed* formula is:	H–C–C–C–C–H (with H atoms shown)	The relative placing of atoms and the bonds between them.
the *skeletal* formula	(zig-zag line)	The carbon skeleton and functional groups only.

Carbon chains

AQA	M1	OCR	M2
EDEXCEL		SALTERS	
CCEA	M2	WJEC	M2

Hydrocarbons

Hydrocarbons are compounds of carbon and hydrogen **only**.

- A saturated hydrocarbon has single bonds only.
- An unsaturated hydrocarbon contains a multiple carbon carbon bond.

saturated – single bonds only

unsaturated – contains a double bond

Alkanes

Carbon atoms can bond with other carbon atoms to form an enormous range of compounds with different carbon-chain lengths. The simplest organic compounds are a family of saturated hydrocarbons called the alkanes, shown below.

meth- C

eth- C_2

prop- C_3

but- C_4

pent- C_5

hex- C_6

hept- C_7

oct- C_8

non- C_9

dec- C_{10}

Number of carbons	Name	Molecular formula	Structural formula
1	methane	CH_4	CH_4
2	ethane	C_2H_6	CH_3CH_3
3	propane	C_3H_8	$CH_3CH_2CH_3$
4	butane	C_4H_{10}	$CH_3CH_2CH_2CH_3$
5	pentane	C_5H_{12}	$CH_3CH_2CH_2CH_2CH_3$
6	hexane	C_6H_{14}	$CH_3CH_2CH_2CH_2CH_2CH_3$

Note the following points.

- The name of the alkane ends with –*ane*.
- The prefixes (*meth-*, *eth-*, …) are used to represent the number of carbon atoms. You will need to use these many times in organic chemistry and they must be learnt.

General formula

Alkanes, C_nH_{2n+2} are saturated.

- The **general formula** is the simplest algebraic representation for any member in a series of organic compounds.
- The general formula for any alkane is C_nH_{2n+2}, where n = the number of carbon atoms.

Homologous series

Members in a homologous series react similarly.

By studying the reactions of one member of the series, you know how all members in the series are likely to react.

The alkanes are an example of a **homologous series** with the following features.

- Each successive member differs by $-CH_2-$. You can see this by comparing the formula of each alkane in the table above.
- Each member in a homologous series has the same general formula. The alkanes have the general formula: C_nH_{2n+2}.
- All members in a homologous series have the same functional group and similar chemical reactions.

Note that physical properties, such as boiling point and density, do gradually change as the length of the carbon chain increases (see also page 127).

Functional groups

AQA	M1	OCR	
EDEXCEL	M1	SALTERS	M1
CCEA	M2	WJEC	M2

Saturated hydrocarbon chains are comparatively unreactive. The reactivity is increased by the presence of a functional group – the reactive part of a carbon compound.

Unreactive carbon chain
alkyl group
C_nH_{2n+1}

Functional group
The part responsible for the reactions

Common functional groups

It is essential that you can instantly identify a functional group within a molecule. You can then apply the relevant chemistry.

If you continue to study Chemistry to A Level, you will meet more functional groups.

The main functional groups for AS Chemistry are shown below.

name	functional group	structural formula	general formula	prefix or suffix (for naming)
alkane	C–H	$CH_3CH_2CH_3$ *propane*	C_nH_{2n+2}	-ane
alkene	$\diagdown C=C \diagup$	$CH_3CH=CH_2$ *propene*	C_nH_{2n}	-ene
halogenoalkane	C–X (X=halogen)	CH_3CH_2Br *bromoethane*	$C_nH_{2n+1}Br$	bromo-
alcohol	C–OH	CH_3CH_2OH *ethanol*	$C_nH_{2n+1}OH$	-ol
aldehyde	$-C \diagup\diagdown^O_H$	CH_3CHO *ethanal*	$C_nH_{2n}O$	-al
ketone	$R-\overset{O}{\overset{\|}{C}}-R$	$CH_3COC_2H_5$ *butanone*	$C_nH_{2n}O$	-one
carboxylic acid	$-C \diagup\diagdown^O_{OH}$	CH_3COOH *ethanoic acid*	$C_nH_{2n+1}COOH$	-oic acid

- When studying reactions of a homologous series, the alkyl group is relatively unimportant – it is the functional group that reacts.
- The alkyl group, C_nH_{2n+1} is often represented simply as R–. This shifts the emphasis in the formula towards the reactive part of the molecule, the functional group.

The alcohols, shown below, is a homologous series with the –OH functional group.

- The alcohols have the general formula:
  $C_nH_{2n+1}OH$
- *Any* alcohol can be represented as:
  R–OH

The alcohols
CH_3OH
CH_2 ▲ C_2H_5OH Same functional group, OH
CH_2 ▲ C_3H_7OH

Structural isomerism

AQA   M1   OCR
EDEXCEL   M1   SALTERS
CCEA   M2   WJEC   M2

In addition to forming long chains, the atoms making up a molecular formula are often arranged differently, forming **isomers**.

Structural isomerism

Structural isomers are molecules with the same molecular formula but with different structural formulae (structural arrangements of atoms).

The two structural isomers of C_4H_{10} are shown below:

See also *E/Z isomerism*, page 129.

$$H-\overset{\overset{H}{|}}{\underset{\underset{H}{|}}{C}}-\overset{\overset{H}{|}}{\underset{\underset{H}{|}}{C}}-\overset{\overset{H}{|}}{\underset{\underset{H}{|}}{C}}-\overset{\overset{H}{|}}{\underset{\underset{H}{|}}{C}}-H$$
butane

$$H-\overset{\overset{H}{|}}{\underset{\underset{H}{|}}{C}}-\overset{\overset{CH_3}{|}}{\underset{\underset{H}{|}}{C}}-\overset{\overset{H}{|}}{\underset{\underset{H}{|}}{C}}-H$$
methylpropane

Butane and methylpropane are chain isomers of C_4H_{10}. Chain isomerism is a type of structural isomerism in which the carbon skeleton is different.

The number of different structural isomers possible from a molecular formula increases dramatically as the carbon-chain length increases. This is shown in the table below.

molecular formula	number of isomers
C_5H_{12}	3
C_6H_{14}	5
C_7H_{16}	9
C_8H_{18}	18
C_9H_{20}	35
$C_{10}H_{22}$	75
$C_{15}H_{32}$	4,347
$C_{20}H_{42}$	366,319

Look at the different names given to the two isomers.

- The branched isomer, methylpropane, is treated as a side chain attached to a straight-chain alkane.
- Side chains, such as the methyl group CH_3-, are called *alkyl* groups with the general formula of C_nH_{2n+1}.
- An alkyl group can be regarded as an alkane with one hydrogen atom removed (to allow another group to be attached).

The naming of organic compounds is discussed in more detail below.

Notice how the alkyl groups are named and how their formulae are linked to the parent alkane.

You must learn these.

Alkanes and alkyl groups

Number of carbons	Alkane C_nH_{2n+2}		Alkyl group C_nH_{2n+1}	
	Formula	Name	Formula	Name
1	CH_4	methane	CH_3-	methyl
2	C_2H_6	ethane	C_2H_5-	ethyl
3	C_3H_8	propane	C_3H_7-	propyl

In position isomerism, a functional group can be at different positions on the chain. For example, there are two position isomers with the molecular formula $C_4H_{10}O$ that have the alcohol, –OH, functional group:

- butan-1-ol with the –OH functional group at the end of the carbon chain
- butan-2-ol with the –OH functional group one carbon in from the end of the carbon chain.

butan-1-ol

butan-2-ol

Naming of organic compounds

AQA   M1   OCR   M2
EDEXCEL   M1   SALTERS   M1
CCEA   M2   WJEC   M2

With so many isomers possible, it is important that each has an individual name. The examples below show the steps needed to name an organic compound.

Rule 1

- The name is based upon the longest carbon chain on an **alkane**.

butane

Rule 2

- Any functional groups and alkyl groups are identified.
- These are then added to the name as a prefix (e.g. chloro-) or suffix (e.g. -ol) (see page 121).

methylbutanol

Rule 3

- If there is more than one possible isomer then the carbon atoms are labelled with numbers. Numbering starts from the end giving the lowest possible numbers for any functional groups and side chains. In this example, numbering from the left would give the incorrect name of 3-methylbutan-4-ol.

2-methylbutan-1-ol

Rule 4

- If there is more than one alkyl or functional group, they are placed in alphabetical order. This rule is not needed for the example above but it does need to be applied to name the compound to the right.

3 ethyl 2 methylpentane

Types of bond fission

AQA   M2   OCR   M2
EDEXCEL   M2   SALTERS   M2
CCEA   M2   WJEC   M2

The breaking of a covalent bond is called **bond fission**. Reactions of organic compounds involve bond fission followed by the formation of new bonds. Two types of bond fission are possible, **homolytic** fission and **heterolytic** fission.

Homolytic fission

These are very important principles, used throughout organic chemistry.

In homolytic fission, bond-breaking produces two species of the same (*homo-*) type:

$$A{:}B \longrightarrow A\bullet + \bullet B$$
radicals

A radical is a species with an unpaired electron (see page 128).

- In the example above, a covalent bond breaks so that one of the bonding electrons goes to each of **A** and **B**.
- Homolytic fission forms **two radicals**.

Heterolytic fission

Homolytic fission
⟶ radicals

Heterolytic fission
⟶ ions

In heterolytic fission, bond-breaking produces two species of different (*hetero-*) type:

$$A{:}B \longrightarrow A{:}^- + B^+ \qquad or \qquad A{:}B \longrightarrow A^+ + {:}B^-$$
ions   *ions*

- In the example above, a covalent bond breaks so that both the bonding electrons go to either **A** or **B**.
- Heterolytic fission forms **oppositely-charged ions**.

Percentage yields

AQA   M1   OCR   M2
EDEXCEL   M1   WJEC   M1
CCEA   M2

Organic reactions typically produce a lower yield than expected from the balanced equation. The yield is usually expressed as a percentage yield:

$$\text{percentage yield} = \frac{\text{actual yield}}{\text{theoretical yield}} \times 100\%$$

KEY POINT

Example

See also page 39.

5.2 g of 1-bromobutane is formed by reacting 5.0 g of butan-1-ol with sodium bromide and concentrated sulfuric acid. Find the percentage yield of 1-bromobutane.

equation	$C_4H_9OH + NaBr + H_2SO_4$	$\longrightarrow$	$C_4H_9Br + H_2O + NaHSO_4$
moles	1 mol	$\longrightarrow$	1 mol
reacting masses	74.0 g	$\longrightarrow$	136.9 g
	1 g	$\longrightarrow$	$\frac{136.9}{74.0}$ g
	5.0 g	$\longrightarrow$	$5 \times \frac{136.9}{74.0}$ g

Factors that can contribute to a low yield:

Organic reactions often do not go to completion.

An organic compound often reacts to produce a mixture of products.

The purification stages result in loss of some of the desired product.

∴ 5.0 g of C_4H_9OH produces a theoretical yield of 9.3 g C_4H_9Br.

Mass of C_4H_9Br that forms = 5.2 g.

$$\text{percentage yield} = \frac{\text{actual yield}}{\text{theoretical yield}} \times 100\% = \frac{5.2}{9.3} \times 100 = 56\%$$

Atom economy

AQA	M1	OCR		
EDEXCEL	M1	SALTERS		
CCEA	M2	WJEC	M1	

We are all becoming far more aware of our environment and the need to conserve resources and produce less waste. Atom economy is used to assess the efficiency of a reaction in terms of atoms:

$$\text{atom economy} = \frac{\text{molecular mass of the desired product}}{\text{sum of molecular masses of all products}} \times 100$$

KEY POINT

Atom economy is calculated from the balanced equation for the reaction.

Example

To improve the atom economy for C_4H_9Br production, a chemical company must either find uses for the other products or use another method for C_4H_9Br production.

equation: $C_4H_9OH + NaBr + H_2SO_4 \longrightarrow C_4H_9Br + H_2O + NaHSO_4$

molecular mass of desired product:

$$C_4H_9Br = 136.9$$

sum of molecular masses of all products:

$$C_4H_9Br + H_2O + NaHSO_4 = 136.9 + 18.0 + 120.1 = 275.0$$

$$\text{atom economy} = \frac{136.9}{275.0} \times 100 = 49.8\%$$

The consequence is that, even if this reaction could be carried out with a 100% percentage yield, converting **all** of the organic starting material, C_4H_9OH, into the organic product, C_4H_9Br, more than half the mass of products is **waste**.

Atom economy and type of reaction

An addition reaction produces a single product: all atoms are used and the reaction has an atom economy of 100%.

Addition reactions are the most atom efficient, giving a single product and an atom economy of 100%.

addition: $CH_2{=}CH_2 + H_2O \longrightarrow C_2H_5OH$

Substitution and elimination reactions produce by-products and will always have less atom economy unless some use is found for by-products.

Substitution and elimination always produce by-products.

These are wasted atoms, unless the by-products can be used.

substitution: $CH_3CH_2CH_2Br + NaOH \longrightarrow CH_3CH_2CH_2OH + NaBr$

elimination: $CH_3CH_2CH_2Br \longrightarrow CH_3CH=CH_2 + HBr$

Progress check

1 C_6H_{14} has 5 structural isomers. Show the structural and displayed formula for each of these isomers and name them.

2 8.52 g of 1-chloropentane was reacted with aqueous sodium hydroxide, producing 4.75 g of pentan-1-ol.
$CH_3CH_2CH_2CH_2CH_2Cl + NaOH \longrightarrow CH_3CH_2CH_2CH_2CH_2OH + NaCl$
Find the percentage yield of 1-bromobutane and the atom economy of the reaction.

2 Percentage yield: 67.5 atom economy: 60.1%.

4.2 Hydrocarbons from oil

After studying this section you should be able to:

- *understand how hydrocarbon fractions are separated from crude oil*
- *describe how low demand fractions are processed into hydrocarbons of higher demand*

LEARNING SUMMARY

Useful products from crude oil

AQA	M1	OCR	
EDEXCEL		SALTERS	
CCEA	M2	WJEC	M2

Crude oil is a fossil fuel, formed from the decay of sea creatures over millions of years. It is a complex mixture of hydrocarbons, containing mainly alkanes. Crude oil cannot be used directly but it is the source of many useful products. The first stages in the processing of crude oil are described below.

Fractional distillation

Crude oil is heated and passed into a fractionating tower, separating the complex mixture into *fractions*. Note that the fractions are not pure and contain mixtures of hydrocarbons within a range of boiling points. The fractionating tower, shown below, shows the temperature gradient which separates the fractions according to their boiling points (see also page 127).

Fractions with the lowest boiling points are collected at the top of the tower.

On descending the tower, the temperature rises and the boiling points increase.

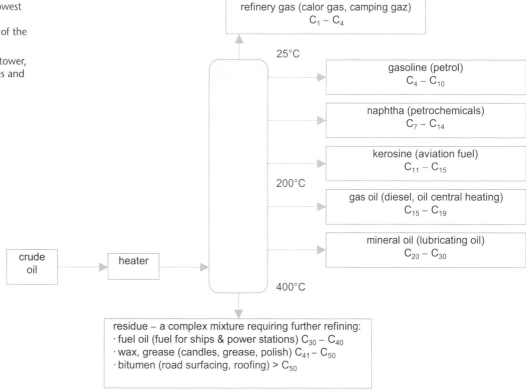

refinery gas (calor gas, camping gaz) $C_1 - C_4$

25°C

gasoline (petrol) $C_4 - C_{10}$

naphtha (petrochemicals) $C_7 - C_{14}$

kerosine (aviation fuel) $C_{11} - C_{15}$

200°C

gas oil (diesel, oil central heating) $C_{15} - C_{19}$

mineral oil (lubricating oil) $C_{20} - C_{30}$

crude oil → heater →

400°C

residue – a complex mixture requiring further refining:
- fuel oil (fuel for ships & power stations) $C_{30} - C_{40}$
- wax, grease (candles, grease, polish) $C_{41} - C_{50}$
- bitumen (road surfacing, roofing) $> C_{50}$

Cracking

Cracking is the starting point for the manufacture of many organics.

Cracking is the breaking down of an unsaturated hydrocarbon into smaller hydrocarbons. Cracking breaks C–C bonds in alkanes.

The purpose of cracking is to produce high demand hydrocarbons:

- short-chain alkanes for use in petrol
- alkenes, as a feedstock for a wide range of organic chemicals, including polymers (see also page 132).

Thermal cracking heats the *naphtha fraction* with steam at a high temperature (about 800°C) and high pressure. This forms a mixture of straight-chain alkanes and alkenes (mainly ethene) with a small proportion of branched and cyclic hydrocarbons. Some hydrogen is also produced.

The following equations summarise the overall process of cracking.

Thermal cracking breaks C–C bonds by homolytic fission forming radicals.

Catalytic cracking breaks C–C bonds by heterolytic fission forming ions.

$$C_{10}H_{22} \longrightarrow C_8H_{18} + C_2H_4$$
$$C_{10}H_{22} \longrightarrow C_6H_{14} + 2C_2H_4$$
$$alkane \longrightarrow alkane + alkene$$

There are many more equations possible but all must produce both an alkane and an alkene.

Catalytic cracking processes heavy long-chain fractions obtained in larger quantities than required. A mixture of the vapourised fraction and a zeolite catalyst are reacted at about 450°C using a slight pressure only. The process forms a higher proportion of branched and cyclic hydrocarbons than thermal cracking (see also reforming and isomerisation below). These are used for petrol.

Reforming

In a car engine, petrol has the tendency to auto-ignite before the spark, causing 'knocking'. Resistance to auto-ignition is measured as the **octane number** of the fuel.

Straight-chain hydrocarbons have lower octane numbers than branched-chain and cyclic hydrocarbons.

Reforming uses heat and pressure to convert unbranched fractions into cycloalkanes (e.g. cyclohexane) and arenes (e.g. benzene). These products are used in petrol and as a feedstock for a wide range of organic chemicals including many pharmaceuticals and dyes.

Examples

cyclohexane

benzene

Unleaded petrol and lead-replacement petrol require a higher proportion of branched and cyclic hydrocarbons for effective combustion. This has increased demand for reformed alkanes.

Isomerisation (not Edexcel)

Chemists are improving fuels by use of oxygenates such as methanol and ethanol. Hydrogen is also being developed as a fuel to replace petrol in the future.

Isomerisation converts unbranched hydrocarbons into branched hydrocarbons, needed for petrol.

Example

Progress check

1 Write down the formula of an alkane present in each of the main fractions obtained from crude oil.
2 Construct equations for two possible reactions taking place during the cracking of $C_{14}H_{30}$.

1 Any alkane with a formula C_nH_{2n+2} (see page 125) to match each fraction.
2 Any 2 equations — alkane + alkene, e.g. $C_{14}H_{30} \longrightarrow C_8H_{18} + C_6H_{12}$

4.3 Alkanes

After studying this section you should be able to:

- *understand why different alkanes have different melting and boiling points*
- *understand that combustion of alkanes provides useful energy*
- *describe the substitution reactions of alkanes with halogens*

Alkanes

AQA	M1	OCR	M2
EDEXCEL	M1	SALTERS	M1
CCEA	M2	WJEC	M2

General formula: C_nH_{2n+2}

Alkanes are generally unreactive. Alkanes contain only C–H and C–C bonds, which are relatively strong and difficult to break. The similar electronegativities of carbon and hydrogen give molecules which are non-polar. Alkanes are the typical 'oils' used in many non-polar solvents and they do not mix with water.

Melting points and boiling points of alkanes

EDEXCEL	M2	OCR	M2
CCEA	M2	SALTERS	M2
		WJEC	M2

Boiling points increase with increasing chain length because of the increasing strength of the intermolecular forces. This allows the alkane mixture in crude oil to be separated by fractional distillation.

The same principle can be applied to any homologous series.

Increasing chain length

m.pt and b.pt increase

more surface of contact between molecules

greater van der Waals' forces between molecules

C_3H_8 → C_5H_{12}

For hydrocarbons of similar molecular mass, branched alkanes have lower boiling points because there are fewer interactions possible between molecules:

The surface contact between molecules is important because it allows interaction between electrons. Remember that van der Waals' forces result from fluctuations in electron density (see page 64).

Isomers of C_4H_{10}

m.pt and b.pt increase

more surface of contact between molecules

greater van der Waals' forces between molecules

Combustion of alkanes

AQA	M1	OCR	M2
EDEXCEL	M1	SALTERS	M1
CCEA	M2	WJEC	M2

Combustion with oxygen is the most important reaction of alkanes giving their immediate use as fuels:

natural gas: $CH_4 + 2O_2 \longrightarrow CO_2 + 2H_2O$
petrol: $C_8H_{18} + 12\frac{1}{2}O_2 \longrightarrow 8CO_2 + 9H_2O$

Although generally unreactive, alkanes do react with oxygen and the halogens.

The burning of fossil fuels such as alkanes provides society with many environmental problems, some of which are listed below.

- Petroleum fractions contain sulfur impurities. Unless removed before combustion, sulfur oxides are formed from which acid rain (H_2SO_3 and H_2SO_4) is produced. The base, calcium oxide, is used to remove sulfur oxides from flue gases in power stations:

Remember that reactions with oxygen produce oxides.

$$CaO(s) + SO_2(g) \longrightarrow CaSO_3(s)$$

The $CaSO_3$ is oxidised to gypsum, $CaSO_4 \bullet 2H_2O$, which is used to make plaster.

- Toxic gases such as carbon monoxide, nitrogen oxides and unburnt hydrocarbons are present in car emissions. The development of catalytic converters has helped to remove these polluting gases (see also page 109).

Carbon dioxide, methane and water vapour are the main greenhouse gases that contribute to global warming (see page 147).

- Excessive combustion of fossil fuels increases carbon dioxide emissions contributing to global warming.

Substitution of alkanes with halogens

AQA	M2	OCR	
EDEXCEL	M1	SALTERS	
CCEA	M2	WJEC	M2

> **KEY POINT**
>
> A *substitution reaction* involves the swapping over of one species for another.

Alkanes are substituted by halogens, such as chlorine and bromine.

$$CH_4 + Cl_2 \longrightarrow CH_3Cl + HCl$$

This reaction only takes place in the presence of ultraviolet radiation, which produces highly reactive radicals.

You may have used the term 'free radical'. Radical and free radical mean the same thing.

> **KEY POINT**
>
> A radical, such as $Cl\bullet$, $CH_3\bullet$,
> - is a highly reactive species with an unpaired electron
> - reacts by pairing of an unpaired electron
> - is often involved in chain reactions.

Mechanism of radical substitution

Initiation

In this stage, the reaction starts.

Make sure that you learn this mechanism – these are easy marks in exams if you do!

Ultraviolet radiation provides energy to break Cl–Cl bonds homolytically producing chlorine radicals, $Cl\bullet$

$$Cl_2 \longrightarrow 2Cl\bullet$$

Propagation

In this stage, the reaction products are made.

Radicals are recycled in a chain reaction

The chlorine radicals catalyse the reaction.

See also the breakdown of ozone, pages 109 and 149.

$$CH_4 + Cl\bullet \longrightarrow CH_3\bullet + HCl$$
$$CH_3\bullet + Cl_2 \longrightarrow CH_3Cl + Cl\bullet$$

Termination

In this stage, radicals react together and are removed from the reaction mixture.

$$Cl\bullet + Cl\bullet \longrightarrow Cl_2$$
$$CH_3\bullet + Cl\bullet \longrightarrow CH_3Cl$$
$$CH_3\bullet + CH_3\bullet \longrightarrow CH_3CH_3$$

Impure products

This is answered poorly in exams.

Further substitution of the reaction products is possible, producing a mixture of products:

$$CH_3Cl \xrightarrow{Cl\bullet} CH_2Cl_2 \xrightarrow{Cl\bullet} CHCl_3 \xrightarrow{Cl\bullet} CCl_4$$

Progress check

1 Write an equation for the complete combustion of butane.
2 Butane reacts with chlorine in a substitution reaction.
  (a) What conditions are essential for this reaction? Explain your answer.
  (b) Write an equation for this reaction.
  (c) Write the structural formula of 2 possible monosubstituted products.
  (d) What is the formula for the final product obtained from complete substitution?

(d) C_4Cl_{10}
(c) Any 2 isomers of C_4H_9Cl, e.g. $CH_3CH_2CH_2CH_2Cl$ and $CH_3CH_2CHClCH_3$
(b) $C_4H_{10} + Cl_2 \longrightarrow C_4H_9Cl + HCl$
2 (a) UV; to generate radicals from Cl_2
1 $C_4H_{10}(g) + 6\frac{1}{2}O_2(g) \longrightarrow 4CO_2(g) + 5H_2O(l)$

4.4 Alkenes

After studying this section you should be able to:

- recognise the nature of *E/Z* isomerism in alkenes
- understand that the double bond in alkenes takes part in addition reactions with electrophiles
- understand the use of alkenes in the production of organic compounds
- understand what is meant by addition polymerisation

Alkenes

General formula: C_nH_{2n}

AQA	M2	OCR	
EDEXCEL		SALTERS	
CCEA	M2	WJEC	M2

Alkenes are unsaturated compounds with a C=C double bond. The high electron density of the double bond makes alkenes more reactive than alkanes.

Naming of alkenes

Position isomerism

A type of structural isomerism in which the functional group is in a different position on the carbon skeleton.

The position of the double bond is indicated using one number only, as shown in the example below for two isomers of C_4H_8.

Only one number is needed. In but-1-ene, a double bond starting at carbon-1 must finish at carbon-2

but-1-ene   but-2-ene

What is a double bond?

The coverage here is greatly simplified but sufficient for AS Chemistry.

In saturated compounds, the C–H and C–C single bonds are σ-bonds. A σ-bond is on the C–H or C–C axis, formed by overlap of s- and p- atomic orbitals.

In unsaturated compounds, the C=C double bond comprises a σ- and a π-bond.

σ- and π-bonds are *molecular orbitals* – these bond together the atoms in a molecule.

A π-bond is above and below the C–C axis, formed by overlap of atomic p-orbitals. The π-bond introduces an area of high electron density in the molecule, shown in the diagram of ethene below.

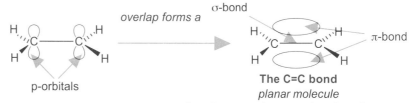

overlap forms a

σ-bond

π-bond

p-orbitals

The C=C bond
planar molecule
3 σ-electron pairs around each carbon atom.
Bond-angle is approximately 120°

E/Z isomerism

AQA	M2	OCR	
EDEXCEL		SALTERS	
CCEA	M2	WJEC	M2

Stereoisomers are molecules with the same structural formula but with a different arrangement in space.

KEY POINT

See also structural isomerism, page 121.

E/Z isomerism is a type of stereoisomerism that is present in some unsaturated hydrocarbons with a C=C bond.

Single C–C bonds can rotate. The C=C double bond, however, restricts rotation and prevents groups from moving from one side of the double bond to the other. This can give rise to *E/Z* isomerism.

For *E/Z* isomerism, there must be:

No rotation possible about a C=C bond.

- a C=C double bond
- two **different** groups attached to **each** carbon end of the double bond.

These are relatively easy exam marks. Make sure you learn this thoroughly.

If two of the groups are the same, this is also referred to as *cis-trans* isomerism:

- the *E* isomer (*trans* isomer) has the same groups on opposite sides
- the *Z* isomer (*cis* isomer) has the same groups on the same side.

The *E* and *Z* isomers of but-2-ene are shown below:

You should be able to label *E* and *Z* isomers in compounds that have *cis* and *trans* isomers.

Cis-trans isomerism is limited to unsaturated compounds in which two of the different groups are the same (as in the example above). The *E/Z* naming system can be applied to compounds with more than 2 different groups attached to either end of the C=C double bond.

The *E/Z* isomers of 3-methylhept-3-ene are shown below:

In exams, you would be expected to show *E/Z* isomers for this example but you would not be expected to label which is which.

It is impossible to label these as *cis* and *trans* isomers. The rules for labelling each stereoisomer in this example as *E* and *Z* go beyond the demands of AS Level Chemistry.

Addition reactions of alkenes

AQA M2 OCR M2
EDEXCEL M1 SALTERS M2
CCEA M2 WJEC M2

An addition reaction of an alkene involves the opening of the double bond with the formation of a saturated addition product.

> **KEY POINT**
>
> An *addition reaction* involves two species adding together to make one.
> Addition usually converts an **unsaturated** compound into a **saturated** compound.

This is made possible by the high electron density of the π-bond which attracts **electrophiles**.

> **KEY POINT**
>
> An *electrophile*, such as Br_2, HBr, H_2SO_4, NO_2^+:
> - is an electron-deficient species
> - 'attacks' an electron-rich carbon atom or double bond by **accepting a pair of electrons**.

This type of reaction is called **electrophilic addition**.

Addition of bromine

The addition reaction between an alkene and bromine, in which the orange colour of bromine is decolourised, is used as a test for unsaturation.

Mechanism of electrophilic addition

Be careful with the direction of the curly arrows.

Don't get confused by the charges and partial charges within this mechanism.

double bond of alkene induces a dipole on Br_2

electrophilic attack

addition product

Remember that a curly arrow shows the movement of an electron pair.

> **KEY POINT**
>
> In mechanisms, *a curly arrow* is used to show the movement of **a pair of electrons**.
> The movement of the electron pair involves either:
> • the formation of a new covalent bond or
> • the heterolytic fission of a covalent bond (see page 123).
> The *curly arrow* always goes **from** an electron pair (or double bond).

Addition using bromine water (Edexcel only)

Most specifications simplify the reaction of alkenes with bromine water as the addition of bromine. However, Edexcel discusses this reaction in more detail in terms of addition by Br (from Br_2) and OH (from H_2O) across the double bond:

$$H_2C=CH_2 + Br_2 + H_2O \longrightarrow BrCH_2CH_2OH + HBr$$

Addition using acidified manganate(VII) (Edexcel only)

Another useful test for unsaturation is the use of cold, dilute acidified manganate(VII). This oxidises the double bond of the alkene with the formation of a diol. The purple colour of manganate(VII) is decolourised.

Note that [O] is used in equations to represent an oxidising agent. Many candidates in exams get this reaction wrong by incorrectly adding Mn. $KMnO_4$ is used as a source of manganate (VII) ions.

ethane-1,2-diol

Reduction with hydrogen

> **KEY POINT**
>
> Reduction of an organic compound involves loss of oxygen OR gain of hydrogen (accompanied by a gain of electrons).

Alkenes are reduced when treated with hydrogen gas in the presence of a nickel catalyst at 150°C. The process of adding hydrogen across a double bond is sometimes referred to as **hydrogenation**.

Hydrogenation of C=C is used in the production of solid margarine from unsaturated liquid vegetable oils. Saturated fats are solid and the amount of hydrogenation can be controlled to give the correct texture of margarine.

Addition of hydrogen across C=C double bond

Further addition reactions of ethene

Addition reactions of alkenes are useful in organic synthesis. Some further addition reactions of ethene are shown in the flowchart below.

BROMOALKANE

ALCOHOL

poly(ethene)

For more details:

polymerisation, see below;

preparation of alcohols, see also page 135.

Although ethanol is made industrially using steam and phosphoric acid H_3PO_4, an alternative synthesis uses:

- addition with concentrated sulfuric acid at room temperature
  $$C_2H_4 + H_2SO_4 \longrightarrow CH_3CH_2OSO_3H$$
- followed by hydrolysis
  $$CH_3CH_2OSO_3H + H_2O \longrightarrow CH_3CH_2OH + H_2SO_4$$

Addition reactions of unsymmetrical alkenes

These structures are shown below.

In the addition of HBr to propene $CH_3CH=CH_2$, two products are possible:

- the secondary bromoalkane $CH_3CHBrCH_3$ as the major product
- the primary bromoalkane $CH_3CH_2CH_2Br$ as the minor product.

Mechanism

The driving force for the formation of $CH_3CHBrCH_3$ is the intermediate carbocation, shown in the mechanism below.

All students should know this mechanism but only AQA and WJEC demand an explanation in terms of carbocation stabilities.

(AQA and WJEC only)
The secondary carbocation is more stable with 2 alkyl groups feeding electrons towards the positive charge.
The alternative primary carbocation has only 1 alkyl group to stabilise the charge.

The reaction with H_2SO_4 proceeds via a similar mechanism. H_2SO_4 can be treated as $H-OSO_2OH$.

Addition polymerisation of alkenes

An addition polymer is a long-chain molecule with a high molecular mass, made by joining together many small molecules called monomers:

MANY MONOMERS $\longrightarrow$ SINGLE POLYMER
n molecules $\longrightarrow$ 1 single molecule

- The **monomer** used is an **unsaturated** alkene containing a double C=C bond.
- The addition **polymer** formed is a **saturated** compound **without** a double bond.

Many different addition polymers can be formed by using different alkene monomer units, based on the ethene molecule.

The equations below show the formation of poly(ethene), poly(chloroethene) (*pvc*) and poly(tetrafluoroethene) (*PTFE*).

You should be able to draw a short section of a polymer given the monomer units (and *vice versa*).

- Notice that the principle is the same for each addition polymerisation.

ethene

poly(ethene)

Make sure that you can show the repeat unit of a polymer.

- Each of the polymer structures shows the *repeat unit* in brackets. This is repeated thousands of times in each polymer molecule.

chloroethene

poly(chloroethene)
(PVC)

Rather than simply disposing of waste polymers, there is now movement towards recycling, combustion for energy production, and use as a feedstock for cracking.

This eliminates problems with disposal and helps to preserve finite energy resources.

Degradable polymers are also being developed from renewable resources such as corn starch.

- It is important to show the repeat unit with 'side-links' to indicate that both sides are also attached to other repeat units.

tetrafluoroethene

poly(tetrafluoroethene) (PTFE)

Properties of monomers and polymers

- The monomers are volatile liquids or gases.
- Polymers are solids.
- This difference in physical properties is explained by increased van der Waals' forces between the much larger polymer molecules (see page 64).

Problems with disposal

- Addition polymers are non-biodegradable and take many years to break down.
- Disposal by burning can produce toxic fumes (e.g. depolymerisation produces monomers; dioxins from combustion of chlorinated polymers).

Different types of polymer

SALTERS

A **thermoplastic** is a polymer that becomes soft and plastic on heating and hard on cooling. This process **can** be repeated.

A **thermoset** is a polymer that sets permanently when heated. This process **cannot** be repeated.

A **co-polymer** is a polymer made by reacting together two different monomers.

Soluble polymers

SALTERS

Hydrocarbon polymers such as poly(ethene) are non-polar and insoluble in water. Polymers with polar side-chains can have very different properties. For example, poly(ethenol), made by polymerisation of ethanol, is soluble in hot water.

Poly(ethenol) is used to manufacture plastic laundry bags for hospitals. The bags dissolve during the washing, protecting hospital workers from the risk of infection from handling dirty bedding.

ethenol

poly(ethenol)

Poly(ethenol) dissolves in water because the hydroxyl groups, –OH, are able to form hydrogen bonds to water molecules.

Progress check

1 Write the structural formula for the organic product for the reaction of but-2-ene with the following:
  (a) Br_2; (b) H_2/Ni; (c) HBr; (d) H_2O/H_3PO_4.

2 Show the repeat unit for the addition polymer formed from propene.

1 (a) $CH_3CHBrCHBrCH_3$ (b) $CH_3CH_2CH_2CH_3$ (c) $CH_3CH_2CHBrCH_3$
  (d) $CH_3CH_2CHOHCH_3$.

4.5 Alcohols

After studying this section you should be able to:

- *understand the polarity and physical properties of alcohols*
- *describe how ethanol is prepared*
- *describe oxidation reactions of alcohols*
- *describe nucleophilic substitution reactions of alcohols*
- *describe esterification and dehydration of alcohols*
- *describe tests for the presence of the hydroxyl group*

LEARNING SUMMARY

Alcohols

General formula: $C_nH_{2n+1}OH$

AQA	M2	OCR	M2
EDEXCEL	M2	SALTERS	M2
CCEA	M2	WJEC	M2

Types and naming of alcohols

The functional group in alcohols is the **hydroxyl** group, C–OH.

Alcohols can be classified as primary, secondary or tertiary, depending on how many alkyl groups are bonded to C–OH. You can see how these types of alcohols are named in the diagrams below.

primary alcohol
1 alkyl group attached to C–OH

propan-1-ol

secondary alcohol
2 alkyl groups attached to C–OH

propan-2-ol

tertiary alcohol
3 alkyl groups attached to C–OH

methylpropan-2-ol

Polarity of alcohols

Carbon, oxygen and hydrogen have different electronegativities, and alcohols have polar molecules:

Oxygen is more electronegative than both carbon and hydrogen – both carbon and hydrogen are electron deficient.

The polarity produces electron-deficient carbon and hydrogen atoms, indicated in the diagram above. Depending upon the reagents used, alcohols can react by breaking either the C–O or O–H bond.

The properties of alcohols are dominated by the hydroxyl group, C–OH.

Physical properties of alcohols

The –OH group dominates the physical properties of short-chain alcohols.

Hydrogen bonding takes place between alcohol molecules, resulting in:

- higher melting and boiling points than alkanes of comparable M_r
- solubility in water.

hydrogen bonding between ethanol molecules

The solubility of alcohols in water decreases with increasing carbon chain length as the non-polar contribution to the molecule becomes more important.

miscible with water
large polar contribution

polar head

non-polar tail

insoluble in water
most of the molecule is a non-polar carbon chain

polar head

non-polar tail

Preparation of ethanol

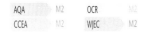

Ethanol finds widespread uses: in alcoholic drinks, as a solvent in the form of methylated spirits, and as a fuel. There are two main methods for its production.

Fermentation of sugars (for alcoholic drinks)

Yeast is added to an aqueous solution containing sugars.

E.g. $$C_6H_{12}O_6(aq) \longrightarrow 2C_2H_5OH(aq) + 2CO_2(g)$$

Hydration of ethene (for industrial alcohol)

Ethene and steam are passed over a phosphoric acid catalyst at 330°C under high pressure (see pages 131–132).

$$C_2H_4(g) + H_2O(g) \longrightarrow C_2H_5OH(l)$$

The preparation method used for ethanol may depend on the raw materials available.

An oil-rich country may use oil to produce ethene, then ethanol. However, this method uses up finite oil reserves.

In a country without oil reserves (especially in hot climates), sugar may provide a renewable source for ethanol production.

Combustion of alcohols

Alcohols such as ethanol and methanol are used as fuels, making use of combustion.

$$C_2H_5OH + 3O_2 \longrightarrow 2CO_2 + 3H_2O$$

Ethanol is used as a petrol substitute in countries with limited oil reserves.

Methanol is an 'oxygenate'. The oxygen in its formula aids combustion.

Methanol is used as a petrol additive in the UK to improve combustion of petrol. It also has increasing importance as a feedstock in the production of organic chemicals.

Oxidation of alcohols

The oxidation of different alcohols is an important reaction in organic chemistry. It links alcohols with aldehydes, ketones and carboxylic acids, shown below.

aldehyde, RCHO   ketone, RCOR'   carboxylic acid, RCOOH

- Aldehydes and ketones both contain the **carbonyl group** C=O.
- Carboxylic acids contain the **carboxyl** group COOH.

The stages of oxidation are shown below:

hydroxyl group C–OH $\longrightarrow$ carbonyl group C=O $\longrightarrow$ carboxyl group COOH

Oxidation of primary alcohols

A primary alcohol can be oxidised to an aldehyde and then to a carboxylic acid.

> **KEY POINT**
>
> Oxidation of an organic compound involves gain of oxygen OR loss of hydrogen (accompanied by loss of electrons from the organic compound).

The stages of oxidation are shown below:

primary alcohol   aldehyde   carboxylic acid

This is carried out using an oxidising agent. The oxidising agent used is a mixture of concentrated sulfuric acid, H_2SO_4 (source of H^+) and potassium dichromate, $K_2Cr_2O_7$ (source of $Cr_2O_7^{2-}$).

For balanced equations, the oxidising agent can be shown simply as [O].

- By heating and distilling the product immediately, oxidation can be stopped at the aldehyde stage.

$$CH_3CH_2OH + [O] \longrightarrow CH_3CHO + H_2O$$

- By refluxing with an excess of the oxidising agent, further oxidation takes place to form the carboxylic acid.

$$CH_3CH_2OH + 2[O] \longrightarrow CH_3COOH + H_2O$$

The orange dichromate ions, $Cr_2O_7^{2-}$, are reduced to green Cr^{3+} ions.

Oxidation of secondary alcohols

By heating with $H^+/Cr_2O_7^{2-}$ a secondary alcohol can be oxidised to form a ketone. No further oxidation normally takes place.

secondary alcohol $\longrightarrow$ ketone

secondary alcohol   ketone

The equation for the oxidation of propan-2-ol is shown below.

$$CH_3(CHOH)CH_3 + [O] \longrightarrow CH_3COCH_3 + H_2O$$

Oxidation of tertiary alcohols

Tertiary alcohols are not oxidised under normal conditions.

tertiary alcohol

Distinguishing between aldehydes and ketones (AQA only)

The presence of an aldehyde can easily be detected using two tests.

Heat aldehyde with Tollens' reagent
- Tollens' reagent $\longrightarrow$ silver mirror

Heat aldehyde with Fehling's (or Benedict's) solution
- Benedict's / Fehling's solution $\longrightarrow$ brick-red precipitate of Cu_2O

Tollens' reagent is a solution of silver nitrate (as a source of Ag^+ ions) in ammonia.
$Ag^+ \longrightarrow Ag$

Fehling's (or Benedict's) solution is an alkaline solution of Cu^{2+} ions.
$Cu^{2+} \longrightarrow Cu^+$

Ketones cannot be oxidised and **do not** react with the above solutions and this provides an easy method for distinguishing between an aldehyde and a ketone.

Identifying alcohols with the CH₃CHOH group (CCEA only)

The presence of an alcohol containing the CH_3CHOH group can be detected by heating an alcohol with iodine, I_2, in aqueous sodium hydroxide, NaOH(aq).

Alcohols containing the CH_3CHOH group produce pale-yellow crystals with I_2/NaOH.

- A pale-yellow crystalline solid called iodoform (triiodomethane), CHI_3, is formed. Iodoform has an antiseptic smell.
- Iodoform can also be confirmed following recrystallisation and determination of its melting point (123°C).

Substitution reactions of alcohols

EDEXCEL M2   SALTERS
CCEA M2

The electron-deficient carbon atom of the polar $C^{\delta+}-O^{\delta-}$ bond of alcohols attracts *nucleophiles*, allowing a **substitution** reaction to take place in which the nucleophile replaces the OH group. This is called **nucleophilic substitution** (see also halogenoalkanes, page 139).

> **KEY POINT**
>
> A nucleophile, such as OH^-, Br^-, H_2O, NH_3, CH_3OH, CN^-
> - is an electron-rich species
> - 'attacks' an electron-deficient carbon atom by **donating a pair of electrons**.

Reaction with bromide ions, Br^- in acid, H^+

$$C_2H_5OH + Br^- + H^+ \longrightarrow C_2H_5Br + H_2O$$

To prepare a chloroalkane, use concentrated HCl.

For a bromoalkane or an iodoalkane, use:

- a metal halide, e.g. NaBr
- and 50% sulfuric acid.

Concentrated acid oxidises bromides and iodides to the halogens (see pages 82–83).

Reaction with phosphorus pentachloride PCl_5 (Edexcel and CCEA only)

Alcohols react with PCl_5, releasing hydrogen chloride gas, which forms misty fumes in air.

$$C_2H_5OH + PCl_5 \longrightarrow C_2H_5Cl + HCl + POCl_3$$

The reaction of alcohols with PCl_5 can be used as a test for an –OH group.

In the presence of ammonia, HCl fumes form dense white fumes of ammonium chloride:

$$NH_3 + HCl \longrightarrow NH_4Cl$$

(CCEA only)

A similar reaction takes place with thionyl chloride, $SOCl_2$, in which SO_2 forms instead of $POCl_3$.

Dehydration of alcohols

AQA M2   SALTERS
OCR   WJEC M2

When an alcohol is refluxed with a concentrated acid catalyst such as sulfuric acid H_2SO_4, or phosphoric acid H_3PO_4, an **elimination** reaction takes place: water is eliminated with formation of an alkene.

$$C_2H_5OH \longrightarrow C_2H_4 + H_2O$$

Use hot concentrated acid for elimination of water from an alcohol.

Elimination (1 $\longrightarrow$ 2) is the opposite process to addition (2 $\longrightarrow$ 1).

> **KEY POINT**
>
> An *elimination* reaction involves one species breaking up into two – the opposite to addition.
>
> Elimination usually converts a **saturated** compound into an **unsaturated** compound.

See page 135 for more details of methods used to prepare ethanol.

Alkenes, such as ethene, can be synthesised from renewable resources rather than from crude oil. Sugars, such as glucose, can be fermented to produce ethanol, C_2H_5OH.

E.g.

$$C_6H_{12}O_6 \longrightarrow 2C_2H_5OH + 2CO_2$$
glucose

Elimination of water from ethanol with an acid catalyst produces ethene.

$$C_2H_5OH \xrightarrow[\text{heat}]{\text{H}^+ \text{ catalyst}} C_2H_4 + H_2O$$

The alkenes can then be polymerised to form plastics.

Formation of esters

CCEA   M2
OCR   M2

Esters are formed by reaction of a carboxylic acid with an alcohol.

Esters are formed by reaction of an alcohol with a carboxylic acid in the presence of an acid catalyst (e.g. concentrated sulfuric acid).

carboxylic acid + alcohol $\longrightarrow$ ester + water

The esterification of ethanol by ethanoic acid is shown below:

$$CH_3COOH + CH_3OH \longrightarrow CH_3COOCH_3 + H_2O$$

(CCEA only)

Esters can also be formed by reacting an alcohol with an acyl chloride at room temperature.

E.g.

$$CH_3COCl + C_2H_5OH \longrightarrow CH_3COOC_2H_5 + HCl$$
ethanoyl chloride

Reaction with sodium

EDEXCEL   M2
CCEA   M2

Sodium releases H_2 from any compound with a hydroxyl group: alcohols, carboxylic acids (and water).

Alcohols react with sodium, releasing hydrogen gas. The reaction involves breaking of the O–H bond of the alcohol.

$$2C_2H_5OH + 2Na \longrightarrow 2C_2H_5O^-Na^+ + H_2$$
sodium ethoxide

Progress check

1 Why does ethanol dissolve in water?

2 Write equations to show possible oxidations for the following:
(a) butan-1-ol; (b) butan-2-ol.

3 What are the 2 possible organic products of the reaction between butan-2-ol and concentrated sulfuric acid at 170°C?

3 $CH_3CH=CHCH_3$ and $CH_3CH_2CH=CH_2$
(b) $CH_3CH_2CHOHCH_3 + [O] \longrightarrow CH_3CH_2COCH_3 + H_2O$
$CH_3CH_2CH_2CH_2OH + 2[O] \longrightarrow CH_3CH_2CH_2COOH + H_2O$
2 (a) $CH_3CH_2CH_2CH_2OH + [O] \longrightarrow CH_3CH_2CH_2CHO + H_2O$
1 Alcohol –OH group forms H-bonds with water molecules.

4.6 Halogenoalkanes (Haloalkanes)

After studying this section you should be able to:

- understand the nature of the polarity in halogenoalkanes
- describe nucleophilic substitution reactions of halogenoalkanes
- explain the relative rates of hydrolysis of different halogenoalkanes
- describe elimination reactions of halogenoalkanes
- understand that reaction conditions can be used to control the products
- understand some common practical techniques used in organic chemistry

Halogenoalkanes General formula: $C_nH_{2n+1}X$, where X = F, Cl, Br or I

AQA	M2	OCR	M2
EDEXCEL	M2	SALTERS	M2
CCEA	M2	WJEC	M2

Uses of halogenoalkanes

Chlorofluorocarbons, CFCs were used as refrigerants, solvents, propellants, dry cleaning and degreasing agents, and for blowing polystyrene.

CFCs are now known to deplete the ozone layer. Their use is now banned in countries that have signed up to the Montreal Protocol.

Chloroethene and tetrafluoroethene are used to produce the plastics PVC and PTFE.

Halogenoalkanes are used as synthetic intermediates in chemistry and solvents.

Polychlorinated compounds are used as herbicides.

Types and naming of halogenoalkanes

1-bromopropane
primary

2-bromopropane
secondary

2-bromo-2-methylpropane
tertiary

Polarity of halogenoalkanes

Carbon and halogens have different electronegativities and halogenoalkanes have polar molecules with a polar C–X bond.

chlorine is more electronegative than carbon
electron flow from carbon to chlorine –
dipole produced

$C^{\delta+}-F^{\delta-}$
$C^{\delta+}-Cl^{\delta-}$ polarity
$C^{\delta+}-Br^{\delta-}$ **decreases**
$C^{\delta+}-I^{\delta-}$

The polarity produces an electron-deficient carbon atom, $C^{\delta+}$ which is important in the reactions of halogenoalkanes. The polarity decreases from fluorine to iodine, reflecting the decrease in electronegativity down the halogen group.

Nucleophilic substitution reactions of halogenoalkanes

AQA	M2	OCR	M2
EDEXCEL	M2	SALTERS	M2
CCEA	M2	WJEC	M2

The electron-deficient carbon atom of the polar $C^{\delta+}-X^{\delta-}$ bond attracts **nucleophiles**, allowing a **nucleophilic substitution** reaction to take place in which the nucleophile replaces the halogen atom (see also alcohols, page 137).

The hydrolysis of halogenoalkanes

> A **hydrolysis** reaction is a reaction in which water breaks a bond.

Reaction with aqueous hydroxide ions, $OH^-(aq)$ produces an alcohol.

$$C_2H_5Br + OH^-(aq) \xrightarrow[\text{reflux}]{OH^-/H_2O} C_2H_5OH + Br^-$$

The hydroxide ion, OH^-, behaves as a nucleophile.

Mechanism of nucleophilic substitution with primary halogenoalkanes

Primary halogenoalkanes react in a **one-step** mechanism.

$$H-\overset{\overset{\displaystyle H}{|}}{\underset{\underset{\displaystyle H}{|}}{C}}-\overset{\overset{\displaystyle H}{|}}{\underset{\underset{\displaystyle H}{|}}{C}}-\overset{\overset{\displaystyle H}{|}}{\underset{\underset{\displaystyle \overset{\bullet\bullet}{O}H^-}{|}}{C^{\delta+}}}-Br^{\delta-} \xrightarrow{\text{single step}} H-\overset{\overset{\displaystyle H}{|}}{\underset{\underset{\displaystyle H}{|}}{C}}-\overset{\overset{\displaystyle H}{|}}{\underset{\underset{\displaystyle H}{|}}{C}}-\overset{\overset{\displaystyle H}{|}}{\underset{\underset{\displaystyle H}{|}}{C}}-OH \; + \; {\bullet\bullet}Br^-$$

Note the usefulness of the curly arrow in showing movement of an electron pair:

- the **:OH⁻** nucleophile donates its electron pair to the electron-deficient $C^{\delta+}$ of the bromoalkane, forming a covalent bond;
- the C–Br bond is broken by heterolytic fission with loss of its electron pair to form a **:Br⁻** ion.

NaOH(aq) is used as a source of OH⁻ (aq).

The OH⁻ ion behaves as a nucleophile by donating an electron pair.

Mechanism of nucleophilic substitution with tertiary halogenoalkanes
(CCEA only)

With tertiary halogenoalkanes, substitution with hydroxide ions takes place in a **two-step** mechanism.

$$H_3C-\overset{\overset{\displaystyle CH_3}{|}}{\underset{\underset{\displaystyle CH_3}{|}}{C}}-Br^{\delta-} \xrightarrow{\text{1st step}} H_3C-\overset{\overset{\displaystyle CH_3}{|}}{\underset{\underset{\displaystyle CH_3}{|}}{C^{+}}} \quad {\bullet\bullet}OH^- \xrightarrow{\text{2nd step}} H_3C-\overset{\overset{\displaystyle CH_3}{|}}{\underset{\underset{\displaystyle CH_3}{|}}{C}}-OH$$

carbocation intermediate

- In the 1st step, a Br⁻ ion is lost.
- The carbocation intermediate is stabilised by the three alkyl groups (see also page 132).
- In the 2nd step, nucleophilic attack by a hydroxide ion forms a tertiary alcohol.

Comparing the hydrolysis reactions of halogenoalkanes

The rates of hydrolysis of different halogenoalkanes can be investigated as follows:

- Alkaline hydrolysis with NaOH(aq)/H₂O, reflux:

$$R-X + OH^- \longrightarrow R-OH + X^-$$

An alternative method uses water itself for the hydrolysis. The halogenoalkane is heated with a mixture of water, ethanol and AgNO₃(aq). The rate of hydrolysis is monitored directly by observing precipitation of replaced halide ions as silver halides.

- Acidification with dilute nitric acid. This removes excess NaOH, which would otherwise form a precipitate with the silver nitrate in the next stage, preventing detection of any silver halides.
- Addition of AgNO₃(aq) shows the presence of aqueous halide ions (see also Testing for halide ions, page 82).

$$Ag^+(aq) + X^-(aq) \longrightarrow AgX(s)$$

The intensity of any precipitate indicates the extent and rate of the hydrolysis.

$C^{\delta+}$–$F^{\delta-}$
$C^{\delta+}$–$Cl^{\delta-}$ polarity
$C^{\delta+}$–$Br^{\delta-}$ **increases**
$C^{\delta+}$–$I^{\delta-}$

bond	bond enthalpy /kJmol⁻¹
C–F	467
C–Cl	364
C–Br	290
C–I	228

Two factors – which is more important?

- **Polarity** predicts that the more polar C–F bond would attract water most easily and would give the fastest reaction.
- **Bond enthalpy** predicts that the C–I bond would be broken most easily and would give the fastest reaction.

For this reaction, bond enthalpy is more important than polarity.
The rate of hydrolysis increases as the C–X bond weakens.

KEY POINT

Further examples of nucleophilic substitution

AQA   M2   CCEA   M2
EDEXCEL   M2   SALTERS   M2

Nucleophilic substitution of halogenoalkanes is used in organic synthesis to introduce different functional groups onto the carbon skeleton. The reaction scheme below shows how halogenoalkanes react with ammonia and cyanide ions.

For hydrolysis, water is used as the solvent. For other nucleophilic substitutions, it is important that water is absent or hydrolysis will take place.

Ethanol is used as the alternative solvent.

(Note that HBr then reacts:
$NH_3 + HBr \longrightarrow NH_4Br$)

The ability of halogenoalkanes to undergo nucleophilic substitution is important in organic synthesis as compounds with different functional groups can be synthesised.

Elimination reactions of halogenoalkanes

AQA   M2   CCEA   M2
EDEXCEL   M2   WJEC   M2

When a halogenoalkane is refluxed with hydroxide ions in **anhydrous** conditions (using NaOH in **ethanol** as a solvent at 78°C), an **elimination** reaction takes place.

The hydroxide OH$^-$ ion behaves as a base, accepting a proton H$^+$ to form H_2O.

Elimination versus hydrolysis

AQA   M2   CCEA   M2
EDEXCEL   M2   WJEC   M2

The reaction of hydroxide ions with halogenoalkanes forms different organic products, depending upon the reaction conditions used.

Use different reaction conditions to control the type of reaction.

Commonly tested in exams.

aqueous conditions $\longrightarrow$	**alcohol**:	OH$^-$ acts as a **nucleophile**
anhydrous conditions $\longrightarrow$	**alkene**:	OH$^-$ acts as a **base**

KEY POINT

The ability to control *which* reaction takes place solely by changing the conditions is an important tool for the synthetic chemist.

Progress check

1 What is meant by the term nucleophilic substitution?

2 Why does 1-bromopropane react with nucleophiles but propane does not?

3 Why do bromoalkanes react more readily than chloroalkanes?

4 What are the organic products for the reactions of 1-bromopropane with:
(a) NaOH/H$_2$O;   (b) NH$_3$/ethanol;   (c) NaOH/ethanol?

4 (a) $CH_3CH_2CH_2OH$ (b) $CH_3CH_2CH_2NH_2$ (c) $CH_3CH=CH_2$.
3 C–Br bond is weaker than C–Cl bond.
2 C–Br bond is polar, C–H bonds are not polar.
1 Replacement of an atom or group in an organic molecule by a nucleophile.

Organic practical techniques

EDEXCEL M2   SALTERS M2
CCEA M2

Common techniques

During your AS Chemistry course, you may have carried out one or more of the following experiments.

- The preparation of a chloroalkane from an alcohol using HCl.
- The oxidation of alcohols to form an aldehyde or a carboxylic acid.
- Preparation of cyclohexene by dehydration of cyclohexanol.

Stages in the preparation of a pure organic liquid product

You will have developed some important practical skills. You may be assessed directly on your practical skills but you may also be asked about some of the following techniques on theory papers.

Important organic practical skills

- Heating under **reflux**.
  - The boiling of a liquid and subsequent condensation back to the same container. This prevents the organic solvent from boiling away and the heated reactants from drying out.
- **Separating** immiscible liquids using a separating funnel.
  - Each liquid can be run off, in turn, through the tap.
- **Removing impurities** using a separating funnel.
  - **Acidic impurities** can be removed by shaking an organic liquid with sodium hydrogencarbonate solution, which reacts with any acid.
- **Drying** an organic liquid.
  - An anhydrous salt is used such as anhydrous sodium sulfate.
- **Distillation** of an organic liquid.
  - Distillation allows collection of a pure liquid product. You can separate organic liquids in a mixture by distilling each off at its boiling point (fractional distillation).

Remember to take the stopper out or nothing will run out of the tap!

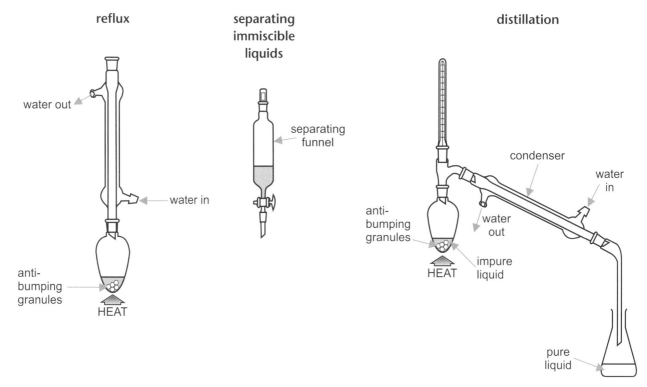

reflux   separating immiscible liquids   distillation

water out

water in

separating funnel

anti-bumping granules

HEAT

condenser

water in

anti-bumping granules

water out

impure liquid

HEAT

pure liquid

4.7 Analysis

After studying this section you should be able to:

- *understand that infra-red spectroscopy can be used to identify functional groups*
- *interpret a simple infra-red spectrum*
- *use a mass spectrum to determine the relative molecular mass of an organic molecule*
- *use the fragmentation pattern of a mass spectrum to deduce likely structures for an unknown compound*

Using infra-red (IR) spectroscopy in analysis

AQA	M2	OCR	M2
EDEXCEL	M2	SALTERS	M2
CCEA	M2	WJEC	M2

AQA (A2)	M4	OCR (A2)	M4
EDEXCEL (A2)	M4 M5	SALTERS (A2)	M4 M5
		WJEC (A2)	M4

The frequency of IR absorption is measured in wavenumbers, units: cm⁻¹.

Basic principles

Bonds in molecules naturally vibrate. Some bonds in molecules increase their vibrations by absorbing energy from IR radiation. Different bonds absorb different frequencies of IR radiation.

An IR spectrum is obtained by passing a range of IR frequencies through a compound. As energy is taken in, **absorption peaks** are produced. The frequencies of the absorption peaks can be matched to those of known bonds to identify structural features in an unknown compound.

IR radiation has less energy than visible light.

> **KEY POINT**
>
> IR spectroscopy is useful for identifying the functional groups in a molecule.

Important IR absorptions

You don't need to learn the absorption frequencies – the data is provided.

bond	functional group	wavenumber/cm⁻¹
O–H	hydrogen bonded in alcohols	3230 – 3550
N–H	amines	3100 – 3500
C–H	organic compound with a C–H bond	2850 – 3100
O–H	hydrogen bonded in carboxylic acids	2500 – 3300 (broad)
C≡N	nitriles	2200 – 2260
C=O	aldehydes, ketones, carboxylic acids, esters	1680 – 1750
C–O	alcohols, esters	1000 – 1300

IR spectroscopy is used in some modern breathalysers for measuring the concentration of blood alcohol. A particular IR absorption identifies the presence of ethanol in the breath and the intensity of the peak is directly related to the ethanol level.

An infra-red spectrum is particularly useful for identifying:

- an **alcohol** from absorption of the O–H bond
- a **carbonyl** compound from absorption of the C=O bond
- a **carboxylic acid** from absorption of the C=O bond **and** broad absorption of the O–H bond.

Interpreting infra-red spectra

AQA	M2	OCR	M2
EDEXCEL	M2	SALTERS	M2
CCEA	M2	WJEC	M2

AQA (A2)	M4	OCR (A2)	M4
EDEXCEL (A2)	M4 M5	SALTERS (A2)	M4 M5
		WJEC (A2)	M4

Carbonyl compounds (aldehydes and ketones)
Butanone, $CH_3COCH_2CH_3$

- C=O absorption 1680 to 1750 cm⁻¹

C=O absorption at 1740 cm⁻¹

IR (liquid film)

IR spectroscopy is most useful for identifying C=O and O–H bonds.

Look for the distinctive patterns.

Alcohols

Ethanol,
C_2H_5OH

- O–H absorption
  3230 to 3500 cm^{-1}
- C=O absorption
  1000 to 1300 cm^{-1}

Note that all these molecules contain C–H bonds, which absorb in the range 2840 to 3045 cm^{-1}.

Carboxylic acids

Propanoic acid,
C_2H_5COOH

- Very broad O–H absorption
  2500 to 3500 cm^{-1}
- C–O absorption
  1680 to 1750 cm^{-1}

There are other organic groups (e.g. N–H, C=C) that absorb IR radiation but the principle of linking the group to the absorption wavenumber is the same.

Ester

Ethyl ethanoate,
$CH_3COOC_2H_5$

- C=O absorption
  1680 to 1750 cm^{-1}
- C–O absorption
  1000 to 1300 cm^{-1}

The fingerprint region is unique for a particular compound.

Fingerprint region

- Between 1000 and 1550 cm^{-1}

Many spectra show a complex pattern of absorption in this range.

- This pattern can allow the compound to be identified by comparing its spectrum with spectra of known compounds.

Progress check

1 Ethanol, CH_3CH_2OH was oxidised to ethanal, CH_3CHO and then to ethanoic acid, CH_3COOH. How could you use IR spectroscopy to follow the course of this reaction?

1 Ethanol absorbs at about 3300 cm^{-1} (OH); ethanal absorbs at about 1700 cm^{-1} (C=O); ethanoic acid absorbs at about 1700 cm^{-1} (C=O) and 2500–3500 cm^{-1} (broad) (OH from COOH group).

Mass spectrometry

AQA	M2	OCR	M2
EDEXCEL	M2	SALTERS	
		WJEC	M2

AQA (A2)	M4	OCR (A2)	
EDEXCEL (A2)	M4 M5	SALTERS (A2)	
		WJEC (A2)	M4

Mass spectrometry can be used to determine relative atomic masses from a mass spectrum (see page 31). Mass spectrometry is also used to determine relative molecular masses and to identify the molecular structures of organic compounds.

Molecular ions

The molecular ion peak is usually given the symbol M.

Organic molecules can be analysed using mass spectrometry.

In the mass spectrometer, organic molecules are bombarded with electrons. This can lead to the formation a **molecular ion**.

The mass spectrum also contains fragments of the molecule ion (see detail below).

The equation below shows the formation of a molecular ion from butanone, $CH_3COCH_2CH_3$.

$$H_3C-\overset{\overset{\displaystyle O}{\|}}{C}-CH_2CH_3 \;+\; e^- \longrightarrow \left[H_3C-\overset{\overset{\displaystyle O}{\|}}{C}-CH_2CH_3\right]^+ \;+\; 2e^-$$

molecular ion, m/z: **72**

- The molecular ion peak, M, provides the relative molecular mass of the compound.

High resolution mass spectrometry

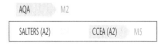

Molecules of similar relative molecular masses can be distinguished using high resolution mass spectrometry.

Using this technique:

- the molecular mass is obtained with a very accurate value
- this is compared with very accurate relative isotopic masses.

Worked Example

isotope	relative isotopic mass
^1H	1.0078
^{12}C	12.0000
^{14}N	14.0031
^{16}O	15.9949

Three gases, CO, C_2H_4 and N_2, each have an approximate relative molecular mass of 28.

The table below shows accurate molecular masses for these gases, calculated from accurate relative isotopic masses (see the margin).

gas	CO	C_2H_4	N_2
molecular mass	27.9949	28.0312	28.0062

The gases can be distinguished by matching the m/z values from high resolution mass spectrometry with the calculated values.

Fragment ions

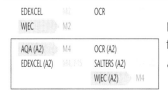

The fragmentation pattern provides clues about the molecular structure of the compound.

In the conditions within the mass spectrometer, some molecular ions are fragmented by bond fission.

- The bond fission that takes place is a fairly random process:
  - different bonds are broken
  - a **mixture** of **fragment ions** is obtained.
- The mass spectrum contains both the molecular ion and the mixture of fragment ions.

Mass spectrometry of organic compounds is useful for identifying:

- the relative molecular mass
- parts of the skeletal formula.

Fragmentation of butane, C_4H_{10}

The mass spectrum of butane would contain peaks for these four ions:

$C_4H_{10}{}^+$: $m/z = 58$
$C_3H_7{}^+$: $m/z = 43$
$C_2H_5{}^+$: $m/z = 29$
$CH_3{}^+$: $m/z = 15$

Bond fission forms a fragment ion and a radical. The diagram below shows that fission of the same C–C bond in a butane molecule can form two different fragment ions.

$$H_3C-CH_2-CH_2 \!\vdots\! CH_3{}^+$$
$m/z = 58$

loss of 15 mass units $\longrightarrow$ $H_3C-CH_2-CH_2{}^+$ + $^\bullet CH_3$
$m/z = 43$

loss of 43 mass units $\longrightarrow$ $H_3C-CH_2-\overset{\bullet}{C}H_2$ + $CH_3{}^+$
$m/z = 15$

- Fragmentation of the molecular ion produces a fragment ion by loss of a radical.
- The mass spectrometer only detects the fragment ion.
- The fragment ion may itself fragment into another ion and radical.

The mass spectrum of butanone

The mass spectrum of butanone, $CH_3CH_2COCH_3$, is shown below. The m/z values of the main peaks have been labelled.

Note that only ions are detected in the mass spectrum.

Uncharged species such as the radical cannot be deflected within the mass spectrometer.

base peak at m/z: 43

m/z: 15

m/z: 29

m/z: 57

M peak at m/z: 72

M+1 peak at 73

m/z

The M+1 peak is a small peak 1 unit higher than the molecular ion peak.

The origin of the M+1 peak is the small proportion of carbon-13 in the carbon atoms of organic molecules.

The table below shows the identities of the main peaks.

A common mistake in exams is to show both the fragmentation products as ions.

m/z	ion	fragment lost
72	$CH_3COCH_2CH_3^+$	–
57	$CH_3CH_2CO^+$	$CH_3\bullet$
43	CH_3CO^+	$CH_3CH_2\bullet$
29	$CH_3CH_2^+$	$CH_3CO\bullet$
15	CH_3^+	$CH_3CH_2CO\bullet$

The acylium (RCO+) ion is particularly stable and is often present as a high dominant peak.

The most abundant peak in the mass spectrum is the base peak.

- The base peak in the mass spectrum of butanone has m/z: 43, formed by the loss of a $CH_3CH_2\bullet$ radical:

The base peak is given a relative abundance of 100. Other peaks are compared with this base peak.

$$\left[\underset{\substack{molecular\ ion \\ m/z:\ 72}}{H_3C-\overset{\displaystyle O}{\overset{\|}{C}}-CH_2CH_3} \right]^+ \longrightarrow \left[\underset{\substack{fragment\ ion \\ m/z:\ 43}}{H_3C-\overset{\displaystyle O}{\overset{\|}{C}}} \right]^+ + \underset{\substack{radical \\ M-29}}{\bullet CH_2CH_3}$$

Common patterns in mass spectra

Different fragmentations are possible depending on the structure of the molecular ion.

- The table above shows common fragment ions and fragmented radicals.
- The skill in interpreting a mass spectrum is in searching for known patterns, evaluating all the evidence and reassembling all the information into a molecular structure.
- You should also look for a peak at m/z 77 or loss of 77 units. This is a giveaway for the presence of a phenyl group, C_6H_5, in a molecule.

Progress check

1 Write an equation for the formation of the fragment ion at $m/z=29$ in the mass spectrum of butanone above.

2 What peaks would you expect in the mass spectrum of pentane?

m/z: 29, $C_2H_5^+$; m/z: CH_3^+, 15.

2 m/z: 72, $C_5H_{12}^+$; m/z: 57, $C_4H_9^+$; m/z: 43, $C_3H_7^+$;

1 $CH_3CH_2COCH_3^+ \longrightarrow CH_3CH_2^+ + CH_3CO\bullet$

4.8 Chemistry in the environment

Global warming

AQA   M2   OCR   M2
EDEXCEL M2   SALTERS M2

The main gases in the air are:

nitrogen 78%

oxygen 21%

argon 0.9%

carbon dioxide 0.04%

Although O_2 and N_2 are the main gases in the atmosphere, their molecules do not change polarity when they vibrate and so they do not absorb IR radiation.

The greenhouse effect

The **greenhouse effect** is a natural process, keeping the Earth at a temperature capable of supporting life. Most IR radiation emitted by the Earth goes back into space. However, greenhouse gases in the atmosphere absorb some of the IR radiation, which is then re-emitted as heat, with some passing back towards the Earth.

- The Earth absorbs energy at the same rate as it radiates energy.
- The equilibrium maintains a steady temperature.

The absorption-emission process keeps the heat close to the surface of the Earth.

Greenhouse gases

Absorption of IR radiation by **greenhouse gases** causes bonds to vibrate. A molecule only absorbs IR radiation if its polarity changes as the bonds vibrate.

The main absorbers of IR radiation are the C=O, O–H and C–H bonds in H_2O, CO_2 and CH_4. Other IR absorbers include the nitrogen–oxygen bonds in nitrogen oxides such as NO and NO_2 (often referred to as NO_x).

The greenhouse effect of a gas depends on:

- its atmospheric concentration
- its ability to absorb IR radiation.

Climate change

Climate change refers to variations in the Earth's climate over time.
- Over hundreds of thousands of years, natural climate changes have been responsible for ice-ages, followed by warming.
- Over shorter periods of time, natural climate change can be the result of factors such as sunspot activity and the presence of volcanic dust in the atmosphere.

The Earth has always experienced **natural climate** change.

Scientific evidence has shown that substantial climate change is now taking place and that the Earth's climate is warming. Scientists agree that the production of greenhouse gases from **human activities** is the primary factor driving **climate change**.

- Human activity is producing **more** greenhouse gases, particularly CO_2.
- Increased proportions of greenhouse gases are upsetting the sensitive, natural balance of nature, resulting in man-made global warming.

Climate change derived from human activities is known as **anthropogenic** climate change, as opposed to natural climate change without human influences.

Computer modelling has predicted significant problems, e.g. melting ice caps, rising sea levels and more extreme storms and flooding.

International treaties, such as the Kyoto Protocol, have set targets for reductions in CO_2 production.

Carbon neutrality

AQA   M2   OCR
SALTERS

Carbon neutrality can be achieved in various ways, for example:

changing from finite fuels, such as oil, to renewable fuels based on plants that can be grown annually, such as ethanol and rape-seed oil

cutting down fewer trees and growing more trees to take in CO_2

using less fuel and more public transport

carbon storage and capture (CSS)

improving the efficiency of engines.

Although biofuels are often assumed to be carbon neutral, this is not strictly true – additional energy is required for transportation of the fuels, production of fertilisers, etc.

Carbon footprint

As with carbon neutrality, 'carbon footprint' is used to quantify CO_2 emissions.

Carbon footprint is the total amount of CO_2 and other greenhouse gases emitted over the full life cycle of a product or service.

The proportion of CO_2 in the atmosphere has increased more in the last hundred years than at any other time in the Earth's history. This increase is blamed on human activities. Carbon neutrality is seen as an important step in curbing global warming by re-balancing the CO_2 in the atmosphere.

> **KEY POINT**
>
> **Carbon neutral** refers to 'an activity that has no net annual carbon (greenhouse gas) emissions to the atmosphere'.

For carbon neutrality, amounts of CO_2 released into the atmosphere by human activities, such as from burning fuels, is balanced by the **same amounts** of CO_2 being taken in **from** the atmosphere during plant growth.

Alternatively, CO_2 produced during combustion could be stored or 'sequestered' by **carbon storage and capture** (CSS). This can achieved by:
- removal of waste carbon dioxide as a liquid, injected deep in the oceans
- storage as a gas in deep geological formations
- reacting carbon dioxide with metal oxides to form stable carbonate minerals.

This *trapping* of CO_2 emissions would substantially reduce CO_2 emissions from power stations. CSS could make the combustion of fossil fuels carbon neutral.

For transport, carbon neutrality can be approached by use of **biofuels**.

> **KEY POINT**
>
> A **biofuel** is a liquid or gas transportation fuel derived from biomass.

- Biofuels are **renewable**. Once used, fresh fuel can be made again from plants, grown in a short length of time.
- **Finite** resources, such as oil and coal take millions of years to form. Once used, fossil fuels are spent – they are non-renewable.

The best sources of biofuels are alcohols, such as ethanol and methanol, and vegetable oils, such as the rape-seed oil. Biofuels can be used with petrol to fuel cars.

The carbon neutrality of bio-ethanol is achieved in the following way:
- Growing plants produces sugars, taking in CO_2 from the atmosphere.
- The sugars are fermented with yeast, making bio-ethanol.
- The bio-ethanol is burned as a fuel, releasing CO_2 into the atmosphere.
- More sugars are made to form fresh bio-ethanol.

Alternative fuels, such as hydrogen, would produce no CO_2 emissions but the processes required to produce H_2 as a fuel are currently energy intensive. Hydrogen fuels remain a goal for the future, particularly as a fuel for cars.

The ozone layer

EDEXCEL   M2   OCR
SALTERS

The ozone layer is present 10–60 km above the Earth's surface in the region of the upper atmosphere known as the **stratosphere**.

The ozone layer forms naturally by the interaction of **ultraviolet** (UV) radiation with **oxygen**.

Ultraviolet radiation breaks O_2 molecules into oxygen radicals.

$$O_2 \longrightarrow 2O$$

The O radicals react with O_2 molecules to form **ozone** molecules, O_3. The bond-forming process releases energy as heat.

$$O_2 + O \longrightarrow O_3 + heat$$

O_3 molecules in turn absorb more UV radiation and are themselves broken into O_2 and O. An equilibrium is set up:

$$O_2 + O \rightleftharpoons O_3$$

The Sun mainly radiates radiation in the visible region of the spectrum, with an appreciable amount of UV.

The Earth mainly radiates IR radiation.

- The process maintains a natural concentration of ozone.
- Overall, **UV radiation is absorbed** and **heat is evolved**. This increases the temperature in the stratosphere.

The high energy of the UV radiation would cause harm to life on Earth. Excess UV can lead to skin cancer. The ozone layer is essential in protecting life on Earth by filtering out the harmful UV radiation.

Depletion of ozone

AQA   M2   OCR
EDEXCEL   M2   SALTERS

The ozone layer is beneficial for life on Earth.

The Cl and NO radicals are recycled in chain reactions, which continue many thousand of times until two radicals collide, terminating the reaction. A single Cl radical can remove thousands of O_3 molecules from the ozone layer.

See also homogeneous catalysis: page 109.

Depletion of ozone is caused by **radicals**, particularly Cl and NO.

Cl and NO radicals have been formed from human activity, affecting the natural balance that maintains the ozone layer.

Cl radicals are formed by the action of **UV radiation** on **CFCs**, which were once used in aerosols and as refrigerants.

e.g. $CClF_3 \longrightarrow Cl + CF_3$

Oxides of nitrogen (NO_x) are formed from thunderstorms and aircraft flying through the upper atmosphere.

e.g. $N_2 + O_2 \longrightarrow 2NO$

Both processes lead to depletion of ozone in the stratosphere:

stage 1 $O_3 + Cl\bullet \longrightarrow ClO\bullet + O_2$
stage 2 $ClO\bullet + O \longrightarrow O_2 + Cl\bullet$
overall: $O_3 + O \longrightarrow 2O_2$

stage 1 $O_3 + NO\bullet \longrightarrow NO_2\bullet + O_2$
stage 2 $NO_2\bullet + O \longrightarrow O_2 + NO\bullet$
overall: $O_3 + O \longrightarrow 2O_2$

In the processes above, Cl and NO act as catalysts.

- They are used up in the first stage.
- They are regenerated in the second stage.

Overall, Cl and NO radicals are **not consumed**.

Benefits of CFC use versus damage to ozone

In the 1930s, CFCs were seen as being the ideal chemicals for refrigerants, aerosols and as blowing agents for 'expanded poly(styrene)'.

CFCs have the ideal properties for these uses: high volatility, non-toxicity, non-flammability, no smell and extreme unreactivity.

This unreactivity causes the problem.

- Most pollutants are broken down naturally in the lower atmosphere. However, unreactive CFCs diffuse over many years up to the upper atmosphere.
- High-energy UV radiation in the stratosphere finally breaks down the CFCs, forming Cl radicals. The highly reactive Cl radicals attack the ozone.

In the 1970s and 1980s, scientists discovered that Cl radicals were responsible for thinning of the ozone layer.

- Most governments have signed up to the Montreal Protocol, which sets targets for reduced use of CFCs.
- Since the early 2000s, the use of CFCs has been banned internationally. Although replacement chemicals are now in use, many old refrigerators contain CFCs and these have to be disposed of carefully by collecting the CFCs.

It may take a century before the ozone layer fully recovers.

Green chemistry

EDEXCEL M2   SALTERS M2
OCR M2   WJEC M1

Exam questions are likely to be set in the context of green chemistry. You need to know the main principles and be able to apply them.

The chemical industry is currently reinventing itself to make it more sustainable ('greener'). Green chemistry encourages the design of products and processes that reduce or eliminate the use and generation of hazardous substances. It seeks to produce more efficient processes with less energy input.

The main principles of green chemistry are listed below.

- Using industrial processes that reduce or eliminate hazardous chemicals and which involve the use of fewer chemicals.
  - Lead has been eliminated from petrol, paints and electrical components.
  - Waste CO_2 has replaced CFCs as the blowing agent in expanded polystyrene.

- Designing processes with a high atom economy (see page 124):
  - reducing waste and preventing pollution of the environment
  - discovering catalysts for improving the atom economies of reactions.

- Ensuring that any waste products produced are non-toxic:
  - producing waste that can be safely biodegraded into harmless substances
  - increasing the use of recycling.

- Consuming less energy:
  - replacing or supplementing fossil fuels with renewable fuels
  - making efficient use of energy or developing renewable energy sources.

Other factors

The apparent benefits may be offset by unexpected and detrimental side-effects.

Although rich countries may benefit from the use of renewable fuels, the population in the poorer countries may get insufficient food, e.g. the production of biodiesel uses grain crops and land.

Progress check

1 State three greenhouse gases and the bonds responsible for IR absorption.

2 Outline how the use of bio-ethanol approaches carbon neutrality.

3 Outline the natural processes responsible for maintaining the ozone layer in the stratosphere.

4 Write equations to show how the presence of CFC in the stratosphere depletes the ozone layer.

1 H_2O (O–H); CO_2 (C=O); CH_4 (C–H).

2 Growing plants produces sugars, taking in CO_2 from the atmosphere.
The sugars are fermented with yeast, making bio-ethanol.
The bio-ethanol is burned as a fuel, releasing CO_2 into the atmosphere
Overall, the CO_2 taken in during growth = CO_2 released to the atmosphere (although there will be some extra CO_2 released from transportation and fertiliser production.)

3 Ultraviolet radiation breaks O_2 molecules into oxygen radicals.
$O_2 \longrightarrow 2O$
The O radicals react with O_2 to form O_3, with excess energy released as heat.
$O_2 + O \longrightarrow O_3 + heat$
O_3 molecules absorb more UV radiation and are broken into O_2 and O.
An equilibrium is set up: $O_2 + O \rightleftharpoons O_3$

4 $CClF_3 \longrightarrow Cl + CF_3$
stage 1 $O_3 + Cl\bullet \longrightarrow ClO\bullet + O_2$
stage 2 $ClO\bullet + O \longrightarrow O_2 + Cl\bullet$
overall: $O_3 + O \longrightarrow 2O_2$

Sample question and model answer

For open-ended questions of this type, there are usually more marking points than the total — you have some breathing space.

1

(a) Explain, using examples, the following terms used in organic chemistry.

(i) *general formula*

This is the simplest algebraic representation for any member of a series of organic compounds, e.g. alkanes: C_nH_{2n+2} ✓

(ii) *homologous series*

This is a series containing compounds with similar chemical properties ✓ with each successive member differing by CH_2 ✓, e.g. 'alkanes' ✓

Do make sure that you choose valid examples. In the actual examination, many candidates managed to confuse homolytic and heterolytic fission – poor preparation!

(iii) *homolytic fission* and *heterolytic fission*

Fission means the breaking of a bond. ✓

In homolytic fission, one electron from the bond is released to each atom with radicals forming. ✓ e.g. $Cl_2 \longrightarrow 2Cl\bullet$ ✓

In heterolytic fission, both electrons in the bonding pair are released to one of the atoms with ions forming ✓ e.g. $Br_2 \longrightarrow Br^+ + Br^-$ ✓

9 marks $\longrightarrow$ [7]

Always explain your working. This type of problem is often easier than it looks.

You should look back at the question many times to check that you have answered all parts.

(b) Cracking of the unbranched compound **A**, C_6H_{14}, produced the saturated compound **B** and the unsaturated hydrocarbon **C** (M_r, 42). Compound **B** reacted with bromine in UV light to form a monobrominated compound **D** and an acidic gas **E**. Compound **C** reacted with hydrogen bromide to form a mixture of two compounds **F** and **G**.

(i) Use this evidence to suggest the identity of each of compounds **A** to **G**. Include equations for the reactions in your answer.

A is unbranched. ∴ **A** = $CH_3CH_2CH_2CH_2CH_2CH_3$ ✓

C has M_r = 42; must have a double bond: must contain 3 carbons ∴ **C** = C_3H_6. ✓

Notice that everything 'works': questions are designed in this way. If you find yourself with 'strange' structures containing functional groups that you haven't seen before, you have probably made a mistake. This is the time to check back through your logic.

Cracking of A produces an alkane and an alkene, B is an alkane: **B** = C_3H_8 ✓

The equation is: $C_6H_{14} \longrightarrow C_3H_6 + C_3H_8$ ✓

Alkanes react with bromine in UV by substitution. ∴ **D** = C_3H_7Br ✓, **E** = HBr ✓

The equation is: $C_3H_8 + Br_2 \longrightarrow C_3H_7Br + HBr$ ✓

Alkenes react with bromine by addition. HBr can be added across a double bond in two ways to give: **F** and **G** as $CH_3CH_2CH_2Br$ ✓ and $CH_3CHBrCH_3$. ✓

The equation is: $C_3H_6 + HBr \longrightarrow C_3H_7Br$ ✓

Any 2 examples here formed from C_6H_{14}. These could be branched hydrocarbons, aromatic or alicyclic – but any chosen **must** contain 6 carbons. (Read the question!)

(ii) Oil companies often process compounds such as **A** into better fuels. Suggest two organic products of this processing of **A**.

Compound **A**, hexane, could be converted into benzene ✓ and 2-methylpentane. ✓

(iii) Predict the structure of the polymer that could be formed from compound **C**.

When drawing a polymer: 'side bonds' are always needed. Always show the repeat unit.

$$\left[\begin{array}{cc} H & CH_3 \\ | & | \\ -C-C- \\ | & | \\ H & H \end{array} \right]_n$$

✓✓(1 for no double bond, 1 for correct skeleton)   [14]

Practice examination questions

1 There are four structural isomers with the molecular formula $C_3H_6Cl_2$.

(a) (i) Explain the term *structural isomers*.

(ii) Show the structural formulae of the structural isomers of $C_3H_6Cl_2$.

[5]

(b) Some propane was reacted with chlorine to form a mixture of isomers.

(i) What conditions are required for this reaction to take place?

(ii) Two structural isomers, **A** and **B**, were separated from the mixture. These isomers had a molar mass of 78.5 g mol^{-1}. Deduce the molecular formula of these two isomers.

(iii) Draw the structural formulae of **A** and **B** and name each compound.

[6]
[Total: 11]

2 1,2-Dichloroethene has *E* and *Z* isomers.

(a) (i) Explain the two conditions required for an organic molecule to have *E* and *Z* isomers.

(ii) What is the difference between *E* and *Z* isomers?

(iii) Draw and label the *E* and *Z* isomers of 1,2-dichloroethene.

[5]

(b) The *E* and *Z* isomers of 1,2-dichloroethene can both be reduced to 1,2-dichloroethane. What are the reagents and conditions required?

[2]

(c) The *E/Z* isomers of $C_2H_2Cl_2$ polymerise to form poly(1,2-dichloroethene).

Draw a short section of poly(1,2-dichloroethene), showing **two** repeat units.

[2]
[Total: 9]

3 The flowchart below shows a reaction sequence starting from but-2-ene.

$$CH_3CH=CHCH_3 \xrightarrow{\text{HBr}} A \xrightarrow[\text{heat}]{\text{OH}^-\text{(aq)}} B \xrightarrow[\text{heat}]{Cr_2O_7{}^{2-}/H^+} C$$

(a) Draw a short section of the addition polymer formed by but-2-ene showing two repeat units.

[2]

(b) Identify compounds **A**–**C** in the flowchart above.

[3]

(c) Name the types of reaction taking place in the flowchart above and state the role of the reagents.

(i) $CH_3CH=CHCH_3 \longrightarrow A$

(ii) $A \longrightarrow B$

(iii) $B \longrightarrow C$

[6]

(d) What reagents and conditions would be needed to convert but-2-ene directly into compound **B**?

[2]
[Total: 13]

4 A branched alkene **A** was reduced with hydrogen to form a compound **B**. Compound **B** had the composition by mass of C, 82.8 %; H, 17.2 % (M_r = 58). Compound **A** was reacted with steam with an acid catalyst to produce a mixture of two structural isomers **C** and **D**.

(a) Calculate the empirical and molecular formula of compound **B**. [3]

(b) Show structural formulae for compounds **A–D**. [4]

(c) Write an equation for each reaction. [2]

[Total: 9]

5 (a) There are four structural isomers of molecular formula C_4H_9Br. The structural formulae of two of these isomers are:

- Compound **A**: $CH_3CH_2CH_2CH_2Br$

- Compound **B**: $(CH_3)_2CHCH_2Br$.

(i) Show the remaining two structural isomers.

(ii) Give the name of **B**.

[3]

(b) All four structural isomers of C_4H_9Br react with hot aqueous sodium hydroxide.

(i) What type of reaction is taking place?

(ii) Draw the structural formula of the organic product formed by the reaction of **B** with hot aqueous sodium hydroxide.

[2]

(c) Ethene, C_2H_4, reacts with bromine.

(i) State the name of the mechanism involved.

(ii) Show the mechanism for this reaction.

[4]

[Total: 9]

Practice examination answers

Chapter 1 Atoms, moles and reactions

1 (a)

particle	relative mass	relative charge	
proton	1	1+	✓
neutron	1	0	✓
electron	1/2000	1–	✓

[3]

(b) (i)   neutron ✓; ^{13}C has 1 more neutron ✓

   (ii)   electron ✓; $^{16}O^{2-}$ has 2 more electrons ✓

   (iii)   proton ✓; $^{24}Mg^{2+}$ has 1 more proton ✓   [6]
[Total: 9]

2 (a) Difference: different numbers of neutrons. ✓

   Similarities: same numbers of protons ✓ and electrons. ✓   [3]

(b) ^{11}B ✓   [1]

(c) (i) $1s^22s^22p^63s^23p^6$ ✓;   (ii) p block ✓   [2]

(d) (i) $Si(g) \longrightarrow Si^+(g) + e^-$; ✓✓   (ii) The fifth electron is lost from a different shell ✓, closer to the nucleus ✓, less shielding ✓, experiencing a greater nuclear attraction ✓.   [6]
[Total: 12]

3 (a) (i)   0.0170 ✓;   (ii) 0.0850 mol dm⁻³ ✓✓;

   (iii)   204 cm³ using 24 dm³; 211 cm³ using $pV=nRT$ (AQA) ✓✓✓   [6]

(b) 0.0741 mol dm⁻³ ✓✓✓   [3]
[Total: 9]

4 (a) (i)   0.0145 mol ✓;   (ii) 0.0290 mol ✓;   (iii) 23.2 cm³ ✓   [3]

(b) $M(Na_2CO_3)$ = 106 g mol⁻¹ ✓

   mass Na_2CO_3 = 0.0145 x 106 = 1.537 g ✓   [2]

(c) Mg is oxidised from 0 to +2 ✓; H is reduced from +1 to 0 ✓   [2]
[Total: 7]

5 (a) (i)   0.0625 ✓;   (ii) $M(NaN_3)$ = 65 g mol⁻¹ ✓;

   moles NaN_3 = 0.0417 mol ✓; mass NaN_3 = 0.0417 x 65 = 2.71 g ✓   [4]

(b) (i)   concentration = $0.0417 \times \dfrac{1000}{50.0}$ ✓ = 0.834 mol dm⁻³ ✓

   (ii)   Na is oxidised from 0 to +1 ✓; H is reduced from +1 to 0 ✓   [4]
[Total: 8]

Chapter 2 Bonding, structure and the Periodic Table

1 (a) A covalent bond is a shared ✓ pair ✓ of electrons. [2]

   (b) In a dative covalent bond, the same atom provides both electrons for the bonded pair ✓.
       The O of a water molecule uses a lone pair of electrons to form a dative covalent bond
       with H^+ to form an H_3O^+ ion ✓. [2]

   (c) H_2O: 104.5° ✓; H_3O^+: 107° ✓. In H_3O^+, there is repulsion between three bonded pairs
       of electrons and one lone pair of electrons ✓. The lone pair repels more than the bonded
       pairs, reducing the tetrahedral bond angle by 2.5° ✓. [4]

   (d) Hydrogen bonding ✓. This arises from dipole-dipole attraction ✓ between $H^{\delta+}$ of
       a water molecule and $O^{\delta-}$ on a different water molecule ✓. [3]

   [Total: 11]

2 Electron pairs repel one another ✓ and assume a position as **far away** from one another
   as possible ✓. Lone pairs repel more than bonded pairs ✓. Diagrams are shown below:

   (a) (b) (c)

   1 mark for each diagram
   1 mark for each bond angle
   1 mark for lone pair on PH_3

   tetrahedral ✓   pyramidal ✓   trigonal planar ✓ [Total: 13]

3 (a) (i) covalent ✓; (ii) van der Waals' forces ✓. [2]

   (b) Weak van der Waals' (or induced dipole-dipole) forces break on vaporising ✓. [1]

   (c) Strong ✓ ionic bonds ✓ hold the lattice together. [2]

   (d) Ions vibrate faster ✓. [1]

   (e) On melting, the ionic bonds remain close together and are just loosened from their
       rigid lattice positions ✓. On vaporising, the ions must be separated, breaking the
       ionic attraction ✓. [2]

   [Total: 8]

4 (a) Water has hydrogen bonding ✓. This is dipole-dipole attraction ✓ between $H^{\delta+}$
       of a water molecule and a lone pair ✓ on $O^{\delta-}$ on a different water molecule ✓. [4]

   (b) HCl has dipole-dipole interactions ✓ between $H^{\delta+}$ of a HCl molecule ✓ and $Cl^{\delta-}$
       on a different HCl molecule ✓. any 2 ⟶ [2]

   (c) Kr has van der Waals' forces only ✓. These are caused by oscillating dipoles ✓
       which cause induced dipoles on another Kr atom ✓.
       H-bonding > dipole-dipole > van der Waals' ✓. The intermolecular forces are
       broken on boiling ✓. The stronger the intermolecular forces, the higher the
       boiling point ✓. any 4 ⟶ [4]

   [Total:10]

5 Across Period 3, the atomic radii decrease ✓ because there are more protons in the nucleus
   across the period ✓ increasing the attraction ✓ on the electrons in the same outer shell ✓.

   Down Group 2, the atomic radii increase ✓ as more shells are being added ✓. The shielding
   effect increases as the number of inner shells increases ✓. The attraction experienced by the
   outer electrons from the nucleus decreases ✓. [Total: 8]

6 (a) (i) $Ba(s) + 2H_2O(l) \longrightarrow Ba(OH)_2(aq) + H_2(g)$ ✓

       (ii) $2Mg(s) + O_2(g) \longrightarrow 2MgO$ ✓

       (iii) $Mg(s) + 2HCl(aq) \longrightarrow MgCl_2(aq) + H_2(g)$ ✓ [3]

   (b) (i)   On descending, outer electrons are further from the nucleus ✓ with greater
             shielding ✓. Therefore less attraction on outer electrons which are lost more easily ✓.
       (ii) Reaction with $H_2O/O_2/Cl_2$/named acid ✓. [4]

   (c) (i)   Very high melting point/strong forces holding lattice together ✓.
       (ii) CaO: cement ✓; $Mg(OH)_2$: indigestion remedy ✓ *many others possible*. [3]

   [Total: 10]

7 (a) On descending Group 7, the atomic radii increase ✓ resulting in less nuclear attraction at the edge of the atom ✓. There are more electron shells between the nucleus and the edge of the atom to shield the nuclear charge ✓. The nuclear attraction at the edge of the atom decreases down the group ✓. ∴ the oxidising power of the Group 7 elements decreases down the group ✓. [5]

(b) (i) $Cl_2 \longrightarrow NaCl$ (ox no: Cl 0 → –1) ✓ reduction ✓;

$Cl_2 \longrightarrow NaClO_3$ (ox no: Cl 0 → +5) ✓ oxidation ✓

(ii) 2.7 mol Cl_2 reacts ✓; 0.90 mol $NaClO_3$ forms ✓; 9.6 g ✓

(iii) $Cl_2 + 2NaOH \longrightarrow NaOCl + NaCl + H_2O$ ✓; bleach ✓ [9]

[Total: 14]

8 (a) (i) Electronegativity is a measure of the attraction of an atom in a molecule for the pair of electrons in a covalent bond ✓.

(ii) Electronegativity decreases from fluorine to iodine ✓. The atomic radii increase down the group ✓. The shielding effect also increases down the group as the number of electron shells increases ✓. The smaller the halogen atom, the greater the nuclear attraction experienced by the bonding electrons and the greater the electronegativity ✓. [5]

(b) The boiling points increase from fluorine to iodine ✓. The number of electrons increases ✓, increasing the van der Waals' forces between molecules ✓. [3]

(c) With aqueous potassium chloride, there is no change ✓. With aqueous potassium iodide, the solution turns dark brown ✓ as iodine is displaced ✓.

$Br_2(aq) + 2I^-(aq) \longrightarrow 2Br^-(aq) + I_2(aq)$ ✓ [4]

(d) Bromine is a weaker oxidising agent than chlorine because it does not oxidise chloride ions to chlorine ✓. Bromine is a stronger oxidising agent than iodine because it oxidises iodide ions to iodine ✓. [2]

[Total: 14]

Chapter 3 Energetics, rates and equilibrium

1 (a) $2Na(s) + C(s) + 1\frac{1}{2}O_2(g) \longrightarrow Na_2CO_3(s)$ ✓✓ [1, equation; 1, state symbols] [2]

(b) 100 kPa and 298 K ✓ [1]

(c) +90 kJ mol^{-1} ✓✓✓ [3]

[Total: 6]

2 (a) (i) The enthalpy change that takes place when one mole of a substance ✓ reacts completely with oxygen ✓ under standard conditions, all reactants and products being in their standard states ✓.

(ii) 298 K ✓ [4]

(b) (i) $CH_3CH_2CH_2OH(l) + 4\frac{1}{2}O_2(g) \longrightarrow 3CO_2(g) + 4H_2O(l)$ ✓

(ii) –2023 kJ mol^{-1} ✓✓✓ [4]

[Total: 8]

3

(a) The enthalpy change that takes place when one mole of a compound ✓ in its standard states is formed from its constituent elements ✓ in their standard states under standard conditions ✓. [3]

(b) $C(s) + 2H_2(g) \longrightarrow CH_4(g)$ ✓✓ [1, equation; 1, state symbols] [2]

(c) –76 kJ mol^{-1} ✓✓✓ [3]

(d) C–H: +412 kJ mol^{-1} ✓; C–C: +348 kJ mol^{-1} ✓✓ [3]

[Total: 11]

4 (a) (i) ✓✓✓; (ii) ✓✓

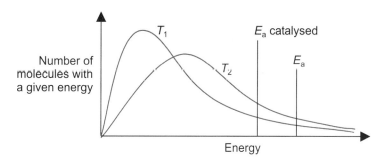

[5]

(b) (i) ✓✓✓; (ii) ✓

(iii) A catalyst provides an alternative reaction route ✓ with a lower activation energy ✓. More molecules exceed the lower activation energy so more are able to react. ✓ [7]

[Total: 12]

5 (a) (i) A homogeneous equilibrium has all equilibrium species in the same phase ✓. All species in this equilibrium are gaseous ✓.

(ii) The forward reaction proceeds at the same rate as the reverse reaction ✓. The concentrations of reactants and products are constant ✓. Equilibrium can only be achieved in a closed system ✓. [5]

(b) (i) Equilibrium position is unaffected ✓. There are equal numbers of gaseous moles on either side of the equilibrium ✓.

(ii) Rate increases ✓. Increasing the pressure also increases the concentration ✓. [4]

(c) Endothermic ✓. With increased temperature, an equilibrium shifts in the endothermic direction ✓. [2]

[Total: 11]

Chapter 4 Organic chemistry, analysis and the environment

1 (a) (i) Same **molecular** formula with a different arrangement/structural formula ✓.

(ii) $ClCH_2CH_2CH_2Cl$, $CH_3CHClCH_2Cl$, $CH_3CH_2CHCl_2$, $CH_3CCl_2CH_3$ ✓✓✓✓ [5]

(b) (i) UV / sunlight ✓; (ii) C_3H_7Cl ✓

(iii) $CH_3CH_2CH_2Cl$ ✓ 1-chloropropane ✓
$CH_3CHClCH_3$ ✓ 2-chloropropane ✓

[6]
[Total: 11]

Organic chemistry, analysis and the environment (continued)

2 (a) (i) There must be a C=C double bond that prevents rotation. ✓
There must be different groups attached to each carbon atom in the C=C double bond. ✓
(ii) *E*: groups are on opposite sides of double bond.
Z: groups are on same side of double bond. ✓
(iii) correct structures ✓✓

Z   E   [5]

(b) H_2 ✓ and a named catalyst/Ni/Pt/Pd ✓.   [2]

(c)

2 repeat units ✓ with side links ✓

[2]
[Total: 9]

3 (a)

no double bond ✓, correct skeleton ✓   [2]

(b) **A** = $CH_3CH_2CHBrCH_3$ ✓; **B** = $CH_3CH_2CHOHCH_3$ ✓; **C** = $CH_3CH_2COCH_3$ ✓   [3]

(c) (i) addition ✓, electrophile ✓;
(ii) substitution ✓, nucleophile ✓;
(iii) oxidation ✓, oxidising agent ✓.   [6]

(d) steam ✓ and H_3PO_4 ✓   [2]

[Total: 13]

4 (a) Empirical formula of **B** = C_2H_5 ✓✓; Molecular formula of **B** = C_4H_{10} ✓   [3]

(b) **A** = $(CH_3)_2C=CH_2$ ✓; **B** = $(CH_3)_2CHCH_3$ ✓;
C and **D** = $(CH_3)_2CHCH_2OH$ ✓ and $(CH_3)_2COHCH_3$ ✓;   [4]

(c) $C_4H_8 + H_2 \longrightarrow C_4H_{10}$ ✓
$C_4H_8 + H_2O \longrightarrow C_4H_9OH$ ✓   [2]

[Total: 9]

5 (a) (i) $CH_3CH_2CHBrCH_3$ ✓; $(CH_3)_3CBr$ ✓.
(ii) 1-bromo-2-methylpropane ✓   [3]

(b) (i) nucleophilic substitution ✓; (ii) $(CH_3)_2CHCH_2OH$ ✓   [2]

(c) (i) electrophilic addition ✓
(ii) arrows stage 1 ✓; arrow stage 2 ✓; carbocation ✓

[4]
[Total: 9]

The Periodic Table

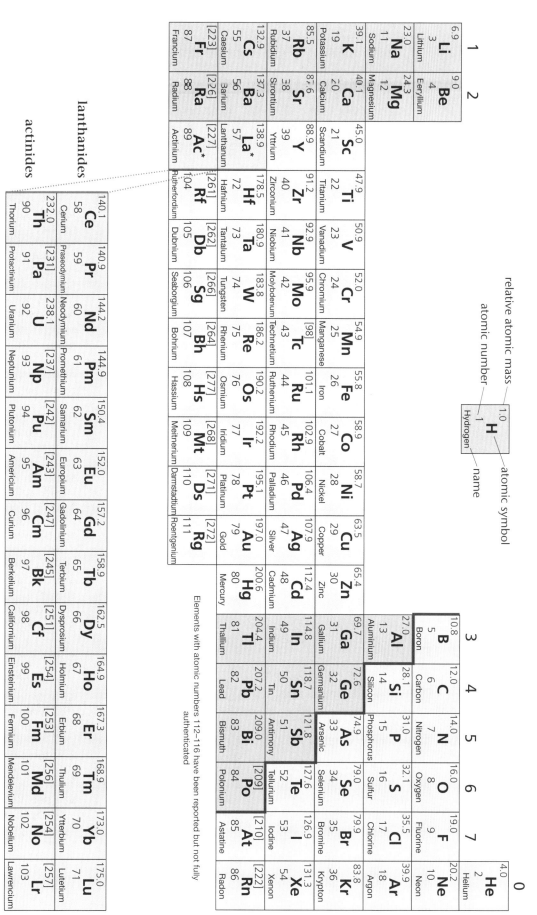

Index

Revise
A2

Chemistry

Rob Ritchie

Contents

Chapter 1 Rates and equilibria

Chapter 2 Energy changes in chemistry

Chapter 3 The Periodic Table

Specification lists

MODULE	SPECIFICATION TOPIC	CHAPTER REFERENCE	STUDIED IN CLASS	REVISED	PRACTICE QUESTIONS
A2 Unit 1 (M4) *Kinetics, equilibria and organic chemistry*	Kinetics	1.1, 1.2			
	Equilibria	1.3			
	Acids and bases	1.5, 1.6, 1.7			
	Nomenclature and isomerism on organic chemistry	4.1			
	Compounds containing the carbonyl group	4.2, 4.3, 4.4, 4.5			
	Aromatic chemistry	4.6, 4.7			
	Amines	4.9			
	Amino acids	5.1, 5.3			
	Polymers	5.2			
	Organic synthesis and analysis	5.6			
	Structural determination	5.3, AS 4.7, 5.4, 5.5			
A2 Unit 2 (M5) *Energetics, redox and inorganic chemistry*	Thermodynamics	2.1, 2.2, AS 3.3			
	Periodicity	3.1			
	Redox equilibria	2.3, 2.4			
	Transition metals	3.3, 3.4, 3.6, 3.7			
	Reactions of inorganic compounds in aqueous solution	3.5, 3.8			

Examination analysis

A2 Chemistry comprises two unit tests. All questions are compulsory.
Synoptic assessment will form part of both units. Practical and investigative skills will also be assessed.

Units 1, 2 and 3 comprise AS Chemistry			*50%*
Unit 4	*Structured questions: short and extended answers*	*1 hr 45 min test*	*20%*
Unit 5	*Structured questions: short and extended answers*	*1 hr 45 min test*	*20%*
Unit 6	*Internal assessment of practical and investigative skills*		*10%*

OCR A2 Chemistry

MODULE	SPECIFICATION TOPIC	CHAPTER REFERENCE	STUDIED IN CLASS	REVISED	PRACTICE QUESTIONS
A2 Unit F324 (M4) Rings, polymers and analysis	Arenes	4.6, 4.7, 4.8			
	Carbonyl compounds	4.2			
	Carboxylic acids and esters	4.3, 4.4, 4.5			
	Amines	4.9			
	Amino acids and proteins	5.1, 5.2			
	Polyesters and polyamides	5.2			
	Synthesis	5.6			
	Chromatography	5.3			
	Spectroscopy	AS 4.7, 5.4, 5.5			
A2 Unit F325 (M5) Equilibria, energy and elements	How fast?	1.1, 1.2			
	How far?	1.3			
	Acids, bases and buffers	1.5, 1.6, 1.7			
	Lattice enthalpy	2.1			
	Enthalpy and entropy	2.2			
	Electrode potentials and fuel cells	2.3, 2.4			
	Transition elements	3.3, 3.4, 3.5, 3.6, 3.7, 3.8			

Examination analysis

A2 Chemistry comprises two unit tests. All questions are compulsory.
Synoptic assessment will form part of both units. Practical and investigative skills will also be assessed.

Units 1, 2 and 3 comprise AS Chemistry			*50%*
Unit F324 *Structured questions: short and extended answers*	*1 hr test*		*15%*
Unit F325 *Structured questions: short and extended answers*	*1 hr 45 min test*		*25%*
Unit F326 *Internal assessment of practical and investigative skills*			*10%*

Salters A2 Chemistry

MODULE	SPECIFICATION TOPIC	CHAPTER REFERENCE	STUDIED IN CLASS	REVISED	PRACTICE QUESTIONS
A2 Unit F334 (M4) Chemistry of materials	Equilibria	1.5			
	Organic functional groups	4.1, 4.2, 4.3, 4.4, 4.5, 4.6, 4.7, 4.8, 4.9, 5.1, 5.2			
	Organic reactions	4.2, 4.3, 4.8, 5.1, 5.2			
	Reaction mechanisms	4.2, 4.9			
	Applications of organic chemistry	5.6			
	Modern analytical techniques	5.3, AS 4.7, 5.4			
	Application	4.3, 4.4, 4.5, 4.9, 5.2			
	Kinetics	1.1, 1.2, 5.2			
	Isomerism	5.3			
	Formulae, equations and amount of substance	1.7, 3.6			
	Redox	2.3, 2.4, 3.4, 3.6			
	Inorganic chemistry and the Periodic Table	3.3, 3.5, 3.7, 3.8			
	Bonding and structure	4.9, 5.2, 5.6			
	Rates of reaction	1.1, 1.2			
A2 Unit F335 (M5) Chemistry by design	Equilibria	1.3, 1.5, 1.6, 1.7, 2.2, 2.4			
	Redox	2.4			
	Inorganic chemistry and the Periodic Table	2.4			
	Organic functional groups	4.4, 4.6			
	Organic reactions	4.7, 4.9, 5.6			
	Modern analytical techniques	3.4, 4.9, 5.3, 5.4, 5.5			
	Energetics	2.1, 2.2			
	Reaction mechanisms	4.2			
	Isomerism	4.1			
	Applications	5.6			

Examination analysis

A2 Chemistry comprises two unit tests. All questions are compulsory.
Synoptic assessment will form part of both units. Practical and investigative skills will also be assessed.

Units 1, 2 and 3 comprise AS Chemistry			*50%*
Unit F334 *Structured questions: short and extended answers*	*1 hr 30 min test*		*15%*
Unit F335 *Structured questions: short and extended answers*	*2 hr test*		*20%*
Unit F336 *Individual investigation*			*15%*

Edexcel A2 Chemistry

MODULE	SPECIFICATION TOPIC	CHAPTER REFERENCE	STUDIED IN CLASS	REVISED	PRACTICE QUESTIONS
	How fast? – rates	1.1, 1.2			
	How far? – entropy	2.1, 2.2			
	Equilibria	1.3, 1.4, 2.2			
	Application of rates and equilibria	2.2			
A2 Unit 1 (M4)	Acid/base equilibria	1.5, 1.6, 1.7			
Rates, equilibria and further organic chemistry	Further organic chemistry – chirality	4.1			
	Further organic chemistry – carbonyl compounds	4.2			
	Further organic chemistry – carboxylic acids	4.3, 4.4, 4.5			
	Further organic chemistry – carboxylic acid derivatives	4.4, 4.5, 5.2			
	Spectroscopy and chromatography	5.3, AS 4.7, 5.4, 5.5			
	Application of redox equilibria	2.3, 2.4, 3.6			
A2 Unit 2 (M5)	Transition metals and their chemistry	3.3, 3.4, 3.5, 3.6 3.7, 3.8			
Transition metals and organic nitrogen chemistry	Organic chemistry – arenes	4.6, 4.7, 4.8			
	Organic chemistry – nitrogen compounds	4.9, 5.1, 5.2			
	Organic chemistry – organic synthesis	5.5, 5.6			

Examination analysis

A2 Chemistry comprises two unit tests. All questions are compulsory.
Synoptic assessment will form part of both units. Practical and investigative skills will also be assessed.

Units 1, 2 and 3 comprise AS Chemistry			*50%*
Unit 4	Objective questions; Structured questions: short and extended answers Data questions with use of a data booklet	*1 hr 40 min test*	*20%*
Unit 5	Objective questions; Structured questions: short and extended answers Contemporary context questions	*1 hr 40 min test*	*20%*
Unit 6	Internal assessment of practical and investigative skills		*10%*

WJEC A2 Chemistry

MODULE	SPECIFICATION TOPIC	CHAPTER REFERENCE	STUDIED IN CLASS	REVISED	PRACTICE QUESTIONS
A2 Unit 1 (M4) Spectroscopy and organic chemistry	Spectroscopy	4.9, AS 4.7, 5.4, 5.5			
	Isomerism and aromaticity	4.6, 4.7			
	Alcohols and phenols	4.8			
	Aldehydes and ketones	4.2			
	Carboxylic acids and derivatives	4.3, 4.4, 4.5, 4.7, 4.8			
	Organic compounds containing nitrogen	4.9, 5.1, 5.2			
	Organic synthesis and analysis	5.2, 5.3, AS 4.7, 5.4, 5.5, 5.6			
	The process of how science works	5.5			
A2 Unit 2 (M5) Physical and inorganic chemistry	Redox and standard electrode potential	2.3, 2.4			
	Redox reactions	3.6			
	Applications of redox	2.3			
	Chemistry of the p-block – general	3.2			
	Chemistry of the p-block – Group 3 (13)	3.2			
	Chemistry of the p-block – Group 4 (14)	3.2			
	Chemistry of the p-block – Group 7 (17)	3.2			
	d-block transition elements	3.3, 3.4, 3.5, 3.7, 3.8			
	Chemical kinetics	1.1, 1.2			
	Enthalpy changes for solids and solutions	2.1			
	Entropy and feasibility of reactions	2.2			
	General equilibria	1.3, 1.4, 2.2			
	Acid-base equilibria	1.5, 1.6, 1.7			

Examination analysis

A2 Chemistry comprises two unit tests. All questions are compulsory.
Synoptic assessment will form part of both units. Practical and investigative skills will also be assessed.

Units 1, 2 and 3 comprise AS Chemistry			50%
Unit 4 Structured questions: short and extended answers		1 hr 45 min test	20%
Unit 5 Structured questions: short and extended answers		1 hr 45 min test	20%
Unit 6 Internal assessment of practical and investigative skills			10%

CCEA A2 Chemistry

MODULE	SPECIFICATION TOPIC	CHAPTER REFERENCE	STUDIED IN CLASS	REVISED	PRACTICE QUESTIONS
A2 Unit 1 (M4) *Periodic trends and further organic, physical and inorganic chemistry*	Lattice enthalpy	2.1			
	Enthalpy, entropy and free energy	2.1, 2.2			
	Kinetics	1.1, 1.2			
	Equilibrium	1.3, 1.4			
	Acid-base equilibria	1.5, 1.6, 1.7			
	Isomerism	4.1			
	Aldehydes and ketones	4.2			
	Carboxylic acids	4.3, 4.5			
	Esters, fats and oils	4.4			
	Periodic trends	3.1			
	Environmental chemistry	2.3, 2.4, 5.2			
A2 Unit 2 (M5) *Analytical, transition metals, electrochemistry and further organic chemistry*	Mass spectrometry	AS 4.7, 5.3, 5.4			
	Nuclear magnetic resonance spectroscopy	5.5			
	Volumetric analysis	3.5, 3.6			
	Colorimetry	3.4			
	Chromatography	5.3			
	Transition metals – general properties	3.3			
	Transition metals – complexes	3.4, 3.5, 3.6			
	Transition metals – catalytic behaviour	3.7			
	Transition metals – oxidation states	2.3, 2.4, 3.6, 3.8			
	Applications of transition metal complexes	3.4			
	Electrode potentials	2.3, 2.4			
	Arenes	4.6, 4.7			
	Amines	4.9			
	Amides	4.5, 4.9			
	Amino acids	5.1, 5.2			
	Polymer chemistry	5.2			

Examination analysis

A2 Chemistry comprises two unit tests. All questions are compulsory.
Synoptic assessment will form part of both units. Practical and investigative skills will also be assessed.

Units 1, 2 and 3 comprise AS Chemistry			*50%*
Unit 4	*Objective questions; structured questions: short and extended answers*	*2 hr test*	*20%*
Unit 5	*Objective questions; structured questions: short and extended answers*	*2 hr test*	*20%*
Unit 6	*Internal assessment of practical and investigative skills*		*10%*

AS/A2 Level Chemistry courses

AS and A2

All Chemistry A Level courses currently studied are in two parts, with three separate modules in each part. Students first study the AS (Advanced Subsidiary) course. Some will then go on to study the second part of the A Level course, called A2. Advanced Subsidiary is assessed at the standard expected halfway through an A Level course: i.e., between GCSE and Advanced GCE. This means that new AS and A2 courses are designed so that difficulty steadily increases:

- AS Chemistry builds from GCSE science
- A2 Chemistry builds from AS Chemistry.

How will you be tested?

Assessment units

For AS Chemistry, you will be tested by three assessment units. For the full A Level in Chemistry, you will take a further three units. AS Chemistry forms 50% of the assessment weighting for the full A Level.

AS 50%   A2 50%

3 units for AS   3 units for A2

A 100%

One of the units in AS and in A2 is practically based. You will take two theory units in each of AS and A2. Each theory unit can normally be taken in either January or June. Alternatively, you can study the whole course before taking any of the unit tests.

If you are disappointed with a module result, you can resit each module. The higher mark counts.

Coursework

Coursework forms part of your A Level Chemistry course in the form of the assessment of practical skills. More details are provided on page 11.

Key skills

It is important that you develop your key skills of Communication, Application of Number and Information Technology throughout your AS and A2 courses. These are important skills that you need whatever you do beyond AS and A Levels. To gain the key skills qualification, you will need to demonstrate your ability to put your ideas across to other people, collect data and use up-to-date technology in your work. You will have many opportunities during A2 Chemistry to develop your key skills.

What skills will I need?

For A2 Chemistry, you will be tested by *assessment objectives*: these are the skills and abilities that you should have acquired by studying the course. The assessment objectives for A2 Chemistry are shown below.

Knowledge and understanding of science and How Science Works

- recognising, recalling and showing understanding of scientific knowledge
- selection, organisation and communication of relevant information in a variety of forms.

Application of knowledge and understanding of science and How Science Works

- analysis and evaluation of scientific knowledge and processes
- applying scientific knowledge and processes to unfamiliar situations including those related to issues
- assessing the validity, reliability and credibility of scientific information.

How Science Works

- demonstrating and describing ethical, safe and skilful practical techniques and processes
- selecting appropriate qualitative and quantitative methods
- making, recording and communicating reliable and valid observations and measurements with appropriate precision and accuracy
- analysis, interpretation, explanation and evaluation of the methodology, results and impact of their own and others' experimental and investigative activities in a variety of ways.

Experimental and investigative skills

Chemistry is a practical subject and part of the assessment of A2 Chemistry will test your practical skills. This will be carried out during your lessons. You will be assessed on three main skills:

- implementing
- analysing evidence and drawing conclusions
- evaluating evidence and procedures.

The skills may be assessed in separate practical exercises. They may also be assessed all together in the context of a single 'whole investigation'.

You will receive guidance about how your practical skills will be assessed from your teacher. This Study Guide concentrates on preparing you for the written examinations testing the subject content of A2 Chemistry.

Synoptic assessment

- bringing together knowledge, principles and concepts from different areas of Chemistry and applying them in a particular context
- using chemical skills in contexts which bring together different areas of the subject.

More details of synoptic assessment in Chemistry are discussed in Chapter 6 of this Study Guide (pages 178–181).

Different types of questions in A2 examinations

In A2 Chemistry examinations, different types of question are used to assess your abilities and skills. Unit tests mainly use structured questions requiring both short answers and more extended answers.

Short-answer questions

A short-answer question may test recall or it may test understanding. Short answer questions normally have space for the answers printed on the question paper.

Here are some examples (the answers are shown in *italics*):

What is meant by isotopes?

Atoms of the same element with different mass numbers.

Calculate the amount (in mol) of H_2O in 4.5 g of H_2O.

1 mol H_2O has a mass of 18 g. $\therefore$ 4.5 g of H_2O contains 4.5/18 = 0.25 mol H_2O.

Structured questions

Structured questions are in several parts. The parts usually have a common context and they often become progressively more difficult and more demanding as you work your way through the question. A structured question may start with simple recall, then test understanding of a familiar or an unfamiliar situation.

Here is an example of a structured question that becomes progressively more demanding.

(a) Write down the atomic structure of the two isotopes of potassium: ^{39}K and ^{41}K.

   (i) ^{39}K ...19... protons; ...20... neutrons; ...19... electrons. ✓
   (ii) ^{41}K ...19... protons; ...22... neutrons; ...19... electrons. ✓   [2]

(b) A sample of potassium has the following percentage composition by mass: ^{39}K: 92%; ^{41}K: 8% Calculate the relative atomic mass of the potassium sample.

   92 x 39/100 + 8 x 41/100 = 39.16 ✓   [1]

(c) What is the electronic configuration of a potassium atom?

   $1s^2 2s^2 2p^6 3s^2 3p^6 4s^1$ ✓   [1]

(d) The second ionisation energy of potassium is much larger than its first ionisation energy.

   (i) Explain what is meant by the *first ionisation energy* of potassium.

      The energy required to remove an electron ✓ *from each atom in 1 mole* ✓ *of gaseous atoms* ✓

   (ii) Why is there a large difference between the values for the first and the second ionisation energies of potassium?

      The 2nd electron removed is from a different shell ✓ *which is closer to the nucleus and experiences more attraction from the nucleus.* ✓ *This outermost electron experiences less shielding from the nucleus because there are fewer inner electron shells than for the 1st ionisation energy.* ✓   [6]

Extended answers

In A2 Level Chemistry, questions requiring more extended answers may form part of structured questions or may form separate questions. They may appear anywhere on the paper and will typically have between 5 and 15 marks allocated to the answers as well as several lines of answer space. These questions are also often used to assess your abilities to communicate ideas and put together a logical argument.

The correct answers to extended questions are often less well-defined than to those requiring short answers. Examiners may have a list of points for which credit is awarded up to the maximum for the question.

An example of a question requiring an extended answer is shown below.

The table below gives data on the oxides of Period 3 in the Periodic Table.

oxide	Na_2O	MgO	Al_2O_3	SiO_2	P_4O_{10}	SO_3
melting point /K	1548	3100	2290	1880	853	306
conductivity of molten compound	good	good	medium	poor	poor	poor

Describe and explain the trends in formula, structure and bonding shown by these data. [11 marks]

Points that the examiners might look for include:

In the formula, the number of oxygen atoms bonded increases across the period. The oxidation number increases steadily by one for each successive element. ✓ This is because the number of electrons in the outer shell increases by one for each element in the period. ✓

The trend in structure is from giant to simple molecular ✓ with a changeover between silicon and phosphorus. ✓ Structure is related to the magnitude of the melting points. ✓

The high melting points (Na_2O ✓ SiO_2) result from strong forces between ions (for Na_2O – Al_2O_3) ✓ and between atoms (for SiO_2). ✓ The low melting points (P_4O_{10} ✓ SO_3) result from weak intermolecular forces or van der Waals' forces. ✓

The trend in bonding is from ionic to covalent ✓ with a changeover between Al_2O_3 and SiO_2. ✓ This is related to the conductivity of the molten oxide. ✓

Good conductivity results from mobile ions (Na_2O ✓ $Al2O_3$. ✓)

Poor conductivity shows that no mobile ions are present (SiO_2 ✓ SO_3. ✓)

13 marking points → 11 marks

In this type of response, there may be an additional mark for a clear, well-organised answer, using specialist terms. In addition, marks may be allocated for legible text with accurate spelling, punctuation and grammar.

Other types of questions

Free-response and open-ended questions allow you to choose the context and to develop your own ideas.

Multiple-choice or objective questions require you to select the correct response to the question from a number of given alternatives.

Stretch and Challenge

Stretch and Challenge is a concept that is applied to the structured questions in Units 4 and 5 of the exam papers for A2. In principle, it means that the sub-questions become progressively harder so as to challenge more able students and help differentiate between A and A* students.

Stretch and Challenge questions are designed to test a variety of different skills and your understanding of the material. They are likely to test your ability to make appropriate connections between different areas and apply your knowledge in unfamiliar contexts (as opposed to basic recall).

Exam technique

Links from AS

A2 Chemistry builds from the knowledge and understanding acquired after studying AS Chemistry. This Study Guide has been written so that you will be able to tackle A2 Chemistry from an AS Chemistry background.

You should not need to search for important chemistry from AS Chemistry because cross-references have been included as 'Key points from AS' to the AS Chemistry Study Guide: Revise AS.

What are examiners looking for?

Examiners use instructions to help you to decide the length and depth of your answer.

If a question does not seem to make sense, you may have misread it – read it again!

State, define or list

This requires a short, concise answer, often recall of material that can be learnt by rote.

Explain, describe or discuss

Some reasoning or some reference to theory is required, depending on the context.

Outline

This implies a short response, almost a list of sentences or bullet points.

Predict or deduce

You are not expected to answer by recall but by making a connection between pieces of information.

Suggest

You are expected to apply your general knowledge to a 'novel' situation, one which you have not directly studied during the A2 Chemistry course.

Calculate

This is used when a numerical answer is required. You should always use units in quantities and significant figures should be used with care.

Look to see how many significant figures have been used for quantities in the question and give your answer to this degree of accuracy.

If the question uses 3 sig figs, then give your answer to 3 sig figs also.

Some dos and don'ts

Dos

Do answer the question.

* No credit can be given for good Chemistry that is irrelevant to the question.

Do use the mark allocation to guide how much you write.

* Two marks are awarded for two valid points - writing more will rarely gain more credit and could mean wasted time or even contradicting earlier valid points.

Do use diagrams, equations and tables in your responses.

* Even in 'essay-type' questions, these offer an excellent way of communicating chemistry.

Do write legibly.

* An examiner cannot give marks if the answer cannot be read.

Do write using correct spelling and grammar. Structure longer essays carefully.

* Marks are now awarded for the quality of your language in exams.

Don'ts

Don't fill up any blank space on a paper.

* In structured questions, the number of dotted lines should guide the length of your answer.
* If you write too much, you waste time and may not finish the exam paper. You also risk contradicting yourself.

Don't write out the question again.

* This wastes time. The marks are for the answer!

Don't contradict yourself.

* The examiner cannot be expected to choose which answer is intended.

Don't spend too much time on a part that you find difficult.

* You may not have enough time to complete the exam. You can always return to a difficult calculation if you have time at the end of the exam.

What grade do you want?

Everyone would like to improve their grades but you will only manage this with a lot of hard work and determination. You should have a fair idea of your natural ability and likely grade in Chemistry and the hints below offer advice on improving that grade.

For a Grade A

You will need to be a very good all-rounder.

- You must go into every exam knowing the work extremely well.
- You must be able to apply your knowledge to new, unfamiliar situations.
- You need to have practised many, many exam questions so that you are ready for the type of question that will appear.

The exams test all areas of the specification and any weaknesses in your Chemistry will be found out. There must be no holes in your knowledge and understanding. For a Grade A, you must be competent in all areas.

For a Grade C

You must have a reasonable grasp of Chemistry but you may have weaknesses in several areas and you will be unsure of some of the reasons for the Chemistry.

- Many Grade C candidates are just as good at answering questions as Grade A students but holes and weaknesses often show up in just some topics.
- To improve, you will need to master your weaknesses and you must prepare thoroughly for the exam. You must become a better all-rounder.

For a Grade E

You cannot afford to miss the easy marks. Even if you find Chemistry difficult to understand and would be happy with a Grade E, there are plenty of questions in which you can gain marks.

- You must memorise all definitions.
- You must practise exam questions to give yourself confidence that you do know some Chemistry. In exams, answer the parts of questions that you know first. You must not waste time on the difficult parts. You can always go back to these later.
- The areas of Chemistry that you find most difficult are going to be hard to score on in exams. Even in the difficult questions, there are still marks to be gained. Show your working in calculations because credit is given for a sound method. You can always gain some marks if you get part of the way towards the solution.

What marks do you need?

The table below shows how your average mark is transferred into a grade.

average	80%	70%	60%	50%	40%
grade	A	B	C	D	E

The new A* grade

A* grades are awarded for the A Level qualification only and not for the AS qualification or individual units.

For an A* grade, you need to achieve a...

- grade A overall (80% or more on uniform mark scale) for the whole A Level qualification
- grade A* (90% or more on the uniform mark scale) across your A2 units.

Four steps to successful revision

Step 1: Understand

- Study the topic to be learned slowly. Make sure you understand the logic or important concepts.
- Mark up the text if necessary – underline, highlight and make notes.
- Re-read each paragraph slowly.

GO TO STEP 2

Step 2: Summarise

- Now make your own revision note summary:
  What is the main idea, theme or concept to be learned?
  What are the main points? How does the logic develop?
  Ask questions: Why? How? What next?
- Use bullet points, mind maps, patterned notes.
- Link ideas with mnemonics, mind maps, crazy stories.
- Note the title and date of the revision notes
  (e.g. Chemistry: Reaction rates, 3rd March).
- Organise your notes carefully and keep them in a file.

This is now in **short-term memory**. You will forget 80% of it if you do not go to Step 3.
GO TO STEP 3, but first take a 10 minute break.

Step 3: Memorise

- Take 25 minute learning 'bites' with 5 minute breaks.
- After each 5 minute break test yourself:
  Cover the original revision note summary
  Write down the main points
  Speak out loud (record on tape)
  Tell someone else
  Repeat many times.

The material is well on its way to **long-term memory**.
You will forget 40% if you do not do step 4. **GO TO STEP 4**

Step 4: Track/Review

- Create a Revision Diary (one A4 page per day).
- Make a revision plan for the topic, e.g. 1 day later, 1 week later, 1 month later.
- Record your revision in your Revision Diary, e.g.
  Chemistry: Reaction rates, 3rd March 25 minutes
  Chemistry: Reaction rates, 5th March 15 minutes
  Chemistry: Reaction rates, 3rd April 15 minutes
  ... and then at monthly intervals.

Chapter 1

Rates and equilibria

The following topics are covered in this chapter:

- Orders and the rate equation
- Determination of reaction mechanisms
- The equilibrium constant, K_c
- The equilibrium constant, K_p

- Acids and bases
- The pH scale
- pH changes

1.1 Orders and the rate equation

After studying this section you should be able to:

- explain the terms 'rate of reaction', 'order', 'rate constant', 'rate equation'
- construct a rate equation and calculate its rate constant
- use the initial rates method to deduce orders
- use graphs to deduce reaction rates and orders
- understand that a temperature change increases the rate constant
- determine activation energy graphically

LEARNING SUMMARY

Key points from AS

- **Reaction rates**
  Revise AS pages 104–107

During the study of reaction rates in AS Chemistry, reaction rates were described in terms of activation energy and the Boltzmann distribution. For A2 Chemistry, you will build upon this knowledge and understanding by measuring and calculating reaction rates using rate equations.

Orders and the rate equation

AQA	M4	SALTERS	M4
EDEXCEL	M4	WJEC	M5
OCR	M5	CCEA	M4

The rate of a reaction is determined from experimental results. Reaction rate **cannot** be determined from a chemical equation.

Suitable experimental techniques used to obtain rate data for a given reaction include:
- titration
- pH measurement
- colorimetry
- measuring volumes of gases evolved
- measuring mass changes.

The rate of a reaction is usually determined by the **change in concentration** of a reaction species **with time**.

- Units of rate = $\underbrace{mol\ dm^{-3}}_{concentration}\ \underbrace{s^{-1}}_{per\ time}$

Key points from AS

- **What is a reaction rate?**
  Revise AS pages 104–105

Orders of reaction

For a reaction: $\mathbf{A + B + C \longrightarrow}$ products,

- the **order** of the reaction shows how the reaction rate is affected by the concentrations of **A**, **B** and **C**.

If the order is 0 (zero order) with respect to reactant **A**,

[A] means the concentration of reactant A in mol dm^{-3}.

- the rate is unaffected by changes in concentration of **A**:
  rate $\propto$ [**A**]^0

If the order is 1 (first order) with respect to a reactant **B**,
- the rate is doubled by doubling of the concentration of reactant **B**:
  rate $\propto$ [**B**]^1

If the order is 2 (second order) with respect to a reactant **C**,
- the rate is quadrupled by doubling the concentration of reactant **C**:
  rate $\propto$ $[\mathbf{C}]^2$

Combining the information above,

rate $\propto$ $[\mathbf{A}]^0[\mathbf{B}]^1[\mathbf{C}]^2$

$\therefore$ *rate* $= k[\mathbf{B}]^1[\mathbf{C}]^2$

- This expression is called the **rate equation** for the reaction.
- k is the rate constant of the reaction.

A number raised to the power of zero = 1 and zero order species can be omitted from the rate equation.

The rate equation

The rate equation of a reaction shows how the rate is affected by the concentration of reactants. A rate equation can only be determined from experiments.

In general, for a reaction: $\mathbf{A} + \mathbf{B} \longrightarrow \mathbf{C}$,
the reaction rate is given by: *rate* $= k[\mathbf{A}]^m[\mathbf{B}]^n$

- m and n are the **orders of reaction** with respect to **A** and **B** respectively.
- The **overall order** of reaction is $m + n$.
- The reaction rate is measured as the change in concentration of a reaction species with time.
- The units of rate are mol dm^{-3} s^{-1}.

The rate of a reaction is determined from experimental results.

It **cannot** be determined from the chemical equation.

The rate constant k indicates the rate of the reaction:

- a large value of $k \longrightarrow$ fast rate of reaction
- a small value of $k \longrightarrow$ slow rate of reaction.

Measuring rates using graphs

EDEXCEL	M4	SALTERS	M4
OCR	M5	WJEC	M5
		CCEA	M4

Concentration/time graphs

It is often possible to measure the concentration of a reactant or product continuously at various times during the course of an experiment.

- From the results, a concentration/time graph can be plotted.
- The shape of this graph can indicate the order of the reaction by measuring the **half-life** of a reactant.

Key points from AS

- **What is a reaction rate?**
  Revise AS pages 104–105

> The half-life of a reactant is the time taken for its concentration to reduce by half.
>
> **A first-order reaction has a constant half-life.**
>
> **KEY POINT**

Example of a first order graph

The reaction between bromine and methanoic acid is shown below.

$Br_2(aq) + HCOOH(aq) \longrightarrow 2Br^-(aq) + 2H^+(aq) + CO_2(g)$

- During the course of the reaction, the orange bromine colour disappears as Br_2 reacts to form colourless Br^- ions. The order with respect to bromine can be determined by monitoring the rate of disappearance of bromine using a colorimeter.
- The concentration of the other reactant, methanoic acid, is kept virtually constant by using an excess of methanoic acid.

The concentration/time graph from this reaction is shown on the next page. Notice that the **half-life is constant** at 200 s, showing that this reaction is first order with respect to $Br_2(aq)$.

The half-life curve of a first-order reaction is concentration independent.

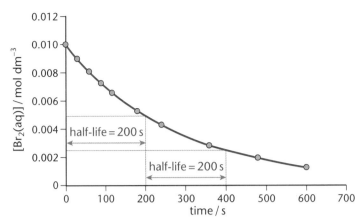

The shapes of concentration/time graphs for zero, first and second order reactions are shown below.

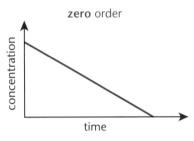

zero order

The concentration falls at a steady rate with time
• half-life **decreases** with time.

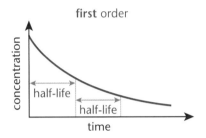

first order

The concentration halves in equal time intervals
• **constant** half-life.

second order

The half-life gets progressively longer as the reaction proceeds
• half-life **increases** with time.

• The first-order relationship can be confirmed by plotting a graph of log[**X**] against time, which gives a straight line.

Measuring rates from tangents

The gradient of the concentration/time graph is a measure of the rate of a reaction.

For the reaction: **A** ⟶ **B**, the graph below shows how the concentration of **A** changes during the course of the reaction.

> At any instant of time during the reaction, the rate can be measured by drawing a tangent to the curve.

KEY POINT

At the start of the reaction ($t = 0$):
• the tangent is steepest
• the concentration of **A** is greatest
• the reaction rate is fastest.

The initial rate of the reaction is the tangent of a concentration/time graph at time=0.

As the reaction proceeds:
• the tangent becomes less steep
• the concentration of **A** decreases
• the reaction rate slows down.

When the reaction is complete:
• the concentration of **A** is very small
• the gradient becomes zero
• the reaction stops.

At an instant of time, t, during the reaction:

• the rate of **change** in concentration of **A** $= -\dfrac{d[A]}{dt}$ *(the tangent)*

A reaction rate is expressed as a positive value.

For the reactant **A**, the **negative** sign shows a **decreasing** concentration with time.

For the product **B**, the **positive** sign shows **increasing** concentration.

The negative sign shows that the concentration of **A** decreases during the reaction.

The rate could also be followed by measuring the rate of **increase** in concentration of the product **B**.

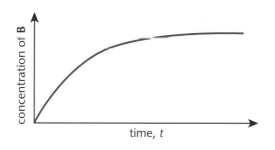

B is formed and its concentration increases during the reaction.

$$\text{rate of } \textbf{change} \text{ in concentration of } \textbf{B} = +\frac{d[\textbf{B}]}{dt}$$

Plotting a rate/concentration graph

- A concentration/time graph is first plotted.
- **Tangents** are drawn at several time values on the concentration/time graph, giving values of reaction rates.
- A second graph is plotted of **rate** against **concentration**.
- The shape of this graph confirms the order of the reaction.

zero order

Rate $\propto [\textbf{A}]^0$ ∴ rate = constant
- Rate unaffected by changes in concentration.

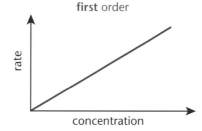

first order

Rate $\propto [\textbf{A}]^1$
- Rate doubles as concentration doubles.

second order

Rate $\propto [\textbf{A}]^2$
- Rate quadruples as concentration doubles.

- The second-order relationship can be confirmed by plotting a graph of rate against $[\textbf{A}]^2$, which gives a straight line.

1 The data below shows how the concentration of reactant **A** changes during the course of a reaction.

Time / 10^4 s	0	0.36	0.72	1.08	1.44
[A] /mol dm^{-3}	0.240	0.156	0.104	0.068	0.045

(a) On graph paper, plot a concentration/time graph and, by drawing tangents, estimate:
   (i) the initial rate
   (ii) the rate 1×10^4 s after the reaction has started.
(b) Determine the half-life of the reaction and show that the reaction is first order with respect to **A**.

(b) Half-life = 5.8 × 10³ s. Successive half-lives are constant.
1 (a) (i) 2.5 × 10⁻⁶ mol dm⁻³ s⁻¹ to 3.5 × 10⁻⁵ mol dm⁻³ s⁻¹ (ii) 8.7 × 10⁻⁶ mol dm⁻³ s⁻¹

Determination of a rate of equation using the initial rates method

AQA	M4	SALTERS	M4
EDEXCEL	M4	WJEC	M5
OCR	M5	CCEA	M4

For a reaction, **X** + **Y** ⟶ **Z**, a series of experiments can be carried out using different **initial** concentrations of the reactants **X** and **Y**.

It is important to change only one variable at a time, so two series of experiments will be required.

- **Series 1** – The concentration of **X** is changed whilst the concentration of **Y** is kept constant.

- **Series 2** – The concentration of **Y** is changed whilst the concentration of **X** is kept constant.

Clock reactions give an approximation to the initial rates method.

Clock reactions are often used to obtain a value for the initial rate of a reaction. A clock reaction measures the **time**, *t*, from the start of the reaction until a visual change – usually a change in colour, appearance of a precipitate or a solid disappearing.

- The clock reaction is repeated several times whilst varying the concentration of one of the reactants. The concentration of any other reactant is kept constant.

- The **initial rate** of the reaction is approximately proportional to $1/t$.

- A **graph** is then plotted of **initial rate** ($1/t$) against **concentration**.

- The **order** with respect to the reactant of which the concentration has been varied, is found from the **shape of the graph**. (See page 21).

A series of clock reactions is repeated for different concentrations of each reactant. A rate/concentration graph is plotted for each reactant.

The results below show the initial rates using different concentrations of **X** and **Y**.

Experiment	[X(aq)] /mol dm^{-3}	[Y(aq)] /mol dm^{-3}	initial rate /mol dm^{-3} s^{-1}
1	1.0×10^{-2}	1.0×10^{-2}	0.5×10^{-3}
2	2.0×10^{-2}	1.0×10^{-2}	2.0×10^{-3}
3	2.0×10^{-2}	2.0×10^{-2}	4.0×10^{-3}

Determine the orders

Comparing experiments 1 and 2: [**Y**(aq)] has been kept constant

- [**X**(aq)] has been doubled,   rate × 4
  ∴ order with respect to **X**(aq) = 2.

Comparing experiments 2 and 3: [**X**(aq)] has been kept constant

- [**Y**(aq)] has been doubled,   rate doubles
  ∴ order with respect to **Y**(aq) = 1.

Use the orders to write the rate equation

- $rate = k[\mathbf{X}(aq)]^2[\mathbf{Y}(aq)]$

- The overall order of this reaction is (2 + 1) = third order

Calculate the rate constant for the reaction

In this example, the results from Experiment 2 have been used.

*You will get the same value of k from the results of **any** of the experimental runs.*

Try calculating k from Experiment 1 and from Experiment 3. All give the same value.

Rearrange the rate equation:

$$\text{The rate constant, } k = \frac{rate}{[\mathbf{X}(aq)]^2[\mathbf{Y}(aq)]}$$

Calculate k using values from one of the experiments.

$$k = \frac{(2.0 \times 10^{-3})}{(2.0 \times 10^{-2})^2 \, (1.0 \times 10^{-2})} = 500 \text{ dm}^6 \text{ mol}^{-2} \text{ s}^{-1}$$

Units of rate constants

Notice how units need to be worked out afresh for reactions with different overall orders.

You do not need to memorise these but it is important that you are able to work these out when needed.

See also page 28 in which the units of the equilibrium constant K_c are discussed.

The units of a rate constant depend upon the rate equation for the reaction. We can determine units of k by substituting units for rate and concentration into the rearranged rate equation. The table below shows how units can be determined.

For a zero-order reaction:	For a first-order reaction:
$rate = k[A]^0 = k$ units of $k = \textbf{mol dm}^{-3}\textbf{ s}^{-1}$	$rate = k[A] \qquad \therefore\ k = \dfrac{rate}{[A]}$ Units of $k = \dfrac{(\text{mol dm}^{-3}\text{ s}^{-1})}{(\text{mol dm}^{-3})} = \textbf{s}^{-1}$
For a second-order reaction:	**For a third-order reaction:**
$rate = k[A]^2 \quad \therefore\ k = \dfrac{rate}{[A]^2}$ Units of $k = \dfrac{(\text{mol dm}^{-3}\text{ s}^{-1})}{(\text{mol dm}^{-3})^2}$ $= \textbf{dm}^3\textbf{ mol}^{-1}\textbf{ s}^{-1}$	$rate = k[A]^2[B] \quad \therefore\ k = \dfrac{rate}{[A]^2[B]}$ Units of $k = \dfrac{(\text{mol dm}^{-3}\text{ s}^{-1})}{(\text{mol dm}^{-3})^2\,(\text{mol dm}^{-3})}$ $= \textbf{dm}^6\textbf{ mol}^{-2}\textbf{ s}^{-1}$

Progress check

1 A chemical reaction is first order with respect to compound **P** and second order with respect to compound **Q**.
(a) Write the rate equation for this reaction.
(b) What is the overall order of this reaction?
(c) By what factor will the rate increase if:
   (i) the concentration of **P only** is doubled
   (ii) the concentration of **Q only** is doubled
   (iii) the concentrations of **P** and **Q** are **both** doubled?
(d) What are the units of the rate constant of this reaction?

2 The reaction of ethanoic anhydride, $(CH_3CO)_2O$, with ethanol, C_2H_5OH, can be represented by the equation:

$$(CH_3CO)_2O + C_2H_5OH \longrightarrow CH_3CO_2C_2H_5 + CH_3CO_2H$$

The table below shows the initial concentrations of the two reactants and the initial rates of reaction.

Experiment	$[(CH_3CO)_2O]$ /mol dm^{-3}	$[C_2H_5OH]$ /mol dm^{-3}	initial rate /mol dm^{-3} s^{-1}
1	0.400	0.200	6.60×10^{-4}
2	0.400	0.400	1.32×10^{-3}
3	0.800	0.400	2.64×10^{-3}

(a) State and explain the order of reaction with respect to:
   (i) ethanoic anhydride   (ii) ethanol.
(b) (i) Write an expression for the overall rate equation.
   (ii) What is the overall order of reaction?
(c) Calculate the value, with units, for the rate constant, k.

(c) 8.25×10^{-3} dm^3 mol^{-1} s^{-1}.
(b) (i) $rate = k\,[(CH_3CO)_2O]\,[C_2H_5OH]$ (ii) second order.
   (ii) first order; concentration doubles; rate doubles.
2 (a) (i) first order; concentration doubles; rate doubles
   (d) dm^6 mol^{-2} s^{-1}
(c) (i) rate doubles (ii) rate quadruples (iii) rate × 8
(b) third order
1 (a) $rate = k[P][Q]^2$.

How does the rate constant, *k*, vary with temperature?

AQA	M4	SALTERS	M4
EDEXCEL	M4	WJEC	M5
OCR	M5	CCEA	M4

Key points from AS

* **Boltzmann distribution**
* **Activation energy**
  Revise AS page 105

During AS Chemistry, you studied how the rate of reaction increases with temperature in terms of collisions, the activation energy of the reaction and the Boltzmann distribution of molecular energies.

The effect of temperature on rate constants

An increase in temperature speeds up the rate of most reactions and the rate constant, *k*, increases.

The table below shows the increase in the rate constant for the decomposition of hydrogen iodide with increasing temperature.

$$2HI(g) \longrightarrow H_2(g) + I_2(g)$$

For many reactions, the rate doubles for each 10°C increase in temperature.

temperature/K	556	629	700	781
rate constant, *k* /dm^3 mol^{-1} s^{-1}	7.04×10^{-7}	6.04×10^{-5}	2.32×10^{-3}	7.90×10^{-2}

A reaction with a large activation energy has a small rate constant. Such a reaction may need a considerable temperature rise to increase the value of the rate constant sufficiently for a reaction to take place at all.

Why are catalysts so effective?

A catalyst is so effective because:

* it **reduces** the **activation energy**
* which **increases** the **rate of reaction**
* which **increases** the **rate constant, *k***.

Measuring the activation energy experimentally

EDEXCEL M5

You do **not** need to remember the Arrhenius equation – it will always be provided in examination papers.

The rate constant, *k*, and the temperature (in K), *T*, are linked mathematically by the Arrhenius equation:

$$\ln k = \text{constant} - \frac{E_a}{R}(1/T)$$

* E_a = activation energy
* R = gas constant, 8.31 J K^{-1} mol^{-1}

If the temperature, *T*, is changed, rate is proportional to *k*.

* A **graph** of **ln (rate)** against 1/*T* will give a **straight line**.
* The **gradient** of the graph is $-\dfrac{E_a}{R}$ from which E_a can be determined.

e^x

$\boxed{\ln}$

This is the key to use on your calculator.

'ln' is 'natural logarithm'.

If you take A level Maths, you will study ln and e^x.

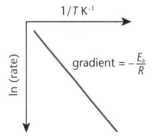

Experimental procedure

Rate is also proportional to 1/time and a value for E_a can be determined by:

* measuring the initial rate for a reaction (as 1/time)
* repeating the experiment at different temperatures, *T*
* plotting a graph of ln (1/time) against 1/*T*
* measuring the gradient, which is $-\dfrac{E_a}{R}$.

1.2 Determination of reaction mechanisms

After studying this section you should be able to:

- *understand what is meant by a rate-determining step*
- *predict a rate equation from a rate-determining step*
- *predict a rate-determining step from a rate equation*
- *use a rate equation and the overall equation for a reaction to predict a possible reaction mechanism*

Predicting reaction mechanisms

EDEXCEL	M4	SALTERS	M3
OCR	M5	WJEC	M5
AQA	M4	CCEA	M4

Key points from AS

- **The hydrolysis of halogenoalkanes**
  Revise AS pages 139–140

You are **not** expected to remember these examples.

You **are** expected to interpret data to identify a rate-determining step and to suggest possible steps in the mechanism for the reaction.

This reaction proceeds via an S$_N$1 mechanism:

1 particle, (CH$_3$)$_3$CBr, is involved in the rate-determining step of a Nucleophilic Substitution mechanism.

Notice that (CH$_3$)$_3$C$^+$ does not feature in the overall equation. It is generated in one step and used up in another step.

The rate-determining step

Chemical reactions often take place in a series of steps. The detail of these steps is the **reaction mechanism**.

> The rate equation can provide clues about a likely reaction mechanism by identifying the **slowest** stage of a reaction sequence, called the **rate-determining step**.
>
> **KEY POINT**

The hydrolysis of tertiary halogenoalkanes

2-bromo-2-methylpropane, (CH$_3$)$_3$C–Br, is hydrolysed by aqueous alkali, OH$^-$.

Experiments show that the rate equation for this reaction is:

$$rate = k[(CH_3)_3C–Br]$$

- The rate equation shows that a slow reaction step must involve **only** (CH$_3$)$_3$C–Br.
- OH$^-$ has **no effect** on the reaction rate.

This supports the slow step below:

$$\underbrace{(CH_3)_3C–Br}_{} \longrightarrow (CH_3)_3C^+ + Br^- \qquad \textbf{SLOW}$$

One particle in rate-determining step

The overall equation for the hydrolyis of 1-bromobutane by aqueous alkali is shown below:

$$(CH_3)_3C–Br + OH^- \longrightarrow (CH_3)_3C–OH + Br^-$$

- This **does not** match the equation for the slow step above.
- The rate-determining step must be followed by further **fast** steps.
- OH$^-$ must be involved as it is in the **overall equation**.

A possible second step is shown below:

$$(CH_3)_3C^+ + OH^- \longrightarrow (CH_3)_3C–OH \qquad \textbf{FAST}$$

The **overall equation** is the sum of the equations from each step in the mechanism.

In the hydrolysis of a tertiary halogenoalkane:

1st step	(CH$_3$)$_3$C–Br	$\longrightarrow$ (CH$_3$)$_3$C$^+$ + Br$^-$	**SLOW**
2nd step	(CH$_3$)$_3$C$^+$ + OH$^-$	$\longrightarrow$ (CH$_3$)$_3$C–OH	**FAST**
Overall equation	(CH$_3$)$_3$C–Br + OH$^-$	$\longrightarrow$ (CH$_3$)$_3$C–OH + Br$^-$	

The hydrolysis of primary halogenoalkanes

1-Bromobutane, $CH_3CH_2CH_2CH_2Br$, is hydrolysed by aqueous alkali, OH^-.

Experiments show that the rate equation for this reaction is:

$rate = k[CH_3CH_2CH_2CH_2Br] [OH^-]$

The **slow** reaction step must involve **both** $CH_3CH_2CH_2CH_2Br$ and OH^-.

This supports the slow step below:

$$\underbrace{CH_3CH_2CH_2CH_2Br + OH^-}_{\text{Two particles in rate-determining step}} \longrightarrow CH_3CH_2CH_2CH_2OH + Br^- \quad \textbf{SLOW}$$

> This reaction proceeds via an S_N2 mechanism:
>
> **2 particles,** $CH_3CH_2CH_2CH_2Br$ **and** OH^-, are involved in the rate-determining step of a nucleophilic Substitution mechanism.

The overall equation for the hydrolyis of 1-bromobutane by aqueous alkali is:

$$CH_3CH_2CH_2CH_2Br + OH^- \longrightarrow CH_3CH_2CH_2CH_2OH + Br^-$$

- This **does** match the equation for the slow step above.
- The reaction **mechanism** must have just a **single step**.

The reaction of iodine and propanone in acid

The reaction of iodine, I_2, and propanone, CH_3COCH_3, is catalysed by acid, H^+.

The overall equation for the acid-catalysed reaction of iodine and propanone is:

$$I_2 + CH_3COCH_3 \longrightarrow CH_3COCH_2I + H^+ + I^- \quad \textit{overall equation}$$

Experiments show that the rate equation for this reaction is:

$rate = k[CH_3COCH_3] [H^+]$

The **slow** rate-determining step must involve **both** CH_3COCH_3 and H^+.

> The slow step is the **rate-determining step**.

A possible equation for the rate-determining step is shown below:

$$\underbrace{CH_3COCH_3 + H^+}_{\text{Two particles in rate-determining step}} \longrightarrow CH_3C(OH^+)CH_3 \quad \textbf{SLOW}$$

> If the species in the rate equation do not match those in the overall equation, the reaction mechanism must have more than one step.

- This does **not** match the overall equation.

Possible further steps are shown below:

$$CH_3C(OH^+)CH_3 \longrightarrow CH_2{=}C(OH)CH_3 + H^+ \quad \textbf{FAST}$$

$$CH_3C(OH){=}CH_2 + I_2 \longrightarrow CH_3(OH)CICH_2I \quad \textbf{FAST}$$

$$CH_3(OH)CICH_2I \longrightarrow CH_3COCH_2I + H^+ + I^- \quad \textbf{FAST}$$

> H^+ is a catalyst.
>
> It is needed for the first step but is regenerated in a further step.
>
> Overall, it is not consumed.

> **KEY POINT**
>
> The overall equation is the sum of the equations from each step in the mechanism.

- The rate of the reaction is controlled mainly by the slowest step of the mechanism – the **rate-determining step**.
- The rate equation only includes reacting species involved in the slow rate-determining step.
- The orders in the rate equation match the number of species involved in the rate-determining step.

Progress check

1 CH_3CH_2Br and cyanide ions, CN^- react as follows:
$$CH_3CH_2Br + CN^- \longrightarrow CH_3CH_2CN + Br^-$$
Show two different possible reaction routes for this reaction and write down the expected rate equation for each route.

1 Single step: $CH_3CH_2Br + CN^- \longrightarrow CH_3CH_2CN + Br^-$ *rate* = $k[CH_3CH_2Br][CN^-]$

Two steps: $CH_3CH_2Br \longrightarrow CH_3CH_2^+ + Br^-$ slow
$CH_3CH_2^+ + CN^- \longrightarrow CH_3CH_2CN$ fast   *rate* = $k[CH_3CH_2Br]$

Using chirality as evidence for mechanisms

EDEXCEL   M4

By studying the optical activity of reactants and products, we can gain further insights into the actual mode of attack during this mechanism.

If we hydrolyse an **optically active** tertiary halogenoalkane that has a **chiral** carbon atom,

- the organic product has **no** optical activity
- a 50:50 racemic mixture of optical isomers is obtained.

This supported the S_N1 mechanism as shown below.

S_N1: no optical activity in organic product

The alternative mechanism would involve an S_N2 mechanism.

S_N2: alternative for tertiary halogenoalkanes

For primary halogenoalkanes, the situation is the other way around:

The S_N2 mechanism does takes place.

The S_N1 mechanism does not take place.

attack on opposite side of molecule to leaving group

organic product would be optically active but with the chiralty swapped over

- In the actual hydrolysis, a single optical isomer product is **not** produced.
- Therefore, mechanism **cannot** take place.

Addition of carbonyl compounds

A similar situation arises during nucleophilic addition to carbonyl compounds.

For details of nucleophilic addition to carbonyl compounds, see page 115.

- The organic product has **no** optical activity.
- A 50:50 racemic mixture of optical isomers is obtained.

1.3 The equilibrium constant, K_c

After studying this section you should be able to:

- deduce, for homogeneous reactions, K_c in terms of concentrations
- calculate values of K_c given appropriate data
- understand that K_c is changed only by changes in temperature
- understand how the magnitude of K_c relates to the equilibrium position
- calculate, from data, the concentrations present at equilibrium

LEARNING SUMMARY

Key points from AS

- **Chemical equilibrium**
  Revise AS pages 110–115

During the study of equilibrium in AS Chemistry, the concept of dynamic equilibrium is introduced and le Chatelier's principle is used to predict how a change in conditions may alter the equilibrium position. For A2 Chemistry, you will build upon this knowledge and understanding to find out the exact position of equilibrium using the Equilibrium Law.

The equilibrium law

AQA	M4	SALTERS	
EDEXCEL		WJEC	M5
OCR		CCEA	M4

The equilibrium law states that, for an equation:

$$a\,A + b\,B \rightleftharpoons c\,C + d\,D,$$

$$K_c = \frac{[C]^c\,[D]^d}{[A]^a\,[B]^b}$$

- $[A]^a$, etc., are the **equilibrium** concentrations of the reactants and products of the reaction.

The overall equation for the reaction is used to determine K_c.

- Each product and reactant has its equilibrium concentration raised to the **power** of its **balancing number** in the equation.

- The equilibrium concentrations of the **products** are multiplied together on **top** of the fraction.

- The equilibrium concentrations of the **reactants** are multiplied together on the **bottom** of the fraction.

Working out K_c

For the equilibrium: $H_2(g) + I_2(g) \rightleftharpoons 2HI(g)$

applying the equilibrium law above: $K_c = \dfrac{[HI(g)]^2}{[H_2(g)]\,[I_2(g)]}$

Equilibrium concentrations of $H_2(g)$, $I_2(g)$ and $HI(g)$ are shown below:

$[H_2(g)]$ /mol dm^{-3}	$[I_2(g)]$ /mol dm^{-3}	$[HI(g)]$ /mol dm^{-3}
0.0114	0.00120	0.0252

For A2 Chemistry, most courses only use **homogeneous equilibria** – all species are in the same phase:

all gaseous (g)

or

all aqueous (aq)

or

all liquid (l).

$$K_c = \frac{[HI(g)]^2}{[H_2(g)]\,[I_2(g)]} = \frac{0.0252^2}{0.0114 \times 0.00120} = 46.4$$

Units of K_c

The units must be worked out afresh for each equilibrium.

See also page 23 in which the units of the rate constant are discussed.

- In the K_c expression, each concentration value is replaced by its units:

$$K_c = \frac{[HI(g)]^2}{[H_2(g)]\,[I_2(g)]} = \frac{(mol\ dm^{-3})^2}{(mol\ dm^{-3})\,(mol\ dm^{-3})}$$

- For this equilibrium, the units cancel.

∴ in the equilibrium: $H_2(g) + I_2(g) \rightleftharpoons 2HI(g)$, K_c has no units.

Properties of K_c

The magnitude of K_c indicates the extent of a chemical equilibrium.

- $K_c = 1$ indicates an equilibrium halfway between reactants and products.
- $K_c = 100$ indicates an equilibrium well in favour of the products.
- $K_c - 1 \times 10^{-?}$ indicates an equilibrium well in favour of the reactants.

K_c indicates how FAR a reaction proceeds; **not** how FAST.
K_c is unaffected by changes in concentration or pressure.
K_c can **only** be changed by altering the temperature.

How does K_c vary with temperature?

It cannot be stressed too strongly that K_c can be changed only by altering the temperature. How K_c changes depends upon whether the reaction gives out or takes in heat energy.

If the forward reaction is **exothermic**, K_c **decreases** with increasing temperature. Raising the temperature reduces the equilibrium yield of products.

$$H_2(g) + I_2(g) \rightleftharpoons 2HI(g): \qquad \Delta H^{\ominus}_{298} = -9.6 \text{ kJ mol}^{-1}$$

temperature /K	500	700	1100
K_c	160	54	25

If the forward reaction is **endothermic**, K_c **increases** with increasing temperature. Raising the temperature increases the equilibrium yield of products.

$$N_2(g) + O_2(g) \rightleftharpoons 2NO(g): \qquad \Delta H^{\ominus}_{298} = +180 \text{ kJ mol}^{-1}$$

temperature /K	500	700	1100
K_c	5×10^{-13}	4×10^{-8}	1×10^{-5}

Although changes in K_c affect the equilibrium yield, the actual conditions used may be different.

If the forward reaction is **exothermic**, K_c increases and the equilibrium yield increases as temperature decreases.

However, a decrease in temperature may slow down the reaction so much that the reaction is stopped.

In practice, a compromise will be needed where equilibrium and rate are considered together – a reasonable equilibrium yield must be obtained in a reasonable length of time.

Progress check

1 For each of the following equilibria, write down the expression for K_c. State the units for K_c for each reaction.
   (a) $N_2O_4(g) \rightleftharpoons 2NO_2(g)$ (c) $H_2(g) + Br_2(g) \rightleftharpoons 2HBr(g)$
   (b) $CO(g) + 2H_2(g) \rightleftharpoons CH_3OH(g)$ (d) $2SO_2(g) + O_2(g) \rightleftharpoons 2SO_3(g)$

2 Explain whether the two reactions, **A** and **B**, are exothermic or endothermic.

temperature /K	numerical value of K_c	
	reaction **A**	reaction **B**
200	5.51×10^{-8}	4.39×10^4
400	1.46	4.03
600	3.62×10^2	3.00×10^{-2}

Reaction **B**: exothermic; K_c decreases with increasing temperature.
2 Reaction **A**: endothermic; K_c increases with increasing temperature.

1 (a) $K_c = \dfrac{[NO_2(g)]^2}{[N_2O_4(g)]}$ mol dm^{-3} (b) $K_c = \dfrac{[CH_3OH(g)]}{[CO(g)][H_2(g)]^2}$ dm^6 mol^{-2} (c) $K_c = \dfrac{[HBr(g)]^2}{[H_2(g)][Br_2(g)]}$ (d) $K_c = \dfrac{[SO_3(g)]^2}{[SO_2(g)]^2[O_2(g)]}$ dm^3 mol^{-1}

Determination of K_c from experiment

AQA	M4	SALTERS	M5
EDEXCEL	M4	WJEC	M5
OCR	M5	CCEA	M4

The equilibrium constant K_c can be calculated using experimental results. The example below shows how to:

- determine the equilibrium concentrations of the components in an equilibrium mixture
- calculate K_c.

The ethyl ethanoate esterification equilibrium

> The most important stage in this calculation is to find the **change** in the number of moles of each species in the equilibrium.

0.200 mol CH_3COOH and 0.100 mol C_2H_5OH were mixed together with a trace of acid catalyst in a total volume of 250 cm³. The mixture was allowed to reach equilibrium:

$$CH_3COOH + C_2H_5OH \rightleftharpoons CH_3COOC_2H_5 + H_2O.$$

Analysis of the mixture showed that 0.115 mol of CH_3COOH were present at equilibrium.

First summarise the results

It is useful to summarise the results as an 'I.C.E.' table: 'Initial/Change/Equilibrium'.

	CH_3COOH	C_2H_5OH	$CH_3COOC_2H_5$	H_2O
Initial no. of moles	0.200	0.100	0	0
Change in moles				
Equilibrium no. of moles	0.115			

From the CH_3COOH values:

- moles of CH_3COOH that reacted = 0.200 – 0.115 = **0.085 mol**

Find the change in moles of each component in the equilibrium

The amount of each component can be determined from the balanced equation. The change in moles of CH_3COOH is already known.

equation:	CH_3COOH	+	C_2H_5OH	$\rightleftharpoons$	$CH_3COOC_2H_5$	+	H_2O
molar quantities:	1 mol		1 mol	$\longrightarrow$	1 mol		1 mol
change/mol:	**–0.085**		–0.085		+0.085		+0.085

> In this experiment, 0.085 mol CH_3COOH has **reacted** with 0.085 mol C_2H_5OH to form 0.085 mol $CH_3COOC_2H_5$ and 0.085 mol H_2O.

Equilibrium concentrations are determined for each component

> Note that the total volume is 250 cm³ (0.250 dm³). The concentration must be expressed as mol dm⁻³.

	CH_3COOH	+	C_2H_5OH	$\rightleftharpoons$	$CH_3COOC_2H_5$	+	H_2O
Initial amount/mol	0.200		0.100		0		0
Change in moles	–0.085		–0.085		+0.085		+0.085
Equilibrium amount/mol	0.115		0.015		0.085		0.085
Equilibrium conc. /mol dm⁻³	$\dfrac{0.115}{0.250}$		$\dfrac{0.015}{0.250}$		$\dfrac{0.085}{0.250}$		$\dfrac{0.085}{0.250}$

Write the expression for K_c and substitute values

$$K_c = \frac{[CH_3COOC_2H_5]\,[H_2O]}{[CH_3COOH]\,[C_2H_5OH]} = \frac{\dfrac{0.085}{0.250} \times \dfrac{0.085}{0.250}}{\dfrac{0.115}{0.250} \times \dfrac{0.015}{0.250}}$$

Calculate K_c

$\therefore K_c = 4.19$ (no units: all units cancel)

Partition coefficients (CCEA only)

When a solute is distributed between two immiscible solvents (usually an organic solvent and water), different amounts of solute dissolve in each layer.

The **partition coefficient**, K_d, is a measure of the relative solubility of a solute between these two solvents.

$$K_d = \frac{\text{concentration in organic solvent}}{\text{concentration in water}}$$

Non-polar organic compounds are more soluble in an organic solvent than in water and have a K_d value >1. By carrying out multiple extractions, virtually all of the organic compound can be extracted from an aqueous layer.

A solvent extraction is a common technique used in organic chemistry to extract an organic solute from water into an organic solvent.

This is used during purification of an organic compound. The impurities remain dissolved in the water, whilst the required organic compound is transferred into the organic solvent.

Progress check

1 Several experiments were set up for the $H_2(g)$, $I_2(g)$ and $HI(g)$ equilibrium. The equilibrium concentrations are shown below.

$[H_2(g)]$ /mol dm^{-3}	$[I_2(g)]$ /mol dm^{-3}	$[HI(g)]$ /mol dm^{-3}
0.0092	0.0020	0.0296
0.0077	0.0031	0.0334
0.0092	0.0022	0.0308
0.0035	0.0035	0.0235

Calculate the value for K_c for each experiment. Hence, show that each experiment has the same value for K_c (allowing for experimental error). Work out an average value for K_c.

2 2 moles of ethanoic acid, CH_3COOH were mixed with 3 moles of ethanol and the mixture was allowed to reach equilibrium.

$$CH_3COOH + C_2H_5OH \rightleftharpoons CH_3COOC_2H_5 + H_2O$$

At equilibrium, 0.5 moles of ethanoic acid remained.
(a) Work out the equilibrium concentrations of each component in the mixture. (You will need to use V to represent the volume but this will cancel out in your calculation.)
(b) Use these values to calculate K_c.

3 When 0.50 moles of $H_2(g)$ and 0.18 moles of $I_2(g)$ were heated at 500°C, the equilibrium mixture contained 0.01 moles of $I_2(g)$.
The equation is:
$H_2(g) + I_2(g) \rightleftharpoons 2HI(g)$
(a) How many moles of $I_2(g)$ reacted?
(b) How many moles of $H_2(g)$ were present at equilibrium?
(c) How many moles of $HI(g)$ were present at equilibrium?
(d) Calculate the equilibrium constant, K_c for this reaction.

3 (a) 0.17 mol (b) 0.33 mol (c) 0.34 mol (d) 35.
(b) $K = 3$.
2 (a) CH_3COOH, 0.5/V mol dm^{-3}; C_2H_5OH, 1.5/V mol dm^{-3}; $CH_3COOC_2H_5$, 1.5/V mol dm^{-3}; H_2O, 1.5/V mol dm^{-3}.
1 values: 47.6; 46.7; 46.9; 45.1. average = 46.6

1.4 The equilibrium constant, K_p

After studying this section you should be able to:

- understand and use the terms mole fraction and partial pressure
- calculate from data the partial pressures present at equilibrium
- deduce, for homogeneous reactions, K_p in terms of partial pressures
- calculate from data the value of K_p, including determination of units

LEARNING SUMMARY

Partial pressure

EDEXCEL   M4   WJEC   M5
CCEA   M4

Equilibria involving gases are usually expressed in terms of K_p, the equilibrium constant in terms of partial pressures.

> **KEY POINT**
>
> In a gas mixture, the *partial pressure* of a gas, p, is the contribution that a gas makes towards the total pressure, P.

The air is a gas mixture with approximate molar proportions of 80% $N_2(g)$ and 20% $O_2(g)$.

The mole fraction of a gas is the same as its proportion by volume.

- The partial pressure of $N_2(g)$ is 80% of the total pressure.
- The partial pressure of $O_2(g)$ is 20% of the total pressure.

As with all fractions, the sum of the mole fractions in a mixture must equal ONE.

> **KEY POINT**
>
> For a gas **A** in a gas mixture:
>
> the mole fraction of **A**, $x_A = \dfrac{\text{number of moles of } \mathbf{A}}{\text{total number of moles in gas mixture}}$
>
> partial pressure of **A**, p_A = mole fraction of **A** × total pressure = $x_A \times P$

The sum of the partial pressures in a mixture must equal the total pressure.

The mole fractions in air are: $x_{N_2} = \dfrac{80}{100} = 0.8$; $x_{O_2} = \dfrac{20}{100} = 0.2$

At normal atmospheric pressure, 100 kPa,

$p_{N_2} = 0.8 \times 100 = 80$ kPa; $p_{O_2} = 0.2 \times 100 = 20$ kPa

Progress check

1 A gas mixture at a total pressure of 300 kPa contains 3 moles of $N_2(g)$ and 1 mole of $O_2(g)$.
   (a) What is the mole fraction of each gas?
   (b) What is the partial pressure of each gas?

1 (a) N_2, 0.75; O_2, 0.25.   (b) N_2, 225 kPa; O_2, 75 kPa.

The equilibrium constant, K_p

EDEXCEL   M4   WJEC   M5
CCEA   M4

K_p is written in a similar way to K_c but with partial pressures replacing concentration terms.

For the equilibrium: $H_2(g) + I_2(g) \rightleftharpoons 2HI(g)$

the equilibrium constant in terms of partial pressures, K_p, is given by:

Don't use [ ] in K_p. This is a common exam mistake.

$$K_p = \frac{p_{HI}{}^2}{p_{H_2} \times p_{I_2}}$$

- p means the equilibrium partial pressure

K_p only includes gases. Ignore any other species.

- suitable units for partial pressures are kilopascals (kPa) or pascals (Pa) – but the same unit must be used for all gases

- the power to which the partial pressures is raised is the *balancing number* in the chemical equation.

Working out K_p

An equilibrium mixture contains 13.5 mol $N_2(g)$, 3.6 mol $H_2(g)$, and 1.0 mol $NH_3(g)$. The total equilibrium pressure is 200 kPa. Calculate K_p.

Find the mole fractions and partial pressures

Total number of gas moles = 13.5 + 3.6 + 1.0 = 18.1 mol

Check that the partial pressures add up to give the total pressure:
149 + 40 + 11 = 200 kPa

$$p_{N_2} = \frac{13.5}{18.1} \times 200 = 149 \text{ kPa} \qquad p_{H_2} = \frac{3.6}{18.1} \times 200 = 40 \text{ kPa}$$

$$p_{NH_3} = \frac{1.0}{18.1} \times 200 = 11 \text{ kPa}$$

Calculate K_p

For the reaction: $N_2(g) + 3H_2(g) \rightleftharpoons 2NH_3(g)$,

$$K_p = \frac{p_{NH_3}^{2}}{p_{N_2} \times p_{H_2}^{3}}$$

$$\therefore K_p = \frac{11^2}{149 \times 40^3} = 1.27 \times 10^{-5} \text{ kPa}^{-2}$$

The units for K_p are found by replacing each partial pressure value in the K_p expression by its units:

substituting units: $K_p = \dfrac{(kPa)^2}{(kPa)\,(kPa)^3}$ $\qquad \therefore$ units of K_p are: kPa^{-2}

Progress check

1 For the following equilibria, write an expression for K_p and work out the value for K_p including units.
   (a) $2HI(g) \rightleftharpoons H_2(g) + I_2(g)$
       partial pressures: $HI(g)$, 56 kPa; $H_2(g)$, 22 kPa; $I_2(g)$, 22 kPa
   (b) $2NO_2(g) \rightleftharpoons 2NO(g) + O_2(g)$
       partial pressures: $NO_2(g)$, 45 kPa; $NO(g)$, 60 kPa; $O_2(g)$, 30kPa

2 In the equilibrium: $2SO_2(g) + O_2(g) \rightleftharpoons 2SO_3(g)$
   2.0 mol of $SO_2(g)$ were mixed with 1.0 mol $O_2(g)$. The mixture was allowed to reach equilibrium at constant temperature and a constant pressure of 900 kPa in the presence of a catalyst. At equilibrium, 0.5 mol of the $SO_2(g)$ had reacted.
   (a) How many moles of SO_2, O_2 and SO_3 were in the equilibrium mixture?
   (b) What were the mole fractions and partial pressures of SO_2, O_2 and SO_3 in the equilibrium mixture?
   (c) Calculate K_p

(c) 4.4×10^{-4} kPa^{-1}

$p(SO_2) = 495$ kPa; $p(O_2) = 243$ kPa; $p(SO_3) = 162$ kPa.

(b) $x(SO_2) = 0.55$; $x(O_2) = 0.27$; $x(SO_3) = 0.18$.

2 (a) 1.5 mol SO_2; 0.75 mol O_2; 0.5 mol SO_3.

1 (a) 0.15 (b) 53 kPa.

1.5 Acids and bases

After studying this section you should be able to:

- describe what is meant by Brønsted–Lowry acids and bases
- understand conjugate acid–base pairs
- understand the difference between a strong and a weak acid
- define the acid dissociation constant, K_a

Key points from AS

- Calculations in acid–base titrations
  Revise AS page 40
- Acids and bases
  Revise AS pages 45–47

During the study of acids and bases at GCSE, you learnt that the pH scale can be used to measure the strength of acids and bases. You also learnt the reactions of acids with metals, carbonates and alkalis. During AS Chemistry, you revisited these reactions and found out how titrations can be used to measure the concentration of an unknown acid or base. For A2 Chemistry, you will study acids in terms of proton transfer, strength, pH and buffers.

Brønsted–Lowry acids and bases

AQA	M4	SALTERS	M4 M5
EDEXCEL	M4	WJEC	M5
OCR	M5	CCEA	M4

A molecule of an acid contains a hydrogen atom that can be released as a positive hydrogen ion or proton, H^+.

- An acid is a substance producing an excess of hydrogen ions in solution – the Arrhenius theory.
- An acid is a proton donor – the Brønsted–Lowry theory.

At A level, we use the Brønsted–Lowry model of acids and bases in terms of proton transfer.

An acid is a proton donor.

A base is a proton acceptor.

KEY POINT

- A Brønsted–Lowry **acid** is a proton donor.
- A Brønsted–Lowry **base** is a proton acceptor.
- An **alkali** is a base that dissolves in water forming $OH^-(aq)$ ions.

Key points from AS

- Acids, bases and alkalis
  Revise AS pages 45–46

Salts are always formed in solution from these typical acid reactions. The salt can be isolated by evaporation of water.

H_2SO_4 forms sulfates

HCl forms chlorides

HNO_3 forms nitrates

With most metals, nitric acid reacts in a different way. If you are asked for the reaction of an acid with a metal, choose a different acid!

The role of $H^+(aq)$ in acid reactions (OCR only)

The H^+ ion is the active part in typical acid reactions. The ionic equations below emphasise this role. The reaction is essentially the same, irrespective of the acid involved.

Acids react with carbonates, forming carbon dioxide:

solid carbonate:
$$2H^+(aq) + CaCO_3(s) \longrightarrow Ca^{2+}(aq) + H_2O(l) + CO_2(g)$$

aqueous carbonate:
$$2H^+(aq) + CO_3^{2-}(aq) \longrightarrow H_2O(l) + CO_2(g)$$

Acids react with bases (metal oxides), forming water:
$$2H^+(aq) + CaO(s) \longrightarrow Ca^{2+}(aq) + H_2O(l)$$

Acids react with alkalis, forming water:
$$H^+(aq) + OH^-(aq) \longrightarrow H_2O(l)$$

Acids react with many metals, forming hydrogen:
$$2H^+(aq) + Mg(s) \longrightarrow Mg^{2+}(aq) + H_2(g)$$

Acid–base pairs

Acids and bases are linked by H^+ as **conjugate pairs**:

- the **conjugate acid** donates H^+
- the **conjugate base** accepts H^+.

$$HCl \rightleftharpoons H^+ + Cl^-$$

conjugate acid   conjugate base

Examples of some conjugate acid–base pairs are shown below.

		acid				base
hydrochloric acid	HCl	$\rightleftharpoons$	H^+	$+$	Cl^-	
sulfuric acid	H_2SO_4	$\rightleftharpoons$	H^+	$+$	HSO_4^-	
ethanoic acid	CH_3COOH	$\rightleftharpoons$	H^+	$+$	CH_3COO^-	

An acid needs a base

An acid can only donate a proton if there is a base to accept it. Most reactions of acids take place in aqueous conditions with water acting as the base. By mixing an acid with a base, an equilibrium is set up comprising **two** acid–base **conjugate pairs**.

The equilibrium system in aqueous ethanoic acid is shown below.

You should be able to identify acid–base pairs in equations such as this.

proton donor H^+ proton acceptor

ACID 1   BASE 1

$$H_2O(l) + CH_3COOH(aq) \rightleftharpoons H_3O^+(aq) + CH_3COO^-(aq)$$

BASE 2   ACID 2

proton acceptor H^+ proton donor

In equations, the oxonium ion H_3O^+ is usually shown as $H^+(aq)$.

- In a reaction involving an aqueous acid such as hydrochloric acid, the 'active' species is the **oxonium** ion, H_3O^+ (ACID 2 above).
- Formation of the oxonium ion requires **both** an acid **and** water.

Progress check

1 Identify the acid–base pairs in the acid–base equilibria below.
(a) $HNO_3 + H_2O \rightleftharpoons H_3O^+ + NO_3^-$
(b) $NH_3 + H_2O \rightleftharpoons NH_4^+ + OH^-$
(c) $H_2SO_4 + H_2O \rightleftharpoons HSO_4^- + H_3O^+$

2 Write equations for the following acid–base equilibria:
(a) hydrochloric acid and hydroxide ions
(b) ethanoic acid, CH_3COOH, and water.

2 (a) $HCl + OH^- \rightleftharpoons H_2O + Cl^-$
(b) $CH_3COOH + H_2O \rightleftharpoons H_3O^+ + CH_3COO^-$

1 (a) acid 1: HNO_3, base 2: H_2O; acid 2: H_3O^+, base 1: NO_3^-
(b) acid 1: H_2O, base 1: OH^-; acid 2: NH_4^+, base 2: NH_3
(c) acid 1: H_2SO_4, base 1: HSO_4^-; acid 2: H_3O^+, base 2: H_2O

Strength of acids and bases

AQA	M4	SALTERS	M3
EDEXCEL		WJEC	M5
OCR		CCEA	M4

The acid–base equilibrium of an acid, HA, in water is shown below.

$$HA(aq) + H_2O(l) \rightleftharpoons H_3O^+(aq) + A^-(aq)$$

To emphasise the loss of H^+, this can be shown more simply as **dissociation** of the acid HA:

$$HA(aq) \rightleftharpoons H^+(aq) + A^-(aq).$$

The **strength** of an acid is the extent to which it dissociates into H^+ and A^-.

A strong acid is completely dissociated.

Strong acids

A **strong** acid, such as nitric acid, HNO_3, is a **good** proton donor.
- There is almost **complete** dissociation.

$$\overset{\text{equilibrium}}{\xrightarrow{\hspace{3cm}}}$$
$$HNO_3(aq) \rightleftharpoons H^+(aq) + NO_3^-(aq)$$

- Virtually all of the potential acidic power has been released as $H^+(aq)$.
- At equilibrium, $[H^+(aq)]$ is much greater than $[HNO_3(aq)]$.

A weak acid only partially dissociates.

Weak acids

A **weak** acid, such as ethanoic acid, CH_3COOH, is a **poor** proton donor.
- There is only **partial** dissociation.

$$\overset{\text{equilibrium}}{\xleftarrow{\hspace{3cm}}}$$
$$CH_3COOH(aq) \rightleftharpoons H^+(aq) + CH_3COO^-(aq)$$

Only a small proportion of the potential acidic power has been released as $H^+(aq)$.
At equilibrium, $[CH_3COOH(aq)]$ is much greater than $[H^+(aq)]$.

K_a is just a special equilibrium constant K_c for equilibria showing the dissociation of acids.

$[HA(aq)]$, $[H^+(aq)]$ and $[A^-(aq)]$ are equilibrium concentrations.

K_a is sometimes called the acidity constant.

The acid dissociation constant, K_a

The extent of acid dissociation is shown by the **acid dissociation constant**, K_a.
For the reaction: $HA(aq) \rightleftharpoons H^+(aq) + A^-(aq)$,

$$K_a = \frac{[H^+(aq)]\,[A^-(aq)]}{[HA(aq)]}$$

units: $K_a = \dfrac{(mol\ dm^{-3})^2}{(mol\ dm^{-3})} = mol\ dm^{-3}$.

- A **large** K_a value shows that the extent of **dissociation is large** – the acid is strong.
- A **small** K_a value shows that the extent of **dissociation is small** – the acid is weak.

Concentrated and *dilute* are terms used to describe the **amount** of dissolved acid in a solution.

Strong and *weak* are terms used to describe the degree of **dissociation** of an acid.

Acid strength and concentration

The distinction between the strength and concentration of an acid is important.

Concentration is the **amount** of an acid dissolved in 1 dm^3 of solution.
- Concentration is measured in $mol\ dm^{-3}$.

Strength is the extent of **dissociation** of an acid.
- Strength is measured as K_a in units determined from the equilibrium.

KEY POINT

Progress check

1 For each of the following acid–base equilibria, write down the expression for K_a. State the units of K_a for each reaction.
   (a) $HCOOH(aq) \rightleftharpoons H^+(aq) + HCOO^-(aq)$
   (b) $C_6H_5COOH(aq) \rightleftharpoons H^+(aq) + C_6H_5COO^-(aq)$

2 Samples of two acids, hydrochloric acid and ethanoic acid, have the same concentration: 0.1 $mol\ dm^{-3}$. Explain why one 'dilute acid' is *strong* whereas the other 'dilute acid' is *weak*.

2 Concentration applies to the amount, in mol, in 1 dm^3 of solution. Both solutions have 0.1 mol dissolved in 1 dm^3 of solution and are dilute. Hydrochloric acid is strong because its dissociation is near to complete. However, ethanoic acid only donates a small proportion of its potential protons, its dissociation is incomplete and it is a weak acid.

1 (a) $K_a = \dfrac{[H^+(aq)]\,[HCOO^-(aq)]}{[HCOOH(aq)]}$ units: mol dm^{-3}

(b) $K_a = \dfrac{[H^+(aq)]\,[C_6H_5COO^-(aq)]}{[C_6H_5COOH(aq)]}$ units: mol dm^{-3}

1.6 The pH scale

After studying this section you should be able to:

- *define the terms pH, pK$_a$ and K$_w$*
- *calculate pH from [H$^+$(aq)]*
- *calculate [H$^+$(aq)] from pH*
- *understand the meaning of the ionisation product of water, K$_w$*
- *calculate pH for strong bases*

pH and [H$^+$(aq)]

AQA	M4	SALTERS	
EDEXCEL	M4	WJEC	M5
OCR		CCEA	M4

The concentrations of H$^+$(aq) ions in aqueous solutions vary widely between about 10 mol dm^{-3} and about 1×10^{-15} mol dm^{-3}.

The **pH scale** is a logarithmic scale used to overcome the problem of using this large range of numbers and to ease the use of negative powers.

The pH scale	
pH	[H$^+$] / mol dm^{-3}
0	1
1	1×10^{-1}
2	1×10^{-2}
3	1×10^{-3}
4	1×10^{-4}
5	1×10^{-5}
6	1×10^{-6}
7	1×10^{-7}
8	1×10^{-8}
9	1×10^{-9}
10	1×10^{-10}
11	1×10^{-11}
12	1×10^{-12}
13	1×10^{-13}
14	1×10^{-14}

pH is defined as: $pH = -\log_{10} [H^+(aq)]$
- where [H$^+$(aq)] is the concentration of hydrogen ions in aqueous solution.
- [H$^+$(aq)] can be calculated from pH using:
  $[H^+(aq)] = 10^{-pH}$

Notice how the value of pH is linked to the power of 10.

pH	2	9	3.6	10.3
[H$^+$(aq)]/mol dm^{-3}	10^{-2}	10^{-9}	$10^{-3.6}$	$10^{-10.3}$

What does a pH value mean?

- A **low** value of [H$^+$(aq)] matches a **high** value of pH.
- A **high** value of [H$^+$(aq)] matches a **low** value of pH.
- A change of pH by 1 changes [H$^+$(aq)] by 10 times.
- An acid of pH 4 contains 10 times the concentration of H$^+$(aq) ions as an acid of pH 5.

Calculating the pH of strong acids

AQA	M4	SALTERS	
EDEXCEL		WJEC	M5
OCR		CCEA	M4

HA is **monoprotic**: each HA molecule can donate **one** H$^+$ ion.

This is the key to use on your calculator when doing pH and [H$^+$] calculations.

For 10^x, press the **SHIFT** or **INV** key first.

For a strong acid, HA:

- we can assume complete dissociation

- the concentration of H$^+$(aq) can be found directly from the acid concentration:
  [H$^+$] = [HA].

You should be able to convert pH into [H$^+$(aq)] and *vice versa*.

Example 1

A strong acid, HA, has a concentration of 0.010 mol dm^{-3}. What is the pH?

Complete dissociation. $\therefore$ [H$^+$(aq)] = 0.010 mol dm^{-3}

$$pH = -\log_{10} [H^+(aq)] = -\log_{10} (0.010) = \mathbf{2.0}$$

Example 2

A strong acid, HA, has a pH of 3.4. What is the concentration of H$^+$(aq)?

Complete dissociation. $\therefore$ [H$^+$(aq)] = 10^{-pH} = 10$^{-3.4}$ mol dm^{-3}

$$\therefore [H^+(aq)] = \mathbf{3.98 \times 10^{-4}} \text{ mol dm}^{-3}$$

Hints for pH calculations

Calculations involving pH are easy once you have learnt how to use your calculator properly.

- Try the examples on the previous page until you can remember the order to press the keys.
- Try reversing each calculation to go back to the original value. Repeat several times until you have mastered how to use **your** calculator for pH calculations.
- **Don't** borrow a calculator or you will get confused. Different calculators may need the keys to be pressed in a different order!
- Look at your answer and decide whether it looks sensible.

> **KEY POINT**
>
> Learn: $pH = -\log_{10}[H^+(aq)]$
>
> $[H^+(aq)] = 10^{-pH}$.

Progress check

1 Calculate the pH of solutions with the following $[H^+(aq)]$ values.
  (a) 0.01 mol dm^{-3} (d) 2.50×10^{-3} mol dm^{-3}
  (b) 0.0001 mol dm^{-3} (e) 8.10×10^{-6} mol dm^{-3}
  (c) 1.0×10^{-13} mol dm^{-3} (f) 4.42×10^{-11} mol dm^{-3}

2 Calculate $[H^+(aq)]$ of solutions with the following pH values.
  (a) pH 3 (c) pH 2.8 (e) pH 12.2
  (b) pH 10 (d) pH 7.9 (f) pH 9.6

3 How many times more hydrogen ions are in an acid of pH 1 than an acid of pH 5?

4 How can a solution have a pH with a negative value?

4 A solution with [H⁺(aq)] > 1 mol dm⁻³ has a negative pH value.

3 pH 1 has 10 000 times more H⁺ ions than pH 5.

2 (a) 1 × 10⁻³ mol dm⁻³ (d) 1.26 × 10⁻⁸ mol dm⁻³
  (b) 1 × 10⁻¹⁰ mol dm⁻³ (e) 6.31 × 10⁻¹³ mol dm⁻³
  (c) 1.58 × 10⁻³ mol dm⁻³ (f) 2.51 × 10⁻¹⁰ mol dm⁻³

1 (a) 2 (b) 4 (c) 13 (d) 2.60 (e) 5.09 (f) 10.4

Calculating the pH of weak acids

AQA	M4	SALTERS	M5
EDEXCEL		WJEC	M5
OCR		CCEA	M4

The pH of a weak acid HA can be calculated from:

- the **concentration** of the acid and
- the value of the acid dissociation constant, K_a.

Assumptions and approximations

Consider the equilibrium of a weak aqueous acid HA(aq):

$$HA(aq) \rightleftharpoons H^+(aq) + A^-(aq)$$

- Assuming that only a very small proportion of HA dissociates, the equilibrium concentration of HA(aq) will be very nearly the same as the concentration of undissociated HA(aq).

$$\therefore [HA(aq)]_{equilibrium} \approx [HA(aq)]_{start}$$

- Assuming that there is a negligible proportion of H$^+$(aq) from ionisation of water:

$$[H^+(aq)] \approx [A^-(aq)]$$

- Using these approximations

For calculations, use
$$K_a \approx \frac{[H^+(aq)]^2}{[HA(aq)]}$$

$$K_a = \frac{[H^+(aq)]\,[A^-(aq)]}{[HA(aq)]} \qquad \therefore K_a \approx \frac{[H^+(aq)]^2}{[HA(aq)]}$$

Take care to learn this method.

Many marks are dropped on exam papers by students who have not done so!

Example

For a weak acid $[HA(aq)] = 0.100$ mol dm^{-3}, $K_a = 1.70 \times 10^{-5}$ mol dm^{-3} at 25°C. Calculate the pH.

$$K_a = \frac{[H^+(aq)] [A^-(aq)]}{[HA(aq)]} \approx \frac{[H^+(aq)]^2}{[HA(aq)]}$$

$$\therefore 1.70 \times 10^{-5} = \frac{[H^+(aq)]^2}{0.100}$$

$$\therefore [H^+(aq)] = \sqrt{0.100 \times 1.70 \times 10^{-5}} = 0.00130 \text{ mol dm}^{-3}$$

$$pH = -\log_{10} [H^+(aq)] = -\log_{10}(0.00130) = \textbf{2.89}$$

K_a and pK_a

AQA M4 SALTERS M5
EDEXCEL M4 WJEC M5
OCR M5 CCEA M4

K_a and pK_a conversions are just like those between pH and H$^+$.

The smaller the pK_a value, the stronger the acid.

Values of K_a can be made more manageable if expressed in a logarithmic form, pK_a (see also page 37: pH and [H$^+$(aq)]).

$$pK_a = -\log_{10}K_a$$
$$K_a = 10^{-pKa}$$

- A **low** value of K_a matches a **high** value of pK_a
- A **high** value of K_a matches a **low** value of pK_a

Comparison of K_a and pK_a

acid		K_a / mol dm^{-3}	pK_a
methanoic acid	HCOOH	1.6×10^{-4}	$-\log_{10}(1.6 \times 10^{-4}) = \textbf{3.8}$
benzoic acid	C$_6$H$_5$COOH	6.3×10^{-5}	$-\log_{10}(6.3 \times 10^{-5}) = \textbf{4.2}$

Progress check

1 Find the pH of solutions of a weak acid HA ($K_a = 1.70 \times 10^{-5}$ mol dm^{-3}), with the following concentrations:
(a) 1.00 mol dm^{-3}; (b) 0.250 mol dm^{-3}; (c) 3.50×10^{-2} mol dm^{-3}

2 Find values of K_a and pK_a for the following weak acids.
(a) 1.0 mol dm^{-3} solution with a pH of 4.5
(b) 0.1 mol dm^{-3} solution with a pH of 2.2
(c) 2.0 mol dm^{-3} solution with a pH of 3.8.

2 (a) $K_a = 1 \times 10^{-9}$ mol dm^{-3}, $pK_a = 9$
(b) $K_a = 3.98 \times 10^{-4}$ mol dm^{-3}, $pK_a = 3.4$
(c) $K_a = 1.26 \times 10^{-8}$ mol dm^{-3}, $pK_a = 7.9$

1 (a) 2.38 (b) 2.69 (c) 3.11

The ionisation of water and K_w

AQA M4 SALTERS M5
EDEXCEL M4 WJEC M5
OCR M5 CCEA M4

In water, a very small proportion of molecules dissociates into H$^+$(aq) and OH$^-$(aq) ions. The position of equilibrium lies well to the left of the equation below, representing this dissociation.

$$\xleftarrow{\text{equilibrium}}$$
$$H_2O(l) \rightleftharpoons H^+(aq) + OH^-(aq)$$

Treating water as a weak acid: $K_a = \dfrac{[H^+(aq)] [OH^-(aq)]}{[H_2O(l)]}$

Rearranging gives:

$$\underbrace{K_a \times [H_2O(l)]}_{\text{constant, } K_w} = [H^+(aq)] [OH^-(aq)]$$

[H$_2$O(l)] is constant and is included within K_w

> The constant K_w is called the **ionic product of water**
> - $K_w = [H^+(aq)] [OH^-(aq)]$
> - At 25°C, $K_w = 1.0 \times 10^{-14}$ mol^2 dm^{-6}.

KEY POINT

In water, the concentrations of $H^+(aq)$ and $OH^-(aq)$ ions are the same.

- $[H^+(aq)] = [OH^-(aq)] = 10^{-7}$ mol dm^{-3} ($10^{-14} = 10^{-7} \times 10^{-7}$)

Hydrogen ion and hydroxide ion concentrations

Linking [H⁺] and [OH⁻]
$10^{-14} = [H^+] [OH^-]$

Water: pH = 7,
$[H^+] = 10^{-7}$ mol dm^{-3}
$10^{-14} = 10^{-7} \times 10^{-7}$

An acid: pH = 3
$[H^+] = 10^{-3}$ mol dm^{-3}
$10^{-14} = 10^{-3} \times 10^{-11}$

An alkali: pH = 10
$[H^+] = 10^{-10}$ mol dm^{-3}
$10^{-14} = 10^{-10} \times 10^{-4}$

All aqueous solutions contain $H^+(aq)$ and $OH^-(aq)$ ions. The proportions of these ions in a solution are determined by the pH.

In water	$[H^+(aq)] = [OH^-(aq)]$
In **acidic** solutions	$[H^+(aq)] > [OH^-(aq)]$
In **alkaline** solutions	$[H^+(aq)] < [OH^-(aq)]$

The concentrations of $H^+(aq)$ and $OH^-(aq)$ are linked by K_w.

- At 25°C $1.0 \times 10^{-14} = [H^+(aq)] [OH^-(aq)]$
- The indices of $[H^+(aq)]$ and $[OH^-(aq)]$ add up to -14.

Calculating the pH of strong alkalis

The pH of a strong alkali can be found using K_w.

For a strong alkali, BOH:

- we can assume complete dissociation
- the concentration of $OH^-(aq)$ can be found directly from the alkali concentration: $[BOH] = [OH^-]$.

To find the pH of an alkali, first find $[H^+]$ using K_w and $[OH^-]$.

Example

A strong alkali, BOH, has a concentration of 0.50 mol dm^{-3}.

What is the pH?

Complete dissociation. $\therefore$ $[OH^-(aq)] = 0.50$ mol dm^{-3}

$$K_w = [H^+(aq)] [OH^-(aq)] = 1 \times 10^{-14} \text{ mol}^2 \text{ dm}^{-6}$$

$$\therefore [H^+(aq)] = \frac{K_w}{[OH^-(aq)]} = \frac{1 \times 10^{-14}}{0.50} = 2 \times 10^{-14} \text{ mol dm}^{-3}$$

$$pH = -\log_{10} [H^+(aq)] = -\log_{10} (2 \times 10^{-14}) = \mathbf{13.7}$$

Progress check

1 Find the $[H^+(aq)]$ and pH of the following alkalis at 25°C:
   (a) 1×10^{-3} mol dm^{-3} $OH^-(aq)$
   (b) 3.5×10^{-2} mol dm^{-3} $OH^-(aq)$.

2 Find the pH of the following solutions of strong bases at 25°C:
   (a) 0.01 mol dm^{-3} KOH (aq)
   (b) 0.20 mol dm^{-3} NaOH (aq).

2 (a) pH = 12; (b) pH = 13.3.
  (b) $[H^+(aq)] = 2.86 \times 10^{-13}$ mol dm^{-3}; pH = 12.5.
1 (a) $[H^+(aq)] = 1 \times 10^{-11}$ mol dm^{-3}; pH = 11

1.7 pH changes

After studying this section you should be able to:

- recognise the shapes of titration curves for acids and bases with different strengths
- explain the choice of suitable indicators for acid–base titrations
- determine K_a from a titration curve
- carry out unstructured titration calculations
- state what is meant by a buffer solution
- explain how pH is controlled by each component in a buffer solution
- calculate the pH of a buffer solution

LEARNING SUMMARY

Titration curves

AQA	M4	WJEC	M5
EDEXCEL	M4	CCEA	M4
OCR	M4		

Choosing an indicator using titration curves

A titration curve shows the changes in pH during a titration.

In the titration curves below, different combinations of strong and weak acids and alkalis have been used.

- The end point of a titration is identified by the colour change of the indicator.
- Different indicators change colour at different pH values.
- The pH values at which the indicators methyl orange (**MO**) and phenolphthalein (**P**) change colour are shown on each diagram.

strong acid/strong alkali

both indicators suitable

strong acid/weak alkali

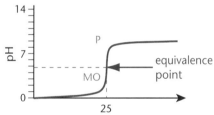

*methyl orange (**MO**) suitable*
*phenolphthalein (**P**) unsuitable*

Choose the correct indicator

Strong acid/strong alkali
phenolphthalein ✓
methyl orange ✓

Strong acid/weak alkali
phenolphthalein ✗
methyl orange ✓

Weak acid/strong alkali
phenolphthalein ✓
methyl orange ✗

Weak acid/weak alkali
phenolphthalein ✗
methyl orange ✗

weak acid/strong alkali

*phenolphthalein (**P**) suitable*
*methyl orange (**MO**) unsuitable*

weak acid/weak alkali

neither indicator suitable

Key features of titration curves

- The pH changes rapidly at the near vertical portion of the titration curve. The mid-point of the vertical portion is the **equivalence point** of the titration.
- The sharp change in pH is brought about by a very small addition of alkali, typically the addition of one drop.
- The indicator is only suitable if its pK_{ind} value is within the pH range of the near vertical portion of the titration curve.

Determination of K_a from a titration curve

EDEXCEL   M4

Titration curves can be used to determine a value for K_a using the procedure below.

- A pH titration curve is plotted and the volume of alkali for neutralisation is read off from the curve (see diagram below).
- The pH at half-neutralisation (i.e. half the alkali required for neutralisation) is read off from the titration curve.

Determination of K_a

The acid dissociation constant, $K_a = \dfrac{[H^+(aq)]\,[A^-(aq)]}{[HA(aq)]}$

At half-neutralisation, **[HA(aq)] = [A⁻(aq)]**, so $K_a = [H^+(aq)]$

$\therefore K_a = \mathbf{10^{-pH}}$, which is easily calculated from the value obtained from the titration curve.

Buffering region of the titration curve

At the start of the titration curve, there is a relatively rapid rise in pH.

As the alkali is added, the salt of the weak acid starts to form. This gives a mixture of a **weak acid** and its **salt** – a **buffer** solution.

The increase in pH in the titration curve slows down in the buffering region.

Enthalpy changes during neutralisation

OCR   M5

The *standard enthalpy change of neutralisation* $\Delta H^{\ominus}_{neut}$ is the energy change that accompanies the neutralisation of an acid by a base to form 1 mole of $H_2O(l)$, under standard conditions.

$$HCl(aq)\ +\ NaOH(aq)\ \rightleftharpoons\ NaCl(aq)\ +\ H_2O(l)\quad \Delta H^{\ominus}_{neut} = -57.9\ kJ\ mol^{-1}$$

acid   base   1 mol

The ionic equation for this reaction is simply:

$$H^+(aq)\ +\ OH^-(aq)\ \longrightarrow\ H_2O(l).$$

- In aqueous solution, strong acids and bases are completely dissociated and $\Delta H^{\ominus}_{neut}$ is approximately equal to the value above.
- For weak acids, this enthalpy change is less exothermic because some input of energy is required to dissociate the acid (see also pages 35–36).

Unstructured titration calculations

AQA	M4, M5	SALTERS	M4
EDEXCEL	M4, M5	WJEC	M4, M5
OCR	M5	CCEA	M5

In AS chemistry examinations, titration problems are usually highly structured. You are helped through calculations with each step being asked in a sequence that leads to the final solution.

In A2 chemistry examinations:

- problems will be set in which you are only provided with a summary of the experimental results
- you will be expected to navigate your own way through the problem; you will have no hints about how to solve the problem.

The key to success is a good understanding of the stages required to solve a titration problem. Most problems follow a common framework.

Acid–base titrations are always likely to be carried out in chemical analysis. In A2, you will also come across redox titrations (see pages 96-98). Although these are different types of titration, the principles for solving any calculations are the same.

AS exam questions would help you through titration calculations by asking each step as a question.

But at A2 you are really on your own! You should now know more chemistry than at AS and also be able to tackle chemistry problems independently.

At least that is what the setters of examination questions expect!

Stage 1

All titrations use a standard solution – one whose concentration is known. In the titration, you measure the volume of this solution that exactly reacts with a second solution.

- You first find the amount, in moles, of the compound in the standard solution.

E.g. In a titration 25.0 cm^3 of 0.100 mol dm^{-3} NaOH reacted exactly with a solution of H_2SO_4.

$$n = c \times \frac{V}{1000} = 0.100 \times \frac{25.0}{1000} = 0.00250 \text{ mol}$$

Remember that the steps shown here are key to solving most titration calculations.

See also redox titrations, pages 96-98.

Stage 2

- The equation is then usually used to find the number of moles of the other substance dissolved in the second solution:

E.g. $2NaOH(aq) + H_2SO_4(aq) \longrightarrow$ *products*

              2 mol   1 mol

amounts: 0.00250   0.00125 mol

Stage 3

- Finally, this value is processed in some way to find out some further information about the second substance. This is the hardest stage and it may contain several steps. This stage may also vary between different problems.

Worked example

Compound **A** is a straight-chain carboxylic acid. A student analysed a sample of acid **A** by the procedure below.

The student dissolved 5.437 g of **A** in water and made the solution up to 250 cm^3.

In a titration, 25.0 cm^3 of 0.200 mol dm^{-3} NaOH were neutralised by exactly 23.45 cm^3 of solution **A**.

Use the results to calculate the molar mass of acid **A** and suggest its identity.

The amount of NaOH used can be calculated from the titration results:

The concentration and volume of NaOH are known. So we work out the amount of NaOH first.

amount of NaOH $= 0.200 \times \frac{25.0}{1000} = 0.00500$ mol

The amount of carboxylic acid that reacts with the NaOH can be determined from the balanced equation for the reaction:

Carboxylic acid **A** must have the formula: RCOOH

The equation and molar reacting quantities are:

$$NaOH(aq) + RCOOH(aq) \longrightarrow RCOONa(aq) + H_2O(l)$$

$$1 \text{ mol} \qquad\qquad 1 \text{ mol}$$

∴ in the titration: 0.00500 mol NaOH reacts with **0.00500 mol RCOOH**

> In this titration there is the same number of moles of each reactant. We work this out from the balanced equation.

You must now look at what the question is asking. You need to think about how you are going to solve the rest of the problem.

- Here, we must first find out the amount, in moles, of **A** that was used to prepare the original 250 cm³ solution.
- In the titration, 23.45 cm³ of this solution of **A** was used – we know the number of moles of **A** dissolved in this volume.
- We need to scale up these quantities to find the number of moles of **A** dissolved in 250 cm³ of solution:

> This step is important but many candidates forget to do this (or don't understand that it must be done!)

23.45 cm³ of RCOOH(aq) in the titration contains 0.00500 mol RCOOH

1 cm³ of RCOOH(aq) in the titration contains $\dfrac{0.00500}{23.45}$ mol RCOOH

250 cm³ of RCOOH(aq) contains $\dfrac{0.00500}{23.45}$ x 250 = 0.0533 mol RCOOH

> This step is easy, provided that you have remembered the following:
> $$n = \frac{\text{mass } m}{\text{Molar mass } M}$$

The molar mass of RCOOH can now be determined and we have solved the first part of the problem:

$$n = \frac{m}{M} \quad \therefore \text{ Molar mass, } M = \frac{m}{n} = \frac{5.437}{0.0533} = \textbf{102.0 g mol}^{-1}$$

*Finally, we need to suggest a possible structure for straight-chain carboxylic acid **A**.*

The formula of the straight-chain carboxylic acid is RCOOH.

COOH has a relative mass of 12.0 + 16.0 x 2 + 1.0 = 45.0.

∴ the alkyl group R has a relative mass of 102.0 – 45.0 = 57.0

∴ the alkyl group must be $CH_3CH_2CH_2CH_2$ (15.0 + 14.0 + 14.0 + 14.0)

and carboxylic acid **A** is $\textbf{CH}_3\textbf{CH}_2\textbf{CH}_2\textbf{CH}_2\textbf{COOH}$ (pentanoic acid).

> These problems may mix together chemistry from different parts of your synoptic assessment. See pages 178–180.

Buffer solutions

A buffer solution minimises changes in pH during the addition of an acid or an alkali to a solution. The buffer solution maintains a near-constant pH by removing most of any added acid or alkali.

A **buffer solution** is a mixture of:

- a **weak acid**, HA, and
- its **conjugate base**, A⁻:

$$HA(aq) \rightleftharpoons H^+(aq) + A^-(aq)$$

$$\text{weak acid} \qquad\qquad\qquad \text{conjugate base}$$

> Buffers are added to foods to prevent deterioration due to pH change (caused by bacterial or fungal activity).

In a buffer solution, the concentration of hydrogen ions, [H⁺(aq)], is very small compared with the concentrations of the weak acid [HA(aq)] or the conjugate base [A⁻(aq)].

$$[H^+(aq)] \ll [HA(aq)] \quad \text{and} \quad [H^+(aq)] \ll [A^-(aq)].$$

How does a buffer act?

Addition of an acid, H⁺(aq), to a buffer

We can explain how a buffer minimises pH changes by using le Chatelier's principle.

On addition of an acid:

- $[H^+(aq)]$ is increased
- the pH change is opposed and the **equilibrium moves to the left**, removing $[H^+(aq)]$ and forming HA(aq)
- the **conjugate base** $A^-(aq)$ removes most of any added $[H^+(aq)]$.

A⁻ removes most of any added acid.

add acid, H⁺(aq)

| HA(aq) | ⇌ | H⁺(aq) | + | A⁻(aq) |
| weak acid | | | | conjugate base |

most added H⁺(aq) is removed

Addition of an alkali, OH⁻(aq), to a buffer

On addition of an alkali:

- the added $OH^-(aq)$ reacts with the small concentration of $H^+(aq)$:

$$H^+(aq) + OH^-(aq) \longrightarrow H_2O(l)$$

- the pH change is opposed – the **equilibrium moves to the right, restoring** $[H^+(aq)]$ as HA(aq) dissociates
- the **weak acid HA restores most** of any $[H^+(aq)]$ that has been removed.

add alkali, OH⁻(aq)

| HA(aq) | ⇌ | H⁺(aq) | + | A⁻(aq) |
| weak acid | | | | conjugate base |

most H⁺(aq) is restored

HA removes added alkali.

> **KEY POINT**
>
> Although the two components in a buffer solution react with added acid and alkali, they **cannot stop** the pH from changing. They do however **minimise** pH changes.

Common buffer solutions

An acidic buffer

Remember these buffers:
CH₃COOH/CH₃COONa
and
NH₄Cl/NH₃

A common acidic buffer is an aqueous solution containing a mixture of ethanoic acid, CH_3COOH, and the ethanoate ion, CH_3COO^-.

- Ethanoic acid acts as the **weak acid**, CH_3COOH.
- Sodium ethanoate, $CH_3COO^-Na^+$, acts as a source of the **conjugate base**, CH_3COO^-.

An alkaline buffer

The pH of healthy blood is maintained at between 7.35 and 7.40 by the carbonic acid–hydrogencarbonate buffer system.

H_2CO_3 is the **weak acid**,

HCO_3^- is the **conjugate base**.

$H_2CO_3(aq) \rightleftharpoons H^+(aq) + HCO_3^-(aq)$

Blood cannot sustain life if its pH lies much outside of this narrow range.

A common alkaline buffer is an aqueous solution containing a mixture of the ammonium ion, NH_4^+, and ammonia, NH_3.

- Ammonium chloride, $NH_4^+Cl^-$, acts as source of the **weak acid**, NH_4^+.
- Ammonia acts as the **conjugate base**, NH_3.

Calculations involving buffer solutions

AQA	M4	SALTERS	M5
CCEA	M4	WJEC	M5
OCR	M5		

For buffer calculations, learn:

$$[H^+(aq)] = K_a \times \frac{[HA(aq)]}{[A^-(aq)]}$$

The pH of a buffer can be controlled by the acid/base ratio.

The pH of a buffer solution depends upon:

- the acid dissociation constant, K_a, of the buffer system
- the **ratio** of the weak **acid** and its conjugate **base**.

For a buffer comprising the weak acid, HA, and its conjugate base, A^-,

$$K_a = \frac{[H^+(aq)]\,[A^-(aq)]}{[HA(aq)]}$$

$$\therefore\ [H^+(aq)] = \underbrace{K_a}_{\text{acid dissociation constant}} \times \underbrace{\frac{[HA(aq)]}{[A^-(aq)]}}_{\begin{array}{l}\text{\textbf{ratio} of the weak acid}\\\text{and its conjugate base}\end{array}}$$

Example

Calculate the pH of a buffer comprising 0.30 mol dm^{-3} CH$_3$COOH(aq) ($K_a = 1.7 \times 10^{-5}$ mol dm^{-3}) and 0.10 mol dm^{-3} CH$_3$COO$^-$(aq).

$$[H^+(aq)] = K_a \times \frac{[HA(aq)]}{[A^-(aq)]}$$

$$\therefore\ [H^+(aq)] = 1.7 \times 10^{-5} \times \frac{0.30}{0.10} = 5.1 \times 10^{-5}\ \text{mol dm}^{-3}$$

$$\therefore\ \text{pH} = -\log_{10}[H^+(aq)] = -\log_{10}(5.1 \times 10^{-5})$$

$$\therefore\ \text{pH of the buffer solution} = \mathbf{4.3}$$

Progress check

1 Three buffer solutions are made from benzoic acid, C$_6$H$_5$COOH and sodium benzoate, C$_6$H$_5$COONa with the following compositions:

> Buffer A: 0.10 mol dm^{-3} C$_6$H$_5$COOH and 0.10 mol dm^{-3} C$_6$H$_5$COONa
> Buffer B: 0.75 mol dm^{-3} C$_6$H$_5$COOH and 0.25 mol dm^{-3} C$_6$H$_5$COONa
> Buffer C: 0.20 mol dm^{-3} C$_6$H$_5$COOH and 0.80 mol dm^{-3} C$_6$H$_5$COONa

(a) Write the equation for the equilibrium in these buffers.
(b) Write an expression for K_a of benzoic acid.
(c) Calculate the pH of each of the buffer solutions
(K_a for C$_6$H$_5$COOH = 6.3 × 10^{-5} mol dm^{-3}.)

1 (a) C$_6$H$_5$COOH $\rightleftharpoons$ H$^+$(aq) + C$_6$H$_5$COO$^-$(aq)
(b) $K_a = \dfrac{[H^+(aq)]\,[C_6H_5COO^-(aq)]}{[C_6H_5COOH(aq)]}$
(c) Buffer A: 4.2 Buffer B: 3.7 Buffer C: 4.8

Sample question and model answer

Many cosmetics contain buffers. Human skin is slightly acidic and has a pH of approximately 5.50. Skin-care products often buffer at a pH 5.50 or below.

(a) What do you understand by the term *buffer*?

A solution that minimises a change in pH. ✓ [1]

(b) Lactic acid, $CH_3CH(OH)COOH$, is used in many cosmetics. Lactic acid is a weak acid with an acid dissociation constant, K_a, of 8.40×10^{-4} mol dm^{-3}.

(i) Write an equation to show the dissociation of lactic acid into its ions.

$CH_3CH(OH)COOH \rightleftharpoons CH_3CH(OH)COO^- + H^+$ ✓ (equilibrium sign essential)

(ii) Write an expression for the acid dissociation constant, K_a, of lactic acid.

$$K_a = \frac{[CH_3CH(OH)COO^-(aq)]\,[H^+(aq)]}{[CH_3CH(OH)COOH(aq)]} ✓$$

(iii) Calculate the pH of a 1.25×10^{-2} mol dm^{-3} solution of lactic acid.

$CH_3CH(OH)COO^- \approx H^+$.

equilibrium conc. of weak acid $\approx$ undissociated concentration of weak acid.

$$\therefore K_a \approx \frac{[H^+(aq)]^2}{[CH_3CH(OH)COOH(aq)]} \qquad 8.4 \times 10^{-4} = \frac{[H^+(aq)]^2}{1.25 \times 10^{-2}} ✓$$

$[H^+(aq)] = \sqrt{(1.25 \times 10^{-2} \times 8.40 \times 10^{-4})} = 3.24 \times 10^{-3}$ ✓ mol dm^{-3}

$\therefore$ pH $= -\log(3.24 \times 10^{-3}) = 2.49.$ ✓ [5]

(c) A buffer solution can be made based on lactic acid and a salt of lactic acid. Explain how this buffer solution acts as a buffer.

In the equilibrium $CH_3CH(OH)COOH \rightleftharpoons CH_3CH(OH)COO^- + H^+$, there are large excesses of $CH_3CH(OH)COOH$ and $CH_3CH(OH)COO^-$. ✓

The equilibrium above shifts in response to added acid and alkali. ✓

On addition of an alkali, OH^- ions react with the small concentration of H^+ present:

$H^+(aq) + OH^-(aq) \longrightarrow H_2O(l)$

The equilibrium shifts to the right, restoring most of any H^+ ions removed: ✓

$CH_3CH(OH)COOH \longrightarrow CH_3CH(OH)COO^- + H^+$ ✓

Lactate ions from sodium lactate remove most of any added acid, shifting the equilibrium to the left: ✓

$CH_3CH(OH)COO^- + H^+ \longrightarrow CH_3CH(OH)COOH$ ✓ [6]

(d) A buffer solution based on lactic acid is required to buffer at a pH 3.80. Calculate the ratio of lactic acid : sodium lactate that will be needed to give this pH.

In the buffer, $[H^+(aq)] = 10^{-3.80} = 1.58 \times 10^{-4}$ ✓ mol dm^{-3}

In the buffer, $[H^+(aq)] = K_a \times \dfrac{[CH_3CH(OH)COOH(aq)]}{[CH_3CH(OH)COO^-(aq)]}$ ✓

$\therefore$ ratio $\dfrac{[CH_3CH(OH)COOH(aq)]}{[CH_3CH(OH)COO^-(aq)]} = \dfrac{[H^+(aq)]}{K_a} = \dfrac{1.58 \times 10^{-4}}{8.4 \times 10^{-4}}$ ✓

ratio lactic acid : sodium lactate $= 0.189 = \dfrac{1}{5.30}$ ✓ [4]

[Total: 16]

Practice examination questions

1 The reaction between hydrogen peroxide and iodide ions in an acid solution can be written as follows.

$$H_2O_2(aq) + 2I^-(aq) + 2H^+(aq) \longrightarrow I_2(aq) + 2H_2O(l)$$

(a) The table below shows initial rates from different initial concentrations of $H_2O_2(aq)$, $I^-(aq)$ and $H^+(aq)$ at constant temperature.

$[H_2O_2(aq)]$ /mol dm^{-3}	$[I^-(aq)]$ /mol dm^{-3}	$[H^+(aq)]$ /mol dm^{-3}	initial rate /10^{-6} mol dm^{-3} s^{-1}
0.00075	0.10	0.10	2.1
0.00150	0.10	0.10	4.2
0.00150	0.10	0.20	4.2
0.00075	0.70	0.10	14.7

(i) Determine, with reasoning, the order with respect to each reactant.

(ii) Write the rate equation and calculate the value of the rate constant, k, for this reaction. [9]

(b) (i) What is meant by the term *rate-determining step*?

(ii) Using the rate equation in (a)(ii), suggest an equation for the rate-determining step of this reaction. [2]

[Total: 11]

2 (a) A chemical reaction is second order with respect to compound **A** and first order with respect to compound **B**.

(i) Write the rate equation for this reaction.

(ii) What is the overall order of this reaction?

(iii) By what factor will the rate increase if the concentrations of **A** and **B** are both tripled?

(iv) When $[A] = 0.40$ mol dm^{-3} and $B = 0.35$ mol dm^{-3}, the rate $= 1.76 \times 10^{-4}$ mol dm^{-3} s^{-1}. Calculate the value of the rate constant, stating its units. [5]

(b) Propanone and iodine react together in acidic solution to give iodopropanone according to the overall equation below.

$$CH_3COCH_3(aq) + I_2(aq) \longrightarrow CH_3COCH_2I(aq) + HI(aq)$$

The rate equation is: *rate* $= k[CH_3COCH_3(aq)] [H^+(aq)]$

(i) State, with an explanation, the role of the $H^+(aq)$ ions in this reaction.

(ii) Suggest a possible equation for the rate-determining step. [3]

[Total: 8]

3 The equilibrium below was set up.

$$CH_3COOCH_2CH_3 + H_2O \rightleftharpoons CH_3COOH + CH_3CH_2OH$$

The equilibrium mixture contains 1.35 mol of ethyl ethanoate, 1.35 mol of water, 0.15 mol of ethanoic acid, and 3.15 mol of ethanol. It has a volume of 353 cm^3.

(a) Calculate K_c for this equilibrium to 2 significant figures. State the units, if any. [3]

(b) More ethanol was to be added to the equilibrium mixture.

(i) What would happen to the equilibrium? Explain your reasoning.

(ii) What would be the effect of this change on the value of K_c? Explain your reasoning. [4]

[Total: 7]

Practice examination questions (continued)

4 When 0.60 moles of $H_2(g)$ and 0.18 moles of $I_2(g)$ were heated to constant temperature in a sealed container with a volume of $1 \ dm^3$, an equilibrium was set up:

$$H_2(g) + I_2(g) \rightleftharpoons 2HI(g)$$

At equilibrium, 0.16 mol $I_2(g)$ had reacted.

(a) (i) Determine the equilibrium concentrations of H_2, I_2, and HI.

(ii) Calculate the equilibrium constant, K_c, for this reaction.   [6]

(b) When the equilibrium mixture was heated by 100°C the value of K_c decreased. What additional information is provided by this observation?   [2]

[Total: 8]

5 Nitric acid, HNO_3 is a strong acid and methanoic acid, HCOOH, is a weak acid ($K_a = 1.6 \times 10^{-4} \ mol \ dm^{-3}$).

(a) Define the term *Brønsted–Lowry acid*.   [1]

(b) Write the acid–base equilibrium that would be set up between methanoic acid and nitric acid. State the role of methanoic acid in this equilibrium.   [2]

(c) Calculate the pH of the following solutions at 25°C:

(i) 0.176 mol dm^{-3} nitric acid

(ii) 0.372 mol dm^{-3} potassium hydroxide

(iii) 0.263 mol dm^{-3} methanoic acid.

(iv) a solution containing 0.255 mol dm^{-3} methanoic acid and 0.325 sodium methanoate.   [10]

[Total: 13]

6 (a) (i) What is meant by the terms *strong acid* and *weak acid*?

(ii) Define the term *Brønsted–Lowry base*.   [2]

(b) 25.0 cm^3 of a solution of sodium hydroxide contained 1.50×10^{-3} mol of NaOH.

(i) Calculate the pH of this solution at 298 K.

(ii) A student added 50.0 cm^3 of 0.250 mol dm^{-3} HCl to the 25.0 cm^3 solution of sodium hydroxide.

Calculate the pH of the solution formed.   [8]

(c) A sample of hydrochloric acid had a pH of 1.52. 25 cm^3 of this acid was diluted with 15 cm^3 of water.

Calculate the pH of the resulting solution.   [4]

[Total: 14]

gy changes in chemistry

topics are covered in this chapter:

- es
- s

- Electrochemical cells
- Predicting redox reactions

2.1 Enthalpy changes

After studying this section you should be able to:

- explain and use the term 'lattice enthalpy'
- construct Born–Haber cycles to calculate the lattice enthalpy of simple ionic compounds
- explain the effect of ionic charge and ionic radius on the numerical magnitude of a lattice enthalpy
- calculate enthalpy changes of solution for ionic compounds from enthalpy changes of hydration and lattice enthalpies

LEARNING SUMMARY

Key points from AS

- **Enthalpy changes**
  Revise AS pages 95–103

During the study of enthalpy changes in AS Chemistry, you learnt how to:

- calculate enthalpy changes directly from experiments using the relationship
  $Q = mc\Delta T$
- calculate enthalpy changes indirectly using Hess' law.

These key principles are built upon by considering the enthalpy changes that bond together an ionic lattice.

Lattice enthalpy

AQA	M5	SALTERS	M5
OCR	M5	WJEC	M5
		CCEA	M4

Ionic bonding is the electrostatic attraction between oppositely charged ions. Lattice enthalpy indicates the strength of the ionic bonds in an ionic lattice.

Key points from AS

- **Ionic bonding**
  Revise AS pages 52–59

Each ion is surrounded by oppositely-charged ions, forming a giant ionic lattice.

KEY POINT

The **lattice enthalpy** ($\Delta H^{\ominus}_{L.E.}$) of an ionic compound is the enthalpy change that accompanies the formation of 1 mole of an ionic compound from its constituent gaseous ions. ($\Delta H^{\ominus}_{L.E.}$ is **exothermic**.)

$$Na^+(g) + Cl^-(g) \longrightarrow Na^+Cl^-(s)$$

Lattice enthalpy is also sometimes called the **enthalpy change of lattice formation**.

The opposite change, to break up the lattice, is called the **enthalpy change of lattice dissociation**.

$$Na^+Cl^-(s) \longrightarrow Na^+(g) + Cl^-(g)$$

Determination of lattice enthalpies

| AQA | M5 | WJEC | M5 |
| OCR | M5 | CCEA | M4 |

Lattice enthalpies cannot be determined directly and must be calculated indirectly using Hess' Law from other enthalpy changes that can be found experimentally. The energy cycle used to calculate a lattice enthalpy is the **Born–Haber cycle**.

The basis of the Born–Haber cycle is the formation of an ionic lattice from its element by two routes.

50

The Born–Haber cycle for sodium chloride is shown below.

Definitions for these enthalpy changes are shown below.

Route 1: **A + B + C + D + E**   Route 2: **F**

Using Hess' Law: **A + B + C + D + E = F**

$\Delta H^{\ominus}_{at}$ Na(g) + $\Delta H^{\ominus}_{at}$ Cl(g) + $\Delta H^{\ominus}_{I.E.}$Na(g) + $\Delta H^{\ominus}_{E.A.}$Cl(g) + **E** = $\Delta H^{\ominus}_{f}$ Na$^+$Cl$^-$(s)

$\therefore$ +107 + 122 + 496 + (−349) + **E** = −411

Hence, the lattice energy of NaCl(s), **E** = **−787** kJ mol⁻¹

- the 1st ionisation energy of Na is **endothermic**
- the 1st electron affinity of Cl is **exothermic**.

Definitions for enthalpy changes

The enthalpy changes involved in these two routes are shown below.

The standard enthalpy change of formation ($\Delta H^{\ominus}_{f}$) is the enthalpy change that takes place when one mole of a compound in its standard state is formed from its constituent elements in their standard states under standard conditions.

$$Na(s) + \tfrac{1}{2}Cl_2(g) \longrightarrow NaCl(s)\qquad \Delta H^{\ominus}_{f} = -411 \text{ kJ mol}^{-1}$$

KEY POINT

Be careful when using $\Delta H^{\ominus}_{at}$ involving diatomic elements.

$\Delta H^{\ominus}_{at}$ relates to the formation of 1 mole of atoms. For chlorine, this involves $\tfrac{1}{2}Cl_2$ only.

The standard enthalpy change of atomisation ($\Delta H^{\ominus}_{at}$) of an element is the enthalpy change that accompanies the formation of 1 mole of gaseous atoms from the element in its standard state.

$$Na(s) \longrightarrow Na(g)\qquad \Delta H^{\ominus}_{at} = +107 \text{ kJ mol}^{-1}$$
$$\tfrac{1}{2}Cl_2(g) \longrightarrow Cl(g)\qquad \Delta H^{\ominus}_{at} = +122 \text{ kJ mol}^{-1}$$

KEY POINT

For gaseous molecules $\Delta H^{\ominus}_{at}$ can be determined from the **bond enthalpy** (B.E.).

Cl–Cl(g) $\longrightarrow$ 2Cl(g)

$\Delta H^{\ominus}_{B.E.}$ = +244 kJ mol⁻¹

$\tfrac{1}{2}$Cl–Cl(g) $\longrightarrow$ Cl(g)

$\Delta H^{\ominus}_{at}$ = +122 kJ mol⁻¹

The first ionisation energy ($\Delta H^{\ominus}_{I.E.}$) of an element is the enthalpy change that accompanies the removal of 1 electron from each atom in 1 mole of gaseous atoms to form 1 mole of gaseous 1+ ions.

$$Na(g) \longrightarrow Na^+(g) + e^-\qquad \Delta H^{\ominus}_{I.E} = +496 \text{ kJ mol}^{-1}$$

KEY POINT

Be careful when using electron affinities. $\Delta H^{\ominus}_{E.A.}$ relates to the formation of 1 mole of 1– ions. For chlorine, this involves Cl(g) only.

The first electron affinity ($\Delta H^{\ominus}_{E.A.}$) of an element is the enthalpy change that accompanies the addition of 1 electron to each atom in 1 mole of gaseous atoms to form 1 mole of gaseous 1– ions.

$$Cl(g) + e^- \longrightarrow Cl^-(g)\qquad \Delta H^{\ominus}_{E.A.} = -349 \text{ kJ mol}^{-1}$$

KEY POINT

Key points from AS

- Polarisation
  Revise AS pages 58–59

Factors affecting the size of lattice enthalpies

The strength of an ionic lattice and the value of its lattice enthalpy depend upon:

- ionic size
- ionic charge.

Effect of ionic size

The effect of increasing ionic size can be seen by comparing the lattice enthalpies of sodium halides as shown below.

Lattice energy has a negative value. You should use the term *'becomes less/more negative'* instead of *'becomes bigger/smaller'* to describe any trend in lattice energy.

compound	lattice enthalpy / kJ mol^{-1}	ions	effect of size of halide ion
NaCl	–787	$+$ $-$	ionic size increases: • charge density decreases • attraction between ions decreases • lattice energy becomes less negative.
NaBr	–751	$+$ $-$	
NaI	–705	$+$ $-$	

Effect of ionic charge

The strongest ionic lattices with the most negative lattice enthalpies contain **small, highly charged ions**.

The diagram below compares the change in ionic size and ionic charge across Period 3 in the Periodic Table.

| Na^+ | Mg^{2+} | Al^{3+} | | P^{3-} | S^{2-} | Cl^- |

With increased charge on cation, two factors increase the magnitude of lattice energy:

- increased charge produces **more** attraction
- decreasing size produces **more** attraction.

With increased charge on anion, there are two competing factors:

- from Cl^- to P^{3-}, the increasing ionic charge produces **more** attraction
- however, the ionic size also increases producing **less** attraction.

Key points from AS

- Polarisation
  Revise AS pages 58–59
- The thermal stability of s-block compounds
  Revise AS pages 78–79

Limitations of lattice enthalpies

Theoretical lattice enthalpies can be calculated by considering each ion as a perfect sphere. Any difference between this theoretical lattice enthalpy and lattice enthalpy calculated using a Born–Haber cycle indicates a degree of covalent bonding caused by polarisation. Polarisation is greatest between:

- a small densely charged cation and
- a large anion.

From the ions shown in the diagram above, we would expect the largest degree of covalency between Al^{3+} and P^{3-}.

Enthalpy change of solution

AQA	M5	SALTERS	
EDEXCEL	M4	WJEC	M5
OCR		CCEA	M4

Key points from AS

- Indirect determination of enthalpy changes
  Revise AS pages 99–100

An ionic lattice dissolves in **polar** solvents (e.g. water). In this process, the giant ionic lattice is broken up by polar water molecules which surround each ion in solution.

When sodium chloride is dissolved in water, the enthalpy change of this process can be measured directly as the enthalpy change of solution.

The energy cycle

The complete energy cycle for dissolving sodium chloride in water is shown below. Definitions of these enthalpy changes are shown below.

Lattice enthalpy **forms** the ionic lattice. Notice that the energy change to break 1 mole of the ionic lattice = – (lattice enthalpy).

Route 1:   **A + B + C**

Route 2:   **D**

The production of hydrated ions comprises two changes:
- hydration of $Na^+(g)$
- hydration of $Cl^-(g)$.

Using Hess' Law: **A + B + C = D**

To determine the enthalpy change of hydration of $Cl^-(aq)$, **C**,

$$-\Delta H^{\ominus}{}_{L.E.}NaCl(s) + \Delta H^{\ominus}{}_{hyd} Na^+(g) + C = \Delta H^{\ominus}{}_{solution} NaCl(s)$$

$$\therefore \ -(-787) + (-390) + C = +13$$

Hence, the enthalpy change of hydration of $Cl^-(aq)$, **C = –384 kJ mol^{-1}**

Definitions for enthalpy changes

The enthalpy changes involved in these two routes are shown below.

> The **standard enthalpy change of solution** ($\Delta H^{\ominus}{}_{solution}$) is the enthalpy change that accompanies the dissolving of 1 mole of a solute in a solvent to form an infinitely dilute solution under standard conditions.
>
> $NaCl(s) + aq \longrightarrow Na^+(aq) + Cl^-(aq)$ $\Delta H^{\ominus}{}_{solution} = +13$ kJmol^{-1}

KEY POINT

> The **standard enthalpy change of hydration** ($\Delta H^{\ominus}{}_{hyd}$) of an ion is the enthalpy change that accompanies the hydration of 1 mole of gaseous ions to form 1 mole of hydrated ions in an infinitely dilute solution under standard conditions.
>
> $Na^+(g) + aq \longrightarrow Na^+(aq)$ $\Delta H^{\ominus}{}_{hyd} = -390$ kJmol^{-1}
> $Cl^-(g) + aq \longrightarrow Cl^-(aq)$ $\Delta H^{\ominus}{}_{hyd} = -384$ kJmol^{-1}

KEY POINT

Factors affecting the magnitude of hydration enthalpy

The values obtained for each ion depend upon the size of the charge and the size of the ion.

ion	Na^+	Mg^{2+}	Al^{3+}	Cl^-	Br^-	I^-
$\Delta H^{\ominus}{}_{hyd}$/kJ mol^{-1}	–390	–1891	–4613	–384	–351	–307
effect of charge	• increasing ionic charge → • greater attraction for water					
ionic radius/nm	0.102	0.072	0.053	0.180	0.195	0.215
effect of size	• decreasing ionic size → • greater attraction for water			• increasing ionic size → • less attraction for water		

Bond enthalpies

AQA ▶ M5

Key points from AS

- **Bond enthalpy**
  Revise AS pages 102–103

For AQA,
- you should be able to use mean bond enthalpies to calculate an approximate value of ΔH for other reactions.

> **KEY POINT**
>
> $\Delta H = \Sigma(\text{bond enthalpies in reactants}) - \Sigma(\text{bond enthalpies in products})$.

- you should be able to explain why values from mean bond enthalpy calculations differ from those determined from enthalpy cycles.

> **KEY POINT**
>
> A **mean bond enthalpy** gives the typical value of a particular bond, having taken into account the same bond in different environments.
>
> In particular environments a bond may have a different value for the bond enthalpy than the mean value.

For full details, you will need to refer to *Revise AS Chemistry*, Chapter 3.3.

Progress check

1 Using the information below:
   (a) name each enthalpy change **A** to **E**
   (b) construct a Born–Haber cycle for sodium bromide and calculate the lattice enthalpy (enthalpy change of lattice formation) of sodium bromide.

enthalpy change	equation	$\Delta H^{\ominus}/kJ\ mol^{-1}$
A	$Na(s) + \tfrac{1}{2}Br_2(l) \longrightarrow NaBr(s)$	−361
B	$Br(g) + e^- \longrightarrow Br^-(g)$	−325
C	$Na(s) \longrightarrow Na(g)$	+107
D	$Na(g) \longrightarrow Na^+(g) + e^-$	+496
E	$\tfrac{1}{2}Br_2(l) \longrightarrow Br(g)$	+112

2 You are provided with the following enthalpy changes.
   $$Na^+(g) + F^-(g) \longrightarrow NaF(s) \qquad \Delta H = -918\ kJ\ mol^{-1}$$
   $$Na^+(g) + aq \longrightarrow Na^+(aq) \qquad \Delta H = -390\ kJ\ mol^{-1}$$
   $$F^-(g) + aq \longrightarrow F^-(aq) \qquad \Delta H = -457\ kJ\ mol^{-1}$$
   (a) Name each of these three enthalpy changes.
   (b) Write an equation, including state symbols, for which the enthalpy change is the enthalpy change of solution of NaF.
   (c) Calculate the enthalpy change of solution of NaF.

1 (a) A enthalpy change of formation of sodium bromide
         B 1st electron affinity of bromine
         C enthalpy change of atomisation of sodium
         D 1st ionisation energy of sodium
         E enthalpy change of atomisation of bromine
   (b) −751 kJ mol⁻¹
2 (a) lattice enthalpy of NaF
         enthalpy change of hydration of Na⁺
         enthalpy change of hydration of F⁻
   (b) NaF(s) + aq ⟶ Na⁺(aq) + F⁻(aq)
   (c) +71 kJ mol⁻¹

Special properties of water arising from hydrogen bonding

Hydrogen bonding is strong enough to have significant effects on physical properties, resulting in some unusual properties for water.

Water has an unexpectedly high boiling point

The boiling points of the Group 6 hydrides are shown below.

hydride	H_2O	H_2S	H_2Se	H_2Te
boiling point/K	373	212	232	271

The boiling points increase from $H_2S \longrightarrow H_2Te$:

- the number of electrons in molecules increase
- van der Waals' forces increase
- more energy is required to break the intermolecular bonds to vaporise the hydrides.

But the **boiling point** of **H_2O** is **higher** than expected. This provides evidence that there are some extra forces acting between the molecules that must be broken to boil each hydride. These extra forces are hydrogen bonds.

This difference is also evident in the unusually high **enthalpy change of vaporisation** of water.

This has consequences for the Earth's climate.

- Energy is **required** to **vaporise** water molecules from the oceans into water vapour but this energy is **released** when water molecules **condense** as rain.
- Heat is transferred from the oceans to the land, stabilising the Earth's temperature across different climate zones.

Water has a relatively large specific heat capacity

Compared with similar compounds, water has an unusually high value specific heat capacity. Hydrogen bonds have to be overcome before energy can be transferred to water molecules to increase the kinetic energy of the molecules.

This has consequences for the Earth's climate.

- Because extra energy is needed to change the temperature of water, the vast stores of water in the oceans help to stabilise the Earth's temperature especially via ocean currents from one region to another.

Water is more dense than ice

- Solids are *usually* denser than liquids – but ice is less dense than water.
- Particles in solids are usually packed closer together than in liquids.
- Hydrogen bonds hold water molecules apart in an open lattice structure.

∴ ice is less dense than water.

The diagram below shows how the open lattice of ice collapses on melting.

hydrogen bonds break

MELTING OF ICE

tetrahedral open lattice
in ice

ice lattice collapses:
molecules move
closer together

When water changes state, the covalent bonds between the H and O atoms in an H_2O molecule are strong and do not break.

It is the intermolecular forces that break.

Without hydrogen bonding, H_2O would be a gas at room temperature and pressure.

Enthalpy change of vaporisation is a measure of the energy required to convert water from a liquid to a gas.

Specific heat capacity, *c*, is a measure of the energy required to heat a substance.

Solids are usually denser than liquids – but ice is less dense than water.

2.2 Entropy changes

After studying this section you should be able to:

- explain that entropy is a measure of disorder
- calculate entropy changes from standard entropies
- understand that entropy, enthalpy and temperature determine the feasibility of a reaction
- apply rates and equilibrium to industrial applications

LEARNING SUMMARY

Entropy

AQA	M5	SALTERS	M5
EDEXCEL	M4	WJEC	M5
OCR	M5	CCEA	M4

What is entropy?

Entropy, S, is a measure of disorder.

Entropy increases whenever particles become more disordered, e.g.

- when a gas spreads spontaneously through a room
- with increasing temperature
- during changes in state: solid $\longrightarrow$ liquid $\longrightarrow$ gas
- when a solid lattice dissolves

$$NaCl(aq) + aq \longrightarrow Na^+(aq) + Cl^-(aq)$$

At 0 K, perfect crystals have zero entropy.

- in a reaction in which there is an increase in the number of gaseous molecules (i.e. when a gas is evolved)

$$NaHCO_3(s) + HCl(aq) \longrightarrow NaCl(aq) + H_2O(l) + CO_2(g)$$

$$MgCO_3(s) \longrightarrow MgO(s) + CO_2(g)$$

Calculating entropy changes

The standard entropy, $S^\ominus$, of a substance is the entropy content of one mole of a substance, under standard conditions.

The entropy change of a reaction can be calculated using standard entropies of the reactants and products.

> $$\Delta S = \Sigma S^\ominus \text{ (products)} - \Sigma S^\ominus \text{ (reactants)}$$

KEY POINT

Example

Notice that standard entropies have units of $J K^{-1} mol^{-1}$.

The equation below is for the carbon reduction of chromium oxide, Cr_2O_3.

$$Cr_2O_3(s) + 3C(s) \longrightarrow 2Cr(s) + 3CO(g)$$

	$Cr_2O_3(s)$	$C(s)$	$Cr(s)$	$CO(g)$
$S^\ominus/J K^{-1} mol^{-1}$	+81	+6	+24	+198

$$\Delta S = \Sigma S^\ominus \text{ (products)} - \Sigma S^\ominus \text{ (reactants)}$$

$$= [(2 \times 24) + (3 \times 198)] - [(+81) + (3 \times 6)] = 543 \text{ J K}^{-1} \text{ mol}^{-1}$$

Spontaneous processes

A **spontaneous** process is one that proceeds of its **own accord**.

Processes lead to **increased stability** if they lead to a **lower energy** state.

This is obviously the situation in an exothermic reaction in which the enthalpy (heat energy) decreases during a reaction. It therefore seems surprising that **some** reactions are **endothermic** and that these can **occur spontaneously** at room temperature.

Enthalpy changes alone cannot explain whether a reaction takes place.

For this, we must consider **both the ΔH and ΔS**.

Different exam boards have taken different approaches to this and these are discussed below. Select the correct approach for your specification.

Free energy

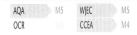

| AQA | M5 | WJEC | M5 |
| OCR | | CCEA | M4 |

The balance between entropy and enthalpy is the feasibility of a reaction.

Whether a chemical or physical process takes place is dependent on

- the temperature, T,
- the entropy change in the system, ΔS,
- the enthalpy change, ΔH, with the surroundings.

The change in *Free Energy*, ΔG, automatically determines the balance between enthalpy change, entropy change and temperature:

$$\Delta G = \Delta H - T\Delta S$$

Feasibility of a reaction

Processes are spontaneous (feasible) when ΔG is negative. The sign of ΔG depends upon ΔH and ΔS, as shown below.

ΔH	ΔS	ΔG
–ve	+ve	always –ve: reaction feasible
+ve	–ve	always +ve: reaction never feasible
–ve	–ve	–ve at low temperatures
+ve	+ve	–ve at high temperatures

ΔG varies with temperature:

when ΔH and ΔS have the same sign, the feasibility of the process depends on the temperature, T.

- $\Delta G = \Delta H - T\Delta S$
- The reaction is feasible when $\Delta G < 0$.

Example

The equation below is for the carbon reduction of chromium oxide, Cr_2O_3.

$Cr_2O_3(s) + 3C(s) \longrightarrow 2Cr(s) + 3CO(g)$

$\Delta H = +807$ kJ mol^{-1}; $\Delta S = 543$ J K^{-1} mol^{-1}

What is the minimum temperature that this reaction takes place spontaneously?

$\Delta S = 543$ J K^{-1} mol^{-1} = 0.543 kJ K^{-1} mol^{-1}

Be careful with units:

ΔH: kJ mol^{-1}

ΔS: J K^{-1} mol^{-1}

$\Delta G = \Delta H - T\Delta S$;

Minimum temperature when $\Delta G = 0$.

$$\therefore 0 = \Delta H - T\Delta S; \quad T = \frac{\Delta H}{\Delta S} = \frac{807}{0.543} = 1486 \text{ K}$$

ΔG and equilibrium constants (WJEC only)

For a reaction with ΔG negative:

- equilibrium constants will have large values as the products predominate.

For a reaction with ΔG positive:

- equilibrium constants will have small values as the reactants predominate.

Total entropy

For a chemical process to take place, there are two types of entropy (disorder) that need to be considered:

- the entropy change of the chemical system, ΔS_{system}
- the entropy change of the surroundings, $\Delta S_{surroundings}$

ΔS_{system}

This is the disorder of the particles making up the chemicals – the chemical 'system'. ΔS_{system} measures the number of ways that the particles can be arranged. It also measures the number of ways that energy can be distributed between the particles.

ΔS_{system} can be calculated using:

$$\Delta S_{system} = \Sigma S^{\ominus} \text{ (products)} - \Sigma S^{\ominus} \text{ (reactants)}$$

A system becomes more stable when ΔS_{system} is positive.

> Entropy is a measure of the number of ways that molecules and their associated energy quanta can be arranged.

$\Delta S_{surroundings}$

This is the disorder of energy in the surroundings.

In terms of energy, the surroundings become more stable when $\Delta S_{surroundings}$ increases and energy spreads from the chemicals into the surroundings in an exothermic process.

$\Delta S_{surroundings}$ is related to the enthalpy change and temperature, in K:

$$\Delta S_{surroundings} = -\frac{\Delta H}{T}$$

The surrounding system becomes more stable during an exothermic reaction when $\Delta S_{surroundings}$ is positive.

Feasibility of a reaction

> ΔS_{total} only tells us the 'thermodynamic stability'. Even if a reaction is thermodynamically stable, it may not seem to take place if the activation energy is great and the rate of reaction is slow ('kinetic stability').

The total entropy, ΔS_{total}, is the overall balance of entropy from the system and from the surroundings:

$$\Delta S_{total} = \Delta S_{system} + \Delta S_{surroundings}$$

or $\quad \Delta S_{total} = \Delta S_{system} - \dfrac{\Delta H}{T}$

For a process to take place, the natural direction of change is towards increasing total entropy when ΔS_{total} is positive, i.e. $\Delta S_{total} > 0$.

The process is then feasible and takes place 'spontaneously'.

Example

The equation below is for the carbon reduction of chromium oxide, Cr_2O_3.

$$Cr_2O_3(s) + 3C(s) \longrightarrow 2Cr(s) + 3CO(g)$$

$\Delta H = +807$ kJ mol^{-1}; $\Delta S = 543$ J K^{-1} mol^{-1}

What is the minimum temperature that this reaction takes place at spontaneously?

$\Delta S_{system} = 543$ J K^{-1} mol^{-1} = 0.543 kJ K^{-1} mol^{-1}

$$\Delta S_{total} = \Delta S_{system} - \frac{\Delta H}{T}$$

Minimum temperature when $\Delta S_{total} = 0$.

> Be careful with units:
> ΔH: kJ mol^{-1}
> ΔS: J K^{-1} mol^{-1}

$$\therefore \Delta S_{system} - \frac{\Delta H}{T} = 0; \; \Delta S_{system} = \frac{\Delta H}{T}; \; T = \frac{\Delta H}{\Delta S_{system}} = \frac{807}{0.543} = 1486 \text{ K}$$

ΔS and equilibrium constants (Edexcel only)

For a reaction in which ΔS_{total} increases, the magnitude of the equilibrium constant also increases:

- $\Delta S = R \ln K$

Progress check (all specifications)

1 State whether the equations below have a positive or negative entropy change.
   (a) $H_2O(g) \longrightarrow H_2O(l)$
   (b) $N_2(g) + 3H_2(g) \longrightarrow 2NH_3(g)$
   (c) $CaCl_2(s) + aq \longrightarrow Ca^{2+}(aq) + 2Cl^-(aq)$

2 The equation below is for the thermal decomposition of calcium carbonate.
   $$CaCO_3(s) \longrightarrow CaO(s) + CO_2(g) \qquad \Delta H = +178 \text{ kJ mol}^{-1}$$

	$CaCO_3(s)$	$CaO(s)$	$CO_2(g)$
$S^\ominus$ / J K^{-1} mol^{-1}	+93	+40	+214

   (a) What is ΔS?
   (b) What is the minimum temperature for decomposition to take place?

2 (a) 161 J K^{-1} mol^{-1}; (b) 1106 K
1 (a) –ve; (b) –ve; (c) +ve.

Applying rates and equilibrium to industrial processes

Industrial processes need to obtain economic yields of the product being manufactured. In the section above, we discussed the importance of the balance between **enthalpy**, **entropy** and **temperature**.

Equilibrium

You do not need to know the actual industrial conditions used in the Haber process.

In the Haber process for the production of ammonia

$$N_2(g) + 3H_2(g) \rightleftharpoons 2NH_3(g) \qquad \Delta H^\ominus = -92 \text{ kJ mol}^{-1}$$

- the right-hand side has **fewer moles of gas** than the left-hand side (reactants), favoured by a **high** pressure
- the forward reaction is **exothermic**, favoured by a **low** temperature.

We can arrive at these conclusions by using le Chatelier's principle from AS Chemistry.

The optimum equilibrium conditions for maximum equilibrium yield are: **high pressure and low temperature**.

The need for compromise

In reality it may not be feasible to use optimum equilibrium conditions.

The rate of a reaction increases with higher temperatures and pressures – but temperature has a much larger effect.

- Compressing gases to **high pressures** has a high **energy cost**. There are also considerable **safety** implications of using very high pressures.
- At **low temperatures**, the rate of reaction is **slow** as few molecules possess the necessary activation energy of the reaction.

There needs to be a compromise.

Compromising
Optimum conditions, reaction rate, feasibility, reality, safety and economics must be considered.

- Use of a **high enough pressure** to **increase** the **equilibrium yield** without increasing **energy costs** in generating the pressure or compromising **safety**.
- Increase the temperature just enough to increase the rate of reaction so that the reaction takes place in a realistic time frame, whilst still producing an **adequate equilibrium yield**.

Equilibrium is never reached

Another factor is that industrial processes do not operate at equilibrium:

- products are removed – we never have a closed system
- there is insufficient time for equilibrium to be reached.

But it is still important to consider the conditions needed to secure a high equilibrium yield.

The search for better processes

The chemical industry takes steps to maximise the **atom economy** of the process. Many factors need to be considered, including:

Hazards can be reduced by developing processes that reduce risks from toxic reactants, explosions, acidic or flammable gases, toxic emissions, etc.

- **recycling** unreacted reagents
- developing **alternative reactions** and processes
- using **more available** raw materials, preferably **renewable**
- using **catalysts** to allow processes to occur at **lower temperatures**, cutting energy needs
- developing **simpler processes** requiring less costly chemical plants
- using **co-products** and **by-products**, rather than disposing of them.

Global warming from energy production

SALTERS   CCEA   M4

Key points from AS

- **Global warming**
  Revise AS pages 147–148

Much of the CO_2 released into the atmosphere dissolves in the oceans.

The concentration of carbon dioxide in the atmosphere depends on natural processes including:

- photosynthesis
- respiration
- the solubility of the gas in surface waters.

Combustion of non-renewable hydrocarbon fuels for energy is causing an increase in the atmospheric concentration of carbon dioxide. Carbon dioxide is a greenhouse gas, implicated as a likely cause for global warming.

The **greenhouse effect** of a gas (e.g. CO_2) depends on:

- its atmospheric concentration
- its ability to absorb infra-red radiation.

Reducing atmospheric CO_2 levels

Various approaches are being considered to reduce atmospheric CO_2 levels. These include:

- more economical use of fuels
- the use of alternative fuels (including hydrogen, see pages 65–66)
- capture and storage of CO_2
- increased photosynthesis.

2.3 Electrochemical cells

After studying this section you should be able to:

- understand the principles of an electrochemical cell
- define the term 'standard electrode (redox) potential', $E^{\ominus}$
- describe how to measure standard electrode potentials
- calculate a standard cell potential from standard electrode potentials
- write cell diagrams
- describe applications of electrochemical cells, including fuel cells
- understand that rusting is an electrochemical process

Key points from AS

- **Redox reactions**
  Revise AS pages 42–44

During the study of redox in AS Chemistry you learnt:

- that oxidation and reduction involve transfer of electrons in a redox reaction
- the rules for assigning oxidation states
- how to construct an overall redox equation from two half-equations.

These key principles are built upon by considering electrochemical cells in which redox reactions provide electrical energy.

Electricity from chemical reactions

AQA	M5	SALTERS	
EDEXCEL		WJEC	M5
OCR		CCEA	M5

In a redox reaction, electrons are transferred between the reacting chemicals.

For example, zinc reacts with aqueous copper(II) ions in a redox reaction.

$$Zn(s) + Cu^{2+}(aq) \longrightarrow Zn^{2+}(aq) + Cu(s)$$

oxidation: $Zn(s) \longrightarrow Zn^{2+}(aq) + 2e^-$
reduction: $Cu^{2+}(aq) + 2e^- \longrightarrow Cu(s)$

Electrochemical cells are the basis of all batteries. Their true value is as a portable form of electricity.

If this reaction is carried out simply by mixing zinc with aqueous copper(II) ions:

- heat energy is produced and, unless used as useful heat, is often wasted.

An electrochemical cell controls the transfer of electrons:

- electrical energy is produced and can be used for electricity.

The zinc-copper electrochemical cell

An electrochemical cell comprises two **half-cells**. In each half-cell, the **oxidation** and **reduction** processes take place **separately**.

A half-cell contains the oxidised and reduced species from a half-equation.

The species can be converted into one another by addition or removal of electrons.

A zinc-copper cell has two half-cells:

- an oxidation half-cell containing reactants and products from the oxidation half-reaction: $Zn^{2+}(aq)$ and $Zn(s)$
- a reduction half-cell containing reactants and products from the reduction half-reaction: $Cu^{2+}(aq)$ and $Cu(s)$.

The circuit is completed using:

- a connecting **wire** which transfers **electrons** between the half-cells
- a **salt bridge** which transfers **ions** between the half-cells.

The salt bridge consists of filter paper soaked in aqueous KNO_3 or NH_4NO_3.

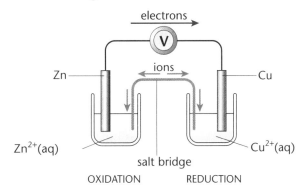

OXIDATION   REDUCTION

Remember these charge carriers:

ELECTRONS through the wire;

IONS through the salt bridge.

oxidation half-cell: $Zn^{2+}(aq)$ and $Zn(s)$
- **Zn is oxidised to Zn^{2+}:** $Zn(s) \longrightarrow Zn^{2+}(aq) + 2e^-$
- electrons are **supplied** to the external circuit
- the polarity is **negative**.

reduction half-cell: $Cu^{2+}(aq)$ and $Cu(s)$
- **Cu^{2+} is reduced to Cu:** $Cu^{2+}(aq) + 2e^- \longrightarrow Cu(s)$
- electrons are **taken** from the external circuit
- the polarity is **positive**.

The measured cell e.m.f. of 1.10 V indicates that there is a potential difference of 1.10 V between the copper and zinc half-cells.

Standard electrode potentials

AQA	M5	SALTERS	M4
EDEXCEL	M5	WJEC	M5
OCR	M5	CCEA	M5

A **hydrogen half-cell** is used as the standard for the measurement of standard electrode potentials.

The standard hydrogen half-cell is based upon the half-reaction below.

$$H^+(aq) + e^- \rightleftharpoons \tfrac{1}{2}H_2(g)$$

A hydrogen half-cell typically comprises

- 1 mol dm^{-3} hydrochloric acid as the source of $H^+(aq)$

- a supply of hydrogen gas, $H_2(g)$ at 100 kPa

- an inert platinum electrode on which the half-reaction above takes place.

> **KEY POINT**
>
> The **standard electrode potential** of a half-cell, $E^\ominus$, is the e.m.f. of a half-cell compared with a standard hydrogen half-cell.
> All measurements are at **298 K** with solution **concentrations of 1 mol dm^{-3}** and gas **pressures of 100 kPa**.

Measuring standard electrode potentials

Half-cells comprising a metal and a metal ion

The diagram below shows how a hydrogen half-cell can be used to measure the standard electrode potential of a copper half-cell.

A high-resistance voltmeter is used to minimise the current that flows.

Hydrogen half-cell

> **KEY POINT**
>
> - The standard electrode potential of a half-cell represents its contribution to the cell e.m.f.
> - The contribution made by the hydrogen half-cell to the cell e.m.f. is defined as 0 V.
> - The **sign** of the standard electrode potential of a half-cell indicates its **polarity** compared with the hydrogen half-cell.

As with the hydrogen half-cell, the platinum electrode provides a surface on which the half-reaction can take place.

Half-cells comprising ions of different oxidation states

The half-reaction in a half-cell can be between aqueous ions of the same element with different oxidation states. To allow electrons to pass into the half-cell, an inert electrode of platinum is used.

A half-cell can contain $Fe^{3+}(aq)$ and $Fe^{2+}(aq)$ ions:

$$Fe^{3+}(aq) \rightleftharpoons Fe^{2+}(aq) + e^-$$

A **standard** $Fe^{3+}(aq)/Fe^{2+}(aq)$ half-cell requires:

- a solution containing **both**
  - 1 mol dm^{-3} $Fe^{3+}(aq)$ **and**
  - 1 mol dm^{-3} $Fe^{2+}(aq)$
- an inert platinum electrode with a connecting wire.

Alternatively the solution can contain **equal** concentrations of Fe^{2+} and Fe^{3+}.

Pt (inert electrode)

$Fe^{2+}(aq)$, 1 mol dm^{-3}
$Fe^{3+}(aq)$, 1 mol dm^{-3}

The electrochemical series

AQA	M5	SALTERS	M4
EDEXCEL	M5	WJEC	M5
OCR	M5	CCEA	M5

A standard electrode potential indicates the availability of electrons from a half-reaction.

Metals are reducing agents.

- Metals typically react by donating electrons in a redox reaction.
- Reactive metals have the greatest tendency to donate electrons and their half-cells have the most negative electrode potentials.

Non-metals are oxidising agents.

Key points from AS

- **Reactivity of s-block elements**
  Revise AS page 77
- **The relative reactivity of the halogens as oxidising agents**
  Revise AS page 81

- Non-metals typically react by accepting electrons in a redox reaction.
- Reactive non-metals have the greatest tendency to accept electrons and their half-cells have the most positive electrode potentials.

The **electrochemical series** shows electrode reactions listed in order of their $E^{\ominus}$ values. The electrochemical series below lists with the **most positive** $E^{\ominus}$ value at the top. This enables oxidising and reducing agents to be easily compared. The reduced form is shown on the right-hand side of the electrode equilibrium.

This version of the electrochemical series shows standard electrode potentials listed with the **most positive** $E^{\ominus}$ value at the top.

You will find versions of the electrochemical series listed with the **most negative** $E^{\ominus}$ value at the top.

This does not matter. The important point is that the potentials are listed **in order** and **can be compared.**

strongest oxidising agent	electrode reaction					$E^{\ominus}$/ V	
	$F_2(g)$	+	$2e^-$	$\rightleftharpoons$	$2F^-(aq)$	+2.87	
	$Cl_2(g)$	+	$2e^-$	$\rightleftharpoons$	$2Cl^-(aq)$	+1.36	
	$Br_2(l)$	+	$2e^-$	$\rightleftharpoons$	$2Br^-(aq)$	+1.09	
	$Ag^+(aq)$	+	e^-	$\rightleftharpoons$	$Ag(s)$	+0.80	
	$Cu^{2+}(aq)$	+	$2e^-$	$\rightleftharpoons$	$Cu(s)$	+0.34	
	$H^+(aq)$	+	e^-	$\rightleftharpoons$	$\frac{1}{2}H_2(g)$	0	
	$Fe^{2+}(aq)$	+	$2e^-$	$\rightleftharpoons$	$Fe(s)$	−0.44	
	$Zn^{2+}(aq)$	+	$2e^-$	$\rightleftharpoons$	$Zn(s)$	−0.76	strongest reducing agent
	$Cr^{3+}(aq)$	+	$3e^-$	$\rightleftharpoons$	$Cr(s)$	−0.77	
	$K^+(aq)$	+	e^-	$\rightleftharpoons$	$K(s)$	−2.92	

KEY POINT

In the electrochemical series:
- the **oxidised** form is on the **left-hand side** of the half-equation
- the **reduced** form is on the **right-hand side** of the half-equation.

Standard cell potentials and cell reactions

AQA	M5	SALTERS	M4
EDEXCEL	M5	WJEC	M5
OCR	M5	CCEA	M5

The **standard cell potential** of a cell is the e.m.f. acting between the two half-cells making up the cell under standard conditions.

The **cell reaction** is:

Key points from AS

- **Combining half-equations**
  Revise AS page 44

- the overall process taking place in the cell
- the sum of the reduction and oxidation half-reactions taking place in each half-cell.

Calculating the standard cell potential of a silver-iron(II) cell

Identify the two relevant half-reactions and the polarity of each electrode.

The more positive of the two systems is the positive terminal of the cell.

$Ag^+(aq) + e^- \rightleftharpoons Ag(s)$ $E^\ominus = +0.80$ V   *positive terminal*
$Fe^{2+}(aq) + 2e^- \rightleftharpoons Fe(s)$ $E^\ominus = -0.44$ V   *negative terminal*

The standard electrode potential is the difference between the $E^\ominus$ values:

$E^\ominus_{cell}$ is the difference between the standard electrode potentials.

Simply subtract the $E^\ominus$ of the negative terminal from the $E^\ominus$ of the positive terminal:

$$E^\ominus_{cell} = E^\ominus \text{ (positive terminal)} - E^\ominus \text{ (negative terminal)}$$
$$\therefore E^\ominus_{cell} = 0.080 - (-0.44) = \textbf{1.24 V}$$

Determination of the cell reaction in a silver-iron(II) cell

Work out the actual direction of the half-equations.

The more negative half-equation is reversed. The negative terminal is then on the left-hand side.

The **half-equation** with the **more negative** $E^\ominus$ value provides the electrons and proceeds to the left. This half-equation is reversed so that the two half-reactions taking place can be clearly seen and compared.

$Ag^+(aq) + e^- \longrightarrow Ag(s)$ *reaction at positive terminal*
$Fe(s) \longrightarrow Fe(aq) + 2e^-$ *reaction at negative terminal*

The overall cell reaction can be found by adding the half-equations.

This is discussed in detail in the AS Chemistry Study Guide, page 44.

The silver half-equation must first be multiplied by '2' to balance the electrons.
The two half-equations are then added.

$$2Ag^+(aq) + Fe(s) \longrightarrow 2Ag(s) + Fe^{2+}(aq)$$

Cell diagrams

AQA   M5

A simple metal electrode comprises a metal in contact with a solution of its ions, e.g. $Cu(s)|Cu^{2+}(aq)$. This is the half-cell.

- The solid vertical line shows a boundary between a solid and a solution.

A cell diagram is a convenient way of representing a cell.

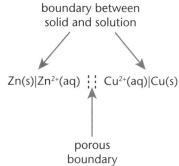

boundary between
solid and solution

$Zn(s)|Zn^{2+}(aq) ⁝ Cu^{2+}(aq)|Cu(s)$

porous
boundary

Two dotted vertical lines show a porous boundary between two half-cells. This is the salt bridge in simple cells.

Notice that the ions in solution are shown in contact with the salt bridge. The metals taking the electric current in and out of the cell are shown at either end of the cell diagram.

The standard cell potential of the cell, as shown in the cell diagram, is defined as:

$$E^\ominus\text{(cell)} = E^\ominus\text{(right-hand side)} - E^\ominus\text{(left-hand side)}$$

The sign of the cell gives the polarity of the right-hand half-cell.

For the cell shown above,

- $E^\ominus\text{(cell)} = 0.34 - (-0.76) = +1.10$ V
- the $Cu(s)|Cu^{2+}(aq)$ half-cell (shown on the right) is positive.

Electrochemical cells

AQA M5 SALTERS
EDEXCEL WJEC M5
OCR

Electrochemical cells can be used as a commercial source of electrical energy.

Cells can be divided into three main types:

- non-rechargeable cells – the cell is used until the chemicals are used up to such a point at which the voltage falls
- rechargeable cells – the cell reaction can readily be reversed when recharging, allowing the cell to be used again
- fuel cells – the cell reaction requires external supplies of a fuel and an oxidant, which are consumed and need to be replenished.

An electrochemical cell comprises two half-cells with different electrode potentials. For example, a simple cell can be set up based on zinc and copper.

The redox systems are shown below.

$$Zn^{2+}(aq) + 2e^- \rightleftharpoons Zn(s) \qquad E^\ominus = -0.74 \text{ V}$$
$$Cu^{2+}(aq) + 2e^- \rightleftharpoons Cu(s) \qquad E^\ominus = +0.34 \text{ V}$$

The more negative zinc system provides the electrons:

Overall: $Zn(s) + Cu^{2+}(aq) \longrightarrow Cu(s) + Zn^{2+}(aq)$

Chargeable and non-rechargeable cells are used as our modern-day cells and batteries.

Other common examples include:
- nickel and cadmium (NiCad) batteries used in rechargeable batteries
- lithium 'button' cells used in watches
- lithium-ion batteries used in laptops.

All these cells work on the same principle, using two redox systems.

Fuel cells for energy

AQA M5 OCR
EDEXCEL WJEC M5

In a fuel cell, energy from the reaction of a fuel with oxygen is used to create a voltage.

- The reactants flow in and products flow out, while the electrolyte remains in the cell.
- Fuel cells can operate virtually continuously as long as the fuel and oxygen continue to flow into the cell. Fuel cells do not have to be recharged.

In the future, a 'hydrogen economy' based on hydrogen fuel cells may well contribute greatly to energy needs.

The diagram below shows a simple hydrogen–oxygen fuel cell.

- At the negative electrode (cathode),

$$H_2(g) \longrightarrow 2H^+(aq) + 2e^-$$

- At the positive electrode (anode),

$$\tfrac{1}{2}O_2(g) + 2H^+(aq) + 2e^- \longrightarrow H_2O(l)$$

- Overall: $H_2(g) + \tfrac{1}{2}O_2(g) \longrightarrow H_2O(l)$

In exams, you are unlikely to have to recall a specific storage cell.

But do take care! Your specification **may** expect you to know the hydrogen fuel cell.

You might also be expected to make predictions on a supplied electrochemical cell. All relevant electrode potentials and other data would be supplied though.

Other hydrogen fuel cells operate with an alkaline electrolyte.

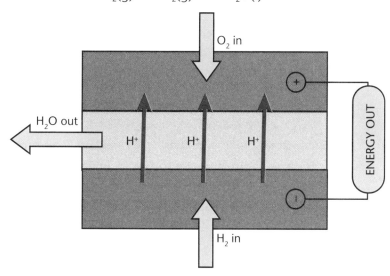

Development of fuel cell vehicles (FCVs)

Fuel cells are being developed as an alternative to finite oil-based fuels in cars.

Scientists in the car industry are developing fuel cell vehicles (FCVs), fuelled by:

- hydrogen gas
- hydrogen-rich fuels.

Hydrogen-rich fuels include methanol, natural gas, or petrol. These are converted into hydrogen gas by an onboard 'reformer'. These fuels are easier to transport and deliver into an FCV than hydrogen gas.

FCVs have advantages over conventional petrol or diesel-powered vehicles.

- Pure hydrogen emits only water.
- Hydrogen-rich fuels produce only small amounts of air pollutants and CO_2.
- Efficiency can be more than twice that of similarly-sized conventional vehicles. Other advanced technologies are being investigated to further increase efficiency.

Hydrogen might be stored in FCVs:

- as a liquid under pressure
- adsorbed on the surface of a solid material
- absorbed within a solid material.

Advantages of hydrogen fuel cells:

less pollution and less CO_2,

greater efficiency.

Limitations of hydrogen fuel cells

There are logistical problems in the development of hydrogen fuel cells:

- storing and transporting hydrogen, in terms of safety, feasibility of a pressurised liquid and a limited life cycle of a solid 'adsorber' or 'absorber'
- limited lifetime (requiring regular replacement and disposal) and high production costs
- use of toxic chemicals in their production
- initial manufacture of hydrogen requires energy
- public and political acceptance of hydrogen as a fuel.

Liquefying hydrogen is costly in terms of energy use.

Safety is a major concern: Hydrogen is very flammable.

Hydrogen is an 'energy carrier' and not an 'energy source'. Hydrogen is first made either by electrolysis of water or by reacting methanol (a finite fuel) with steam.

Future energy needs

Political and social desire to move to a hydrogen economy has many obstacles. Many people do not realise that energy is needed to produce hydrogen and that fuel cells have a finite life.

Fuel cells as breathalysers

EDEXCEL   M5

Ethanol fuel cells are being used in some modern breathalysers to measure the concentration of blood alcohol.

- A platinum catalyst in the fuel cell oxidises ethanol to ethanoic acid, releasing electrons. The electrons increase the current in the fuel cell – the greater the current, the more ethanol in the breath and the greater the concentration of blood alcohol.

Other modern methods are based on IR absorption.

- A particular IR absorption identifies the presence of ethanol in the breath and the intensity of this absorption peak is directly related to the ethanol level.

For IR absorption, see AS pages 143–144.

The oldest method used a chemical reaction based on the reduction of orange chromium(VI) to green chromium(III) by ethanol. The degree of colour change is directly related to the level of alcohol in the expired air.

For reduction of chromium compounds, see page 98.

Progress check

1 Use the standard electrode potentials on page 63, to answer the questions that follow.
   (a) Calculate the standard cell potential for the following cells.
      (i) F_2/F^- and Ag^+/Ag
      (ii) Ag^+/Ag and Cr^{2+}/Cr
      (iii) Cu^{2+}/Cu and Cr^{3+}/Cr.
   (b) For each cell in (a) determine which half-reactions proceed, and hence construct, the cell reaction.

Corrosion as an electrochemical process

SALTERS   M5

Rusting

Rusting is an electrochemical process involving **iron**, **oxygen** and **water**.

The key redox systems involved are shown below:

$$Fe^{2+} + 2e^- \rightleftharpoons Fe \qquad E^\ominus = -0.44 \text{ V}$$
$$\tfrac{1}{2}O_2 + H_2O + 2e^- \rightleftharpoons 2OH^- \qquad E^\ominus = +0.44 \text{ V}$$

- The **Fe^{2+}/Fe** system has the **more negative** $E^\ominus$ value and its half-reaction will **supply electrons**.
- This equation is therefore reversed.

oxidation:	$Fe \longrightarrow$	$Fe^{2+} + 2e^-$
reduction:	$\tfrac{1}{2}O_2 + H_2O + 2e^- \longrightarrow$	$2OH^-$
overall:	$Fe + \tfrac{1}{2}O_2 + H_2O \longrightarrow$	$Fe^{2+} + 2OH^-$

- The Fe^{2+} and OH^- ions now combine to form iron(II) hydroxide.
- Finally $Fe(OH)_2$ is **oxidised** by **oxygen** in the presence of **water**, forming **rust**:

$$Fe(OH)_2 \xrightarrow{O_2/H_2O} Fe_2O_3 \bullet xH_2O$$
RUST

Corrosion prevention

The main methods for protecting iron and steel from corrosion are:
- sacrificial protection by galvanising and use of zinc blocks
- barrier protection using oil, grease, paint or a polymer coating.

In **sacrificial protection**,
- the iron is in contact with a metal with a more negative $E^\ominus$ value
- Mg or Zn are usually used.

Sacrificial protection is used to protect cars (a zinc coating) and North Sea oil platforms (zinc blocks).

$$Mg^{2+} + 2e^- \rightleftharpoons Mg \qquad E^\ominus = -2.36 \text{ V}$$
$$Zn^{2+} + 2e^- \rightleftharpoons Zn \qquad E^\ominus = -0.76 \text{ V}$$
$$Fe^{2+} + 2e^- \rightleftharpoons Fe \qquad E^\ominus = -0.44 \text{ V}$$

more negative $E^\ominus$

When an electrochemical cell is set up the iron is protected:
- the **metal** with a **more negative** $E^\ominus$ value **supplies the electrons** and is sacrificed.

Once reacted, the Zn or Mg can be replaced but the iron is still there.

Recycling of scrap iron and steel

All steel packaging except aerosols can be recycled.
- Incineration is used to clean the metal.
- Iron and steel are separated using magnets.
- The composition of new steel is easily adjusted.
- Scrap is used to adjust the temperature of the furnace.

2.4 Predicting redox reactions

After studying this section you should be able to:

- *use standard cell potentials to predict the feasibility of a reaction*
- *recognise the limitations of such predictions*
- *understand redox processes in the nitrogen cycle*

Key points from AS

- **Redox reactions**
  Revise AS pages 42–44

During AS Chemistry, you learnt how to combine two half-equations to construct a full redox equation.

Standard electrode potentials can be used to predict the direction of any aqueous redox reaction from the relevant half-reactions.

Using standard electrode potentials to make predictions

AQA   M5   SALTERS
EDEXCEL   M5   WJEC   M5
OCR   CCEA   M5

By studying standard electrode potentials, predictions can be made about the likely feasibility of a reaction.

The table below shows three redox systems.

	redox system	$E^{\ominus}$ /V
A	$H_2O_2(aq) + 2H^+ + 2e^- \rightleftharpoons 2H_2O(l)$	+1.77
B	$I_2(aq) + 2e^- \rightleftharpoons 2I^-(aq)$	+0.54
C	$Fe^{3+}(aq) + 3e^- \rightleftharpoons Fe(s)$	−0.04

We can predict the redox reactions that may take place between these species by treating the redox systems as if they are half-cells.

Comparing the redox systems A and B

- The I_2/I^- system has the less positive $E^{\ominus}$ value and its half-reaction will proceed to the left, donating electrons:

$$H_2O_2(aq) + 2H^+(aq) + 2e^- \longrightarrow 2H_2O(l) \qquad E^{\ominus} = +1.77 \text{ V}$$
$$I_2(aq) + 2e^- \longleftarrow 2I^-(aq) \qquad E^{\ominus} = +0.54 \text{ V}$$

We would therefore predict that **H_2O_2 and H^+** would react with **I^-**.

Notice that a reaction takes place between reactants from different sides of each half equation.

Comparing the redox systems B and C

- The Fe^{3+}/Fe system has the more negative $E^{\ominus}$ value and its half-reaction will proceed to the left, donating electrons:

$$I_2(aq) + 2e^- \longrightarrow 2I^-(aq) \qquad E^{\ominus} = +0.54 \text{ V}$$
$$Fe^{3+}(aq) + 3e^- \longleftarrow Fe(s) \qquad E^{\ominus} = -0.04 \text{ V}$$

We would therefore predict that **I_2** would react with **Fe**.

Comparing the redox systems A and C

- The Fe^{3+}/Fe system has the more negative $E^{\ominus}$ value and its half-reaction will proceed to the left, donating electrons:

$$H_2O_2(aq) + 2H^+(aq) + 2e^- \longrightarrow 2H_2O(l) \qquad E^{\ominus} = +1.77 \text{ V}$$
$$Fe^{3+}(aq) + 3e^- \longleftarrow Fe(s) \qquad E^{\ominus} = -0.04 \text{ V}$$

We would therefore predict that **H_2O_2 and H^+** would react with **Fe**.

When comparing two redox systems, we can predict that a reaction may take place between:
- the stronger reducing agent with the more negative $E^{\ominus}$, on the right-hand side of the redox system
- the stronger oxidising agent with the more positive $E^{\ominus}$, on the left-hand side of the redox system.

Limitations of standard electrode potentials

AQA   M5   SALTERS
EDEXCEL   WJEC   M5
OCR   CCEA   M4

A change in electrode potential resulting from concentration changes means that predictions made on the basis of the **standard** value may not be valid.

Electrode potentials and ionic concentration

Non-standard conditions alter the value of an electrode potential.

The half-equation for the copper half-cell is shown below.

$$Cu^{2+}(aq) + 2e^- \rightleftharpoons Cu(s)$$

Using le Chatelier's principle, by increasing the concentration of $Cu^{2+}(aq)$:

- the equilibrium opposes the change by moving to the right
- electrons are removed from the equilibrium system
- the electrode potential becomes more positive.

Will a reaction actually take place?

Remember that these are equilibrium processes.

- Predictions can be made but these give no indication of the reaction rate which may be extremely slow, caused by a large activation energy.
- The actual conditions used may be different from the standard conditions used to record $E^\ominus$ values. This will affect the value of the electrode potential (see above).
- Standard electrode potentials apply to aqueous equilibria. Many reactions take place that are not aqueous.

As a general working rule:

- the larger the difference between $E^\ominus$ values, the more likely that a reaction will take place
- if the difference between $E^\ominus$ values is less than 0.4 V, then a reaction is unlikely to go to completion.

E(cell), ΔS and equilibrium constants

EDEXCEL   M5

For the link between ΔS and K, see page 59.

For the zinc-copper electrochemical cell, we would predict the overall cell reaction as:

$$Zn(s) + Cu^{2+}(aq) \longrightarrow Cu(s) + Zn^{2+}(aq) \qquad E^\ominus(cell) = +1.10 \text{ V}$$

For redox reactions, E(cell), entropy and equilibrium are linked.

A reaction is likely to proceed if:

- E(cell) is positive or
- ΔS_{total} is positive or
- $K > 1$

Remember that these predictions are based upon just the starting and finishing points. There may be a large energy barrier (activation energy) between these extremes and the rate of reaction may be very slow.

> **KEY POINT**
>
> For an overall cell reaction as written above with a positive E(cell),
> $$E(\text{cell}) \quad \propto \quad \Delta S_{total} \quad \propto \quad \ln K$$

Progress check

1 For each of the three predicted reactions on page 68, construct full redox equations.

2 Use the standard electrode potentials below to answer the questions that follow.

$$Fe^{3+}(aq) + e^- \rightleftharpoons Fe^{2+}(aq) \qquad E^\ominus = +0.77 \text{ V}$$
$$Br_2(l) + 2e^- \rightleftharpoons 2Br^-(aq) \qquad E^\ominus = +1.07 \text{ V}$$
$$Ni^{2+}(aq) + 2e^- \rightleftharpoons Ni(s) \qquad E^\ominus = -0.25 \text{ V}$$
$$O_2(g) + 4H^+(aq) + 4e^- \rightleftharpoons 2H_2O(l) \qquad E^\ominus = +1.23 \text{ V}$$

(a) Arrange the reducing agents in order with the most powerful first.

(b) Predict the reactions that could take place.

b) Br_2 with Fe^{2+}; Fe^{3+} with Ni; O_2 and H^+ with Br^-; O_2 and H^+ with Ni.

2 a) Ni(s), Fe^{2+}(aq), Br^-(l), H_2O(l).

A and C: $3H_2O_2$(aq) + 6H⁺(aq) + 2Fe(s) ⟶ 2Fe³⁺(aq) + 6H₂O(l)

1 A and B: H_2O_2(aq) + 2H⁺(aq) + 2I⁻(aq) ⟶ I₂(aq) + 2H₂O(l); B and C: 2Fe(s) + 3I₂(aq) ⟶ 2Fe³⁺(aq) + 6I⁻(aq)

Redox processes in the nitrogen cycle

SALTERS NIS

In the nitrogen cycle, nitrogen is present in many forms and oxidation states, e.g.

gases:			aqueous ions:		
nitrogen	N_2	0	ammonium	NH_4^+	-3
ammonia	NH_3	-3	nitrate(III)	NO_2^-	$+3$
nitrogen(I) oxide	N_2O	$+1$	nitrate(V)	NO_3^-	$+5$
nitrogen(II) oxide	NO	$+2$			
nitrogen(IV) oxide	NO_2	$+4$			

In the environment, many redox processes involve nitrogen.

Chemists benefit agriculture by developing chemicals to:

• provide extra nutrients (fertilisers)

• control soil pH

• control pests (insecticides and pesticides).

Fixation and nitrification

Conversion of nitrogen gas into soluble nitrogen-containing ions:

$$N_2 \rightarrow NH_4^+ \rightarrow NO_2^- \rightarrow NO_3^-$$
$$0 \qquad -3 \qquad +3 \qquad +5$$

Denitrification

Conversion of soluble nitrogen-containing ions into nitrogen gas:

$$NO_3^- \rightarrow NO_2^- \rightarrow NO \rightarrow N_2O \rightarrow N_2$$
$$+5 \qquad +3 \qquad +2 \qquad +1 \qquad 0$$

Advantages and disadvantages of fertilisers (Salters and CCEA only)

Advantages

The soluble nitrogen content of the soil can be enhanced by adding 'artificial' fertilisers containing ammonium and nitrate ions.

• Extra nitrogen nutrients are provided, improving food production.
• Large quantities of fertilisers are produced by the chemical industry.
• Natural fertilisers, based on manure, or crop rotation, can improve soil structure.

Disadvantages

Nitrate fertilisers are soluble and can be leached out of the soil.

• Excessive use of artificial fertilisers can reduce the quality of soil structure.
• Nitrates can run off fields into rivers and ponds – a source of water pollution. Strategies, such as preventing the run off, can be put into place to reduce and manage water pollution.
• Excess nitrogen in watercourses can lead to eutrophication – excessive plant growth and decay causing depletion of dissolved oxygen needed for the survival of other aquatic life, such as fish.

Sample question and model answer

Take care to include correct
state symbols – essential
when showing the equations
representing these enthalpy
changes.

(a) The lattice enthalpy of magnesium oxide can be found using a Born–Haber cycle. Write equations, including state symbols, for the process that takes place during each enthalpy change.

   (i) The enthalpy change of formation of magnesium oxide.

$$Mg(s) + \tfrac{1}{2}O_2(g) \longrightarrow MgO(s)$$ *formulae* ✓ *state symbols* ✓

   (ii) The lattice enthalpy of magnesium oxide.

$$Mg^{2+}(g) + O^{2-}(g) \longrightarrow MgO(s)$$ *formulae* ✓ *state symbols* ✓

   (iii) The first ionisation energy of magnesium.

$$Mg(g) \longrightarrow Mg^+(g) + e^-$$ *formulae* ✓ *state symbols* ✓ [6]

(b) Use the data below to construct a Born–Haber cycle and hence calculate a value for the lattice enthalpy of magnesium oxide.

enthalpy change	*/ kJ mol^{-1}*
atomisation of magnesium	+148
1st ionisation energy of magnesium	+738
2nd ionisation energy of magnesium	+1451
atomisation of oxygen	+248
1st electron affinity of oxygen	−141
2nd electron affinity of oxygen	+791
formation of magnesium oxide	−602

This Born–Haber cycle
includes all the enthalpy
changes that you may
encounter in a Born–Haber
cycle.

Most cycles in exams are
simpler.

Note the different signs for
the 1st and 2nd electron
affinities.

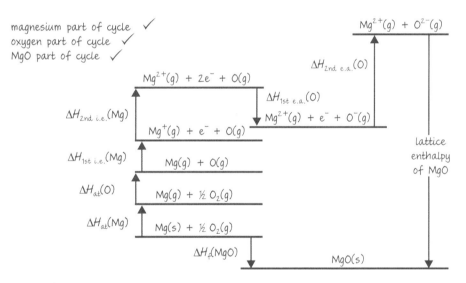

magnesium part of cycle ✓
oxygen part of cycle ✓
MgO part of cycle ✓

By Hess' Law.

$$\Delta H_{at}(Mg) + \Delta H_{at}(O) + \Delta H_{1st\,i.e.}(Mg) + \Delta H_{2nd\,i.e.}(Mg) + \Delta H_{1st\,e.a.}(O) + \Delta H_{2nd\,e.a.}(O) + L.E.(MgO) = \Delta H_f(MgO)$$

∴ $(+148) + (+248) + (+738) + (+1451) + (−141) + (+791) + L.E.(MgO)$
    $= (−602)$

∴ Lattice energy MgO $= −3837 \text{ kJ mol}^{-1}$ *correct value* ✓ *correct sign* ✓

[5]

[Total: 11]

Practice examination questions

1 The Born–Haber cycle for the formation of potassium bromide from potassium and bromine may be represented by a series of stages labelled **A** to **F** as shown.

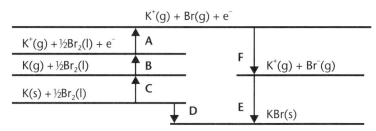

(a) (i) Write the letters **A** to **F** next to the corresponding definition in the table below.

definition	letter	ΔH/kJ mol^{-1}
1st ionisation energy of potassium		+419
1st electron affinity of bromine		−325
the enthalpy of atomisation of potassium		+89
the enthalpy of atomisation of bromine		+112
the lattice enthalpy of potassium bromide		
the enthalpy of formation of potassium bromide		−394

(ii) Calculate the lattice enthalpy of potassium bromide from the data given. [5]

(b) Explain whether you would expect the lattice enthalpies of the following compounds to be more or less exothermic than that of potassium bromide.

(i) KI; (ii) CaBr$_2$ [4]

[Total: 9]

2 (a) Write equations, including state symbols, for the following:

(i) the lattice enthalpy of magnesium chloride

(ii) the enthalpy change of hydration of a Mg^{2+} ion

(iii) the enthalpy change of solution of magnesium chloride. [3]

(b) The lattice enthalpy of MgCl$_2$ is −2526 kJ mol^{-1}.

The enthalpy change of hydration of Mg^{2+} is −1891 kJ mol−1.

The enthalpy change of hydration of Cl$^-$ is −384 kJ mol−1.

Calculate the enthalpy change of solution of magnesium chloride. [3]

[Total: 6]

3 Chromium can be prepared by the reduction of chromium(III) oxide with aluminium.

$$Cr_2O_3(s) + 2Al(s) \longrightarrow 2Cr(s) + Al_2O_3(s)$$

The table below gives information about the reactants and products.

substance	$Cr_2O_3(s)$	$Al(s)$	$Cr(s)$	$Al_2O_3(s)$
ΔH_f/kJ mol^{-1}	−602	0	0	−1676
S/J K^{-1} mol^{-1}	27	28	81	51

Calculate the free energy change for this reaction at 298 K and comment on its feasibility at this temperature.

[Total: 8]

Practice examination questions *(continued)*

4 A standard cell was set up as shown below.

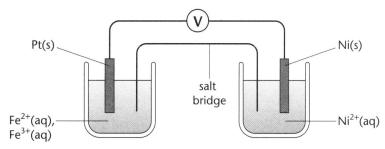

(a) What conditions are necessary for this cell to be a standard cell? [2]

(b) What is the purpose of the salt bridge? [1]

(c) The standard electrode potentials for the half-cells above are:

$$Ni^{2+} + 2e^- \rightleftharpoons Ni \qquad E^\ominus = -0.25 \text{ V}$$
$$Fe^{3+} + e^- \rightleftharpoons Fe^{2+} \qquad E^\ominus = +0.77 \text{ V}$$

   (i) Calculate the standard cell potential of the complete cell.

   (ii) Draw an arrow on the diagram above to show the direction of electron flow in the external circuit.

   (iii) What is the polarity of each electrode? [3]

(d) (i) Write equations for the reaction at each electrode.

   (ii) Hence write an equation to show the overall reaction. [3]

(e) The solution in the nickel half-cell was replaced by a solution that had a higher concentration of Ni^{2+}(aq) ions. Predict the effect on the cell potential. Explain your reasoning. [2]

[Total: 11]

5 (a) Use the standard electrode potentials given below to answer the questions that follow.

				$E^\ominus$/ V
Mg^{2+}(aq) + 2e$^-$	$\rightleftharpoons$	Mg(s)		−2.37
I_2(aq) + 2e$^-$	$\rightleftharpoons$	2I$^-$(aq)		+0.54
Mn^{3+}(aq) + e$^-$	$\rightleftharpoons$	Mn^{2+}(aq)		+1.49

   (i) Which of the six species above is the strongest oxidising agent?

   (ii) Predict, with an equation and a reason, which species above would react with I$^-$(aq) ions to form iodine, I_2(aq). [4]

(b) The standard electrode potentials for two redox equilibria are given below.

$$O_2(g) + 2H_2O(l) + 4e^- \rightleftharpoons 4OH^-(aq) \qquad E^\ominus = +0.40 \text{ V}$$
$$V^{3+}(aq) + e^- \rightleftharpoons V^{2+}(aq) \qquad E^\ominus = -0.26 \text{ V}$$

Two half-cells based on these redox equilibria are connected to make a cell. Determine the cell potential, $E^\ominus$, for this cell and write an equation for the cell reaction. [3]

[Total: 7]

Chapter 3

The Periodic Table

The following topics are covered in this chapter:

- Oxides and chlorides of the Period 3 elements
- Chemistry of the p-block elements
- The transition elements
- Transition element complexes
- Ligand substitution of complex ions
- Redox reactions of transition metal ions
- Catalysis
- Reactions of metal aqua-ions

Key points from AS

- **Bonding and structure**
  Revise AS pages 51–71

During AS Chemistry, you learnt about physical trends in properties of **elements** across Period 3. Across Period 3, **compounds** of the elements also show characteristic trends. For A2 Chemistry, physical and chemical trends in the properties of Period 3 oxides and chlorides are explained in terms of bonding and structure.

3.1 Oxides and chlorides of the Period 3 elements

After studying this section you should be able to:

- describe and explain the periodic variation of the formulae and boiling points of the main oxides and chlorides of Period 3
- describe the reactions of Period 3 elements with oxygen
- describe the action of water on Period 3 oxides and chlorides
- describe these reactions in terms of structure and bonding
- describe the reaction of Period 3 oxides with acids and bases

LEARNING SUMMARY

The formulae of the Period 3 oxides

AQA   M5   CCEA   M4

Across Period 3, the formulae of compounds show a regular pattern which depends upon the **number of electrons** in the outer shell.

The **oxidation state** of an element indicates the **number of electrons** involved in **bonding**. The relationship of the formulae of the Period 3 oxides with electronic structure and oxidation state is shown below.

Key points from AS

- **The modern Periodic Table**
  Revise AS pages 72–75
- **Redox reactions**
  Revise AS pages 42–44

The highest oxidation state of an element is usually the group number.

Chlorine forms several oxides. The highest is Cl_2O_7.

element	Na	Mg	Al	Si	P	S	Cl
electrons in outer shell	1	2	3	4	5	6	7
formula of oxide	Na_2O	MgO	Al_2O_3	SiO_2	P_4O_6 P_4O_{10}	SO_2 SO_3	Cl_2O_7
oxidation state of element in oxide	+1	+2	+3	+4	+3 +5	+4 +6	+7

increase in highest oxidation state

The amount of oxygen atoms per mole of the element increases steadily across Period 3.

For each element in the Period, there is an increase of 0.5 mole of O per mole of element.

$Na_1O_{0.5}$ Mg_1O_1 $Al_1O_{1.5}$ Si_1O_2 $P_1O_{2.5}$ S_1O_3 $Cl_1O_{3.5}$

Periodicity means a repeating trend. This is one of the important ideas of chemistry – a trend across one period is repeated across other periods.

This periodicity is also seen across Periods 2 and 4:

Period 2	Li_2O	BeO	B_2O_3	CO_2
Period 3	Na_2O	MgO	Al_2O_3	SiO_2
Period 4	K_2O	CaO	Ga_2O_3	GeO_2

Preparation of oxides from the elements

AQA ▸ M5   CCEA ▸ M4

Oxides of the Period 3 elements Na to Cl can be prepared by heating each element in oxygen. These are redox reactions:

- the Period 3 element is oxidised
- oxygen is reduced.

Metal oxides

Equations for the reactions of oxygen with the metals sodium, magnesium and aluminium are shown below.

At room temperature and pressure:
Na_2O, MgO and Al_2O_3 are white solids.

$$4Na(s) + O_2(g) \xrightarrow{\textit{yellow flame}} 2Na_2O(s)$$

$$2Mg(s) + O_2(g) \xrightarrow{\textit{white flame}} 2MgO(s)$$

$$4Al(s) + 3O_2(g) \xrightarrow{\textit{white flame}} 2Al_2O_3(s)$$

- Na_2O and MgO are **ionic compounds**.
- Al_2O_3 has bonding that is **intermediate** between ionic and covalent.

Non-metal oxides

Phosphorus and sulfur exist as P_4 and S_8 molecules.

Sulfur, S_8, is usually simplified in equations as S.

Equations for the reactions of phosphorus and sulfur with oxygen are shown below.

$$P_4(s) + 5O_2(g) \xrightarrow{\textit{white flame}} P_4O_{10}(s)$$

$$S(s) + O_2(g) \xrightarrow{\textit{blue flame}} SO_2(g)$$

SO₂ reacts further with oxygen in the presence of a catalyst (V_2O_5):

At room temperature and pressure:
- P_4O_{10} is a white solid
- SO_2 is a colourless gas
- SO_3 is a colourless liquid.

$$2SO_2(g) + O_2(g) \longrightarrow 2SO_3(l)$$

- P_4O_{10}, SO_2 and SO_3 are **covalent compounds**.

Trend in the boiling points of the Period 3 oxides

AQA ▸ M5   WJEC ▸ M5
CCEA ▸ M4

The table below compares the boiling points of some Period 3 oxides with their structure and bonding.

Key points from AS

- **Bonding, structure and properties**
  Revise AS pages 67–71
- **Trends in boiling points of Period 3 elements**
  Revise AS pages 74–75

oxide	Na_2O	MgO	Al_2O_3	SiO_2	P_4O_{10}	SO_3	Cl_2O_7
boiling point /°C	1275	3600	2980	2230	605	45	83

structure	giant lattice				simple molecules	
bonding	ionic		mixed	covalent	covalent and van der Waals'	
	strong forces between ions			strong forces between atoms	weak van der Waals' forces between molecules	

Giant structures and boiling point

During your study of AS Chemistry, you learnt that the boiling points of the elements and compounds are dependent upon bonding and structure.

The same principles can be applied many times in chemistry.

The oxides with **giant** structures ($Na_2O \rightarrow SiO_2$):

- have **high boiling points**
- require a large amount of energy to break the **strong forces** between their particles.

The particles making up a giant lattice can be different.

- Na_2O and MgO have ionic bonding and the **strong** forces are electrostatic attractions acting between positive and negative **ions**.
- Al_2O_3 has bonding intermediate between ionic and covalent. **Strong** forces act between particles that are **intermediate** between ions and atoms.
- SiO_2 has covalent bonding and the **strong** forces are 'shared pairs of electrons' acting between the **atoms**.

P_4O_{10}, SO_2, SO_3 and Cl_2O_7 all exist as covalently bonded molecules. The forces between the molecules though are van der Waals' forces.

Simple molecular structures and boiling point

The oxides with simple molecular structures ($P_4O_{10} \rightarrow Cl_2O_7$):

- have low boiling points
- require a small amount of energy to break the weak van der Waals' forces acting between the molecules.

The action of water on the Period 3 oxides

AQA ▶ M5   CCEA ▶ M4

Sodium and magnesium metals react with water forming the metal hydroxide and hydrogen gas. The reaction with sodium is rapid but magnesium reacts very slowly.

Key points from AS

- **The Group 1 elements**
  Revise AS pages 76–77
- **The Group 2 elements**
  Revise AS pages 77–78

Reactions of sodium and magnesium oxides

Metal oxides form alkalis in water.

$$Na_2O(s) + 2H_2O(l) \longrightarrow 2NaOH(aq) \qquad pH = 14$$
$$MgO(s) + 2H_2O(l) \longrightarrow Mg(OH)_2(aq) \qquad pH = 11$$

Each hydroxide dissociates in water, releasing OH^- ions into solution:

$$NaOH(aq) \longrightarrow Na^+(aq) + OH^-(aq)$$

Although MgO reacts with water, the magnesium hydroxide formed is only slightly soluble in water forming a weak alkaline solution only.

Oxides of aluminium and silicon

Al_2O_3 and SiO_2 have very strong lattices which cannot be broken down by water. Consequently these compounds are insoluble in water.

Reactions of non-metal oxides

Non-metal oxides form acids in water. Equations for the reactions of $P_4O_{10}(s)$, $SO_2(g)$ and $SO_3(l)$ with water are shown below.

$$P_4O_{10}(s) + 6H_2O(l) \longrightarrow 4H_3PO_4(aq) \qquad \textit{phosphoric acid}$$
$$SO_2(g) + H_2O(l) \longrightarrow H_2SO_3(aq) \qquad \textit{sulfurous acid}$$
$$SO_3(l) + H_2O(l) \longrightarrow H_2SO_4(aq) \qquad \textit{sulfuric acid}$$
$$Cl_2O_7(s) + H_2O(l) \longrightarrow 2HClO_4(aq) \qquad \textit{chloric(VII) acid}$$

Each acid dissociates in the water, releasing H^+ ions into solution.

E.g. $H_2SO_4(aq) \longrightarrow H^+(aq) + HSO_4^-(aq)$

An important general rule is:

- **metal oxides** are **basic**
  When soluble, metal oxides form alkaline solutions in water.
- **non-metal oxides** are **acidic**
  When soluble, non-metal oxides form acidic solutions in water.

KEY POINT

The action of water on Period 3 oxides is summarised below.

oxide	Na_2O	MgO	Al_2O_3	SiO_2	P_4O_{10}	SO_3	Cl_2O_7
action of water	alkali		insoluble		acid		

The reactions of oxides with acids and bases

AQA ▶ M4   CCEA ▶ M5

Basic oxides, Na_2O and MgO

Basic oxides are neutralised by acids to form a salt and water.

e.g. $Na_2O(s) + 2HCl(aq) \longrightarrow 2NaCl(aq) + H_2O(l)$
$$MgO(s) + H_2SO_4(aq) \longrightarrow MgSO_4(aq) + H_2O(l)$$

Acidic oxides, P_4O_{10}, SO_2, SO_3 and Cl_2O_7

Acidic oxides are **neutralised** by bases to form a salt and water.

e.g. $P_4O_{10}(s) + 12NaOH(aq) \longrightarrow 4Na_3PO_4(aq) + 6H_2O(l)$

$SO_2(g) + 2KOH(aq) \longrightarrow K_2SO_3(aq) + H_2O(l)$

$SO_3(s) + 2KOH(aq) \longrightarrow K_2SO_4(aq) + H_2O(l)$

$Cl_2O_7(s) + 2NaOH(aq) \longrightarrow 2NaClO_4(aq) + H_2O(l)$

Aluminium oxide, Al_2O_3

Al_2O_3 is **neutralised** by both acids and bases and is **amphoteric**.

> An **amphoteric** oxide is one that is neutralised by both acids and bases.

KEY POINT

With hot dilute $HCl(aq)$,

$Al_2O_3(s) + 6HCl(aq) \longrightarrow 2AlCl_3(aq) + 3H_2O(l)$

With hot concentrated $NaOH(aq)$,

$Al_2O_3(s) + 2NaOH(aq) + 3H_2O(l) \longrightarrow 2NaAl(OH)_4(aq)$

Silicon(IV) oxide, SiO_2

SiO_2 is **neutralised** by hot concentrated bases:

$2KOH(aq) + SiO_2(s) \longrightarrow K_2SiO_3(aq) + H_2O(l)$

Progress check

1 (a) Across the Periodic Table, what pattern is shown in the formulae of the Period 3 oxides?

(b) Why do elements in the **same** group form compounds with **similar** formulae?

2 The reactions of elements with oxygen are redox reactions. Use oxidation numbers to show that the reaction of aluminium with oxygen is a redox reaction.

O changes oxidation state from 0 to –2 (reduction)
Al changes oxidation number from 0 to +3 (oxidation)
2 $4Al(s) + 3O_2(g) \longrightarrow 2Al_2O_3(s)$
(b) Although the outer shell is different, there is the same number of electrons in the outer shell.
1 (a) For each element in the period, there is an increase of 0.5 mole of O per mole of the element.

The formulae of the Period 3 chlorides

CCEA   M4

The relationship of the formulae of some Period 3 chlorides with electron structure and oxidation state is shown below.

Key points from AS

- **The modern Periodic Table**
  Revise AS pages 72–75
- **Redox reactions**
  Revise AS pages 42–44

element	Na	Mg	Al	Si	P
electrons in outer shell	1	2	3	4	5
formula of chloride	NaCl	$MgCl_2$	Al_2Cl_6	$SiCl_4$	PCl_3 PCl_5
oxidation state of element in chloride	+1	+2	+3	+4	+3 +5

increase in highest oxidation state ⟶

The oxidation state of the element in its chloride is equal to the number of chlorine atoms bonded to the element.

The amount of chlorine atoms per mole of the element increases steadily across Period 3.

For each element in the Period, there is an increase of 1 mole of Cl per mole of element.

$NaCl \quad MgCl_2 \quad Al_2Cl_6 \quad SiCl_4 \quad PCl_5$

The same trend is seen across each Period of the Periodic Table.

The reason is the same – the number of electrons in the outer shell increases steadily across each period.

As with the Period 3 oxides, this periodicity in formulae is also seen as a repeating trend across Periods 2 and 4. The table below shows the formulae of the chlorides from Period 2 to Period 4 of the Periodic Table.

Period 2	$LiCl$	$BeCl_2$	BCl_3	CCl_4
Period 3	$NaCl$	$MgCl_2$	$AlCl_3$	$SiCl_4$
Period 4	KCl	$CaCl_2$	$GaCl_3$	$GeCl_4$

Trend in the boiling points of the Period 3 chlorides

CCEA   M4

Al_2Cl_6, $SiCl_4$ and PCl_5 exist as covalently bonded molecules. The forces between the molecules though are van der Waals' forces.

The table below compares the boiling points of some Period 3 chlorides with structure and bonding.

chloride	$NaCl$	$MgCl_2$	Al_2Cl_6	$SiCl_4$	PCl_5
boiling point /°C	1465	1412	178	58	160
structure	giant lattice		simple molecules		
bonding	ionic		covalent and van der Waals'		
forces	strong forces between ions		weak van der Waals' forces between molecules		

Key points from AS

- **Bonding, structure and properties**
  Revise AS pages 67–71
- **Trends in boiling points of Period 3 elements**
  Revise AS pages 74–75

Giant structures and boiling point

The chlorides with **giant** structures (NaCl $\longrightarrow$ MgCl$_2$)

- have **high boiling points**
- require a large amount of energy to break the **strong forces** between their particles
- NaCl and MgCl$_2$ have ionic bonding and the **strong** forces are electrostatic attractions acting between positive and negative **ions**.

The chlorides with **simple molecular** structures (Al$_2$Cl$_6$ $\longrightarrow$ PCl$_5$)

- have low boiling points
- require a small amount of energy to break the weak **van der Waals' forces** acting between the **molecules**.

Phosphorus pentachloride: ionic or covalent?

Phosphorus pentachloride, PCl$_5$, exists as a covalent molecule in the gaseous state. However phosphorus pentachloride exists in the solid state in the **ionic** form, $[PCl_6]^- [PCl_4]^+$.

Thus, solid PCl$_3$ and PCl$_5$ have different structures with PCl$_5$ having a much higher melting point and boiling point than PCl$_3$.

Comparison between the oxides and chlorides of Period 3

The structure and bonding of the Period 3 oxides and chlorides are compared below.

For **oxides**, bonding and structure change at **different** points.

- Bonding changes from ionic to covalent between MgO and SiO$_2$ with Al$_2$O$_3$ having intermediate bonding.
- Structure changes from giant lattice to simple molecular between SiO$_2$ and P$_4$O$_{10}$.

For **chlorides**, both bonding and structure change at the **same** point.

Between $MgCl_2$ and Al_2Cl_6:

- bonding changes from ionic to covalent **and**
- structure changes from giant to simple molecular.

The action of water on the chlorides of Period 3

CCEA   M4

Dissolving ionic chlorides

The ionic chlorides **dissolve** in water forming a **neutral** or very weak acidic solution.

$$NaCl(s) \;\; + aq \longrightarrow Na^+(aq) \;\; + Cl^-(aq) \qquad pH = 7$$
$$MgCl_2(s) \;\; + aq \longrightarrow Mg^{2+}(aq) + 2Cl^-(aq) \qquad pH = 6$$

This periodicity of properties is repeated across each period.

Reactions of covalent chlorides

The covalent chlorides are **hydrolysed** by water in **a vigorous reaction**. Strong acid solutions are formed containing hydrochloric acid.

Equations for the reactions of $Al_2Cl_6(s)$, $SiCl_4(l)$ and $PCl_5(s)$ with water are shown below.

$$Al_2Cl_6(s) \;\; + 6H_2O(l) \longrightarrow 2Al(OH)_3(s) \;\; + 6HCl(aq)$$
$$SiCl_4(l) \;\; + 4H_2O(l) \longrightarrow Si(OH)_4(s) \;\; + 4HCl(aq)$$
$$PCl_5(s) \;\; + 4H_2O(l) \longrightarrow H_3PO_4(aq) \;\; + 5HCl(aq)$$

> An important general rule is:
>
> - **ionic chlorides** form **neutral** solutions in water
> - **covalent chlorides** form **acidic** solutions in water.

KEY POINT

The action of water on the Period 3 oxides is summarised below.

chloride	NaCl	MgCl₂	AlCl₃	SiCl₄	PCl₅
action of water	neutral	very weak acid	strong acid		

Progress check

1 (a) What is the trend in pH shown by the chlorides of Period 3?
 (b) Water is added to sodium chloride and phosphorus pentachloride to form aqueous solutions with different pH values. Explain, with equations, why these solutions have different pH values.

3.2 Chemistry of the p-block elements

After studying this section you should be able to:

- *understand general properties of the p-block elements*
- *understand the electron deficiency of Group 3 compounds*
- *describe the relative stabilities of the oxidation states in Group 4*
- *explain the trend in oxidising power of the Group 7 elements*
- *explain the trend in reducing power of the Group 7 halides*

LEARNING SUMMARY

General properties

WJEC ▶ M5

Electron configurations of the p-block elements

group	3	4	5	6	7	0
outer shell electrons	s^2p^1	s^2p^2	s^2p^3	s^2p^4	s^2p^5	s^2p^6
Period 2: 2nd shell	B	C	N	O	F	Ne
Period 3: 3rd shell	Al	Si	P	S	Cl	Ar

The inert pair effect

On descending Groups 3, 4 and 5, the two electrons in the outer s-orbital (the s^2 pair) are less involved in bonding – the **inert pair effect**.

- At the **top** of the group, the inert pair effect is minor. **All outer shell electrons are involved in covalent bonding.**
- At the **bottom** of the group, the inert pair effect is significant. **Only p-electrons are involved in ionic bonding.**

Expansion of the octet

The maximum number of electron pairs that can surround a central atom is greater in Period 3 (Na–Ar) than in Period 2 (Li–Ne).

- In Period 2, N forms halides with the general formula NX_3 only: NF_3, NCl_3, NBr_3
- In Period 3, P forms halides with the general formulae PX_3 and PX_5: PF_3 and PF_5, PCl_3 and PCl_5, PBr_3 and PBr_5

PF_5, PCl_5 and PBr_5 have 'expanded their octets' with 10 electrons around the central P atom.

Some compounds and ions in the p-block are **amphoteric**.

> An **amphoteric** substance is one that reacts as either an acid or base.
>
> **KEY POINT**

Al_2O_3 is an amphoteric oxide.

Amphoteric behaviour of Al^{3+}.

- *Acidic nature*: With hot dilute HCl(aq), a salt forms:

$$Al_2O_3(s) + 6HCl(aq) \longrightarrow 2AlCl_3(aq) + 2H_2O(l)$$

- *Basic nature*: With hot concentrated NaOH(aq), a salt forms:

$$Al_2O_3(s) + 2NaOH(aq) + 3H_2O(l) \longrightarrow 2NaAl(OH)_4(aq)$$

PbO is an **amphoteric** oxide.

Amphoteric behaviour of Pb^{2+}.

- *Acidic nature*: PbO reacts with bases to form salts:

$$PbO(s) + 2OH^-(aq) \longrightarrow PbO_2^{2-}(aq) + H_2O(l)$$

- *Basic nature*: PbO reacts with acids to form salts:

$$PbO(s) + 2H^+(aq) \longrightarrow Pb^{2+}(aq) + H_2O(l)$$

Group 3

WJEC   MS

Many Group 3 compounds (e.g. BF_3, BCl_3 and $AlCl_3$) are **electron deficient – one of the p-orbitals in B and Al is unused.**

- BF_3, BCl_3 and $AlCl_3$ all react to form **donor-acceptor compounds** by accepting an electron pair.
- The donor-acceptor compounds below are **not** electron-deficient.

The Al_2Cl_6 dimer

At room temperature, aluminium chloride exists as a dimer, with a dative covalent bond bridging from one molecule to the other.

Notice that **all** of the **p-orbitals** in B and Al are **used** and that Al now has a stable octet.

Formation of BF_3NH_3

The electron deficiency of BF_3, BCl_3 and $AlCl_3$ allows the formation of donor-acceptor compounds such as BF_3NH_3.

These ionic liquids are being developed as 'clean technology' solvents and catalysts for processes such as the polymerisation of alkenes.

The affinity of $AlCl_3$ for chlorine species (as in the formation of Al_2Cl_6) finds a use in industrially important catalysts:

- $AlCl_3$ as a halogen carrier in the chlorination of benzene
- low melting temperature ionic liquids, containing the chloroaluminate(III) ion, $AlCl_4^-$.

Boron nitride, BN

Boron nitride, BN, forms different structures.

Cubic 3D structure

BN is finding increasing use as lubricants, wear-resistant coatings and nanotubes for wire sleeving, catalyst support and semiconduction.

- There is a tetrahedral arrangement around each atom in the structure (as in diamond).

A hexagonal structure

- Each layer is bonded weakly to the next (as in graphite) using the two outer shell electrons 'left over' from each N atom.

Boron nitride has hardness, chemical inertness, high melting temperature and semiconductor properties.

Group 4

C   +4 stable
Si
Ge
Sn
Pb   +2 stable

As Group 4 is descended, the obvious trend is from non-metal to metal.

Group 4 compounds have two main oxidation states +2 and +4.

- At the top of the group (C), the +4 oxidation state is more stable.
- At the bottom of the group, the +2 oxidation state is more stable.

Carbon monoxide is the only common example of the +2 oxidation state in a carbon compound. CO is a strong reducing agent and is readily oxidised to CO_2; the more stable +4 oxide of carbon.

- In the extraction of iron from iron(III) oxide in the blast furnace, CO is the reducing agent:

$$Fe_2O_3 + 3CO \longrightarrow 2Fe + 3CO_2$$

Note that the reducing agent (CO) gets oxidised.

$+2 \longrightarrow +4$ oxidation

At the bottom of the group, Pb(IV) compounds act as oxidising agents and can readily be reduced to the more stable +2 oxidation state of lead.

- PbO_2 oxidises concentrated hydrochloric acid:

$$PbO_2 + 4HCl \longrightarrow PbCl_2 + Cl_2 + 2H_2O$$

Note that the oxidising agent (PbO_2) gets reduced

$+4 \longrightarrow +2$ reduction

Properties of oxides: CO_2 and PbO

Oxide	$CO_2(g)$	$PbO(g)$
bonding	covalent	ionic/intermediate
structure	simple molecular	giant ionic

Acid–base properties

- **CO_2** is an **acidic** gas, that forms a weak acidic solution with water:

$$CO_2(g) + aq \longrightarrow CO_2(aq)$$
$$CO_2(aq) + H_2O(l) \rightleftharpoons H^+(aq) + HCO_3^-(aq)$$

- **PbO** is an **amphoteric** oxide.

    Basic nature: PbO reacts with acids to form salts:

$$PbO(s) + 2H^+(aq) \longrightarrow Pb^{2+}(aq) + H_2O(l)$$

    Acidic nature: PbO reacts with bases to form salts:

$$PbO(s) + 2OH^-(aq) \longrightarrow PbO_2^{2-}(aq) + H_2O(l)$$

Properties of chlorides: CCl_4, $SiCl_4$, $PbCl_2$

Chloride	$CCl_4(l)$	$SiCl_4(l)$	$PbCl_2(s)$
bonding	covalent	covalent	mainly ionic
structure	simple molecular	simple molecular	giant ionic

Reaction with water

- CCl_4 does **not react** with water; two layers are formed.
- $SiCl_4$ is **hydrolysed** in a violent reaction with water, forming a strong acidic solution.

$$SiCl_4(l) + 4H_2O(l) \longrightarrow Si(OH)_4(s) + 4HCl(g)$$

- $PbCl_2$ is **slightly soluble** in cold water; more soluble in hot water.

Reactions of $Pb^{2+}(aq)$ with aqueous NaOH, Cl^- and I^-

$Pb^{2+}(aq)$ reacts with NaOH(aq) in two stages.

A precipitation reaction takes place:

$$Pb^{2+}(aq) + 2OH^-(aq) \longrightarrow Pb(OH)_2(s) \qquad \text{white precipitate}$$

With an **excess of NaOH(aq)**, a **colourless solution** forms:

$$Pb(OH)_2(s) + 2OH^-(aq) \longrightarrow Pb(OH)_4^{2-}(aq)$$

Reactions of $Pb^{2+}(aq)$ with aqueous Cl^-

A precipitation reaction takes place:

$$Pb^{2+}(aq) + 2Cl^-(aq) \longrightarrow PbCl_2(s) \qquad \text{white precipitate}$$

With an **excess of $Cl^-(aq)$**, a **colourless solution** forms:

$$PbCl_2(s) + 2Cl^-(aq) \longrightarrow PbCl_4^{2-}(aq)$$

Reactions of $Pb^{2+}(aq)$ with aqueous I^-

A precipitation reaction takes place:

$$Pb^{2+}(aq) + 2I^-(aq) \longrightarrow PbI_2(s) \qquad \text{yellow precipitate}$$

Group 7

WJEC   M5

Key points from AS

- Group 7 and their compounds
  Revise AS pages 80–85

Disproportionation of chlorine in dilute aqueous alkalis

In dilute alkaline solutions, chlorine undergoes disproportionation.

$$Cl_2(aq) + 2NaOH(aq) \longrightarrow NaCl(aq) + NaClO(aq) + H_2O(l)$$

$$0 \xrightarrow{\quad -1 \quad} -1 \qquad \textit{chlorine reduced}$$

$$0 \xrightarrow{\quad +1 \quad} +1 \qquad \textit{chlorine oxidised}$$

Oxidising power of halogens

The oxidising power of the halogens decreases down the group. This can be demonstrated by **displacement reactions** of aqueous halides using Cl_2, Br_2 and I_2.

The $E^\ominus$ values below show that **halogens** gain electrons more readily and become **stronger oxidising agents** as the group is **ascended**.

F_2
Cl_2
Br_2
I_2

Increased strength as oxidising agent

	redox system				$E^\ominus$/ V	
	$F_2(g)$	+ $2e^-$	$\rightleftharpoons$	$2F^-(aq)$	+2.87	
strongest oxidising agent	$Cl_2(g)$	+ $2e^-$	$\rightleftharpoons$	$2Cl^-(aq)$	+1.36	*strongest reducing agent*
	$Br_2(g)$	+ $2e^-$	$\rightleftharpoons$	$2Br^-(aq)$	+1.09	
	$I_2(g)$	+ $2e^-$	$\rightleftharpoons$	$2I^-(aq)$	+0.54	

Reducing power of halides

HCl

HBr

Hi

Increased strength as reducing agent

Halide ions react with concentrated sulfuric acid producing a range of products depending on the halide used.

- HCl does not reduce H_2SO_4
- HBr does not reduce H_2SO_4 to SO_2 (S reduced from +6 to +4)
- HI does not reduce H_2SO_4 to H_2S (S reduced from +6 to −2)

This difference is caused by the **increasing reducing power** of the **hydrogen halides** as the group is **descended**.

The $E^\ominus$ values below show that **halides** lose electrons more readily and become **stronger reducing agents** as the group is **descended**.

3.3 The transition elements

After studying this section you should be able to:

- *explain what is meant by the terms d-block element and transition element*
- *deduce the electronic configurations of atoms and ions of the d-block elements Sc → Zn*
- *know the typical properties of a transition element*

LEARNING SUMMARY

Key points from AS

- **The Periodic Table**
  Revise AS page 72–85

During AS Chemistry, you studied the metals in the s-block of the Periodic Table. For A2 Chemistry, this understanding is extended with a detailed study of the elements scandium to zinc in the d-block of the Periodic Table. Many of the principles introduced during AS Chemistry will be revisited, especially redox chemistry, and structure and bonding.

The d-block elements

AQA	M5	SALTERS	
EDEXCEL		WJEC	M5
OCR		CCEA	M5

The d-block is found in the centre of the Periodic Table. For A2 Chemistry, you will be studying mainly the d-block elements in Period 4 of the Periodic Table: Sc, Ti, V, Cr, Mn, Fe, Co, Ni, Cu and Zn.

Electronic configurations of d-block elements

The diagram below shows the sub-shell being filled across Period 4 of the Periodic Table.

4s		3d									4p					
K	Ca	Sc	Ti	V	Cr	Mn	Fe	Co	Ni	Cu	Zn	Ga	Ge	As	Se	Br	Kr

Key points from AS

- **Sub-shells and orbitals**
  Revise AS pages 21–23
- **Filling the sub-shells**
  Revise AS pages 23–24
- **Sub-shells and the Periodic Table**
  Revise AS page 24

The 3d sub-shell is at a higher energy and fills **after** the 4s sub-shell.

See also the stability from a half-full p sub-shell.

- The elements Sc ⟶ Zn have their highest energy electrons in the 4s and 3d sub-shells.
- Across the d-block, electrons are filling the 3d sub-shell. The diagram below shows the outermost electron shell for Sc ⟶ Zn.

Notice how the orbitals are filled:

- the orbitals in the 3d sub-shell are first occupied singly to prevent any repulsion caused by pairing.

Notice chromium and copper:

- chromium has one electron in each orbital of the 4s and 3d sub-shells ⟶ extra stability
- copper has a full 3d sub-shell ⟶ extra stability.

		4s	3d				
Sc	[Ar] $3d^1 4s^2$	↑↓	↑				
Ti	[Ar] $3d^2 4s^2$	↑↓	↑	↑			
V	[Ar] $3d^3 4s^2$	↑↓	↑	↑	↑		
Cr	[Ar] $3d^5 4s^1$	↑	↑	↑	↑	↑	↑
Mn	[Ar] $3d^5 4s^2$	↑↓	↑	↑	↑	↑	↑
Fe	[Ar] $3d^6 4s^2$	↑↓	↑↓	↑	↑	↑	↑
Co	[Ar] $3d^7 4s^2$	↑↓	↑↓	↑↓	↑	↑	↑
Ni	[Ar] $3d^8 4s^2$	↑↓	↑↓	↑↓	↑↓	↑	↑
Cu	[Ar] $3d^{10} 4s^1$	↑	↑↓	↑↓	↑↓	↑↓	↑↓
Zn	[Ar] $3d^{10} 4s^2$	↑↓	↑↓	↑↓	↑↓	↑↓	↑↓

Transition elements

Successive ionisation energies provide evidence for electronic configurations of the d-block elements Sc to Zn.

This is most evident by the presence of peaks matching the $3d^5$ half-filled shell.

The elements in the d-block, Ti→Cu, are called transition elements.

KEY POINT

- A transition element has **at least one ion** with a **partially-filled d sub-shell**.

Although scandium and zinc are d-block elements, they are **not** *transition elements*. Neither element forms an ion with a partially filled d sub-shell.

Scandium forms one ion only, Sc^{3+} with the 3d sub-shell empty.

$$Sc\ [Ar]3d^14s^2 \xrightarrow{\textit{loss of 3 electrons}} Sc^{3+}\ \text{only:}\quad [Ar]$$

Zinc forms one ion only, Zn^{2+} with the 3d sub-shell full.

$$Zn\ [Ar]3d^{10}4s^2 \xrightarrow{\textit{loss of 2 electrons}} Zn^{2+}\ \text{only:}\quad [Ar]3d^{10}$$

d–block elements

| Sc | Ti | V | Cr | Mn | Fe | Co | Ni | Cu | Zn |

transition elements

Key points from AS

• **The s-block elements**
Revise AS pages 76–79

Properties of transition elements

1 high density

2 high m. and b. pt.

3 moderate to low reactivity

4 two or more oxidation states

5 coloured ions

6 complex ions with ligands

7 catalysts

Typical properties of transition elements

The transition elements are all metals – they are good conductors of heat and electricity. In addition, some general properties distinguish the transition elements from the s-block metals.

Density

• The transition elements are more dense than other metals.
  • They have smaller atoms than the metals in Groups 1 and 2. The small atoms are able to pack closely together ⟶ high density.

Melting point and boiling point

• The transition elements have higher melting and boiling points than other metals.
  • Within the metallic lattice, ions are smaller than those of the s-block metals – the metallic bonding between the metal ions and the delocalised electrons is strong.

Reactivity

• The transition elements have moderate to low reactivity.
  • Unlike the s-block metals, they do not react with cold water. Many transition metals react with dilute acids although some, such as gold, silver and platinum, are extremely unreactive.

Oxidation states

• Most transition elements have compounds with two or more oxidation states.

Colour of compounds

• The transition elements have coloured compounds and ions.
  • Most transition metals have at least one oxidation state that is coloured.

Complex ions

• The ions of transition elements form complex ions with ligands.

Catalysis

• Many transition elements can act as catalysts.

Formation of ions

AQA   M5   SALTERS
EDEXCEL   WJEC   M5
OCR   CCEA   M5

The energy levels of the 4s and 3d sub-shells are very close together.

> • Transition elements are able to form **more than one ion**, each with a different oxidation state, by losing the **4s** electrons and different numbers of **3d** electrons.
> • When forming ions, the 4s electrons are **lost first**, before the 3d electrons.

KEY POINT

The Fe^{3+} ion, $[Ar]3d^5$ is more stable than the Fe^{2+} ion, $[Ar]3d^54s^1$.

Fe^{3+} has a half-full 3d sub-shell $\longrightarrow$ stability.

The 4s and 3d energy levels are very close together and both are involved in bonding.

When forming ions, the 4s electrons are **lost before** the 3d electrons.

Note that this is different from the order of **filling** – 4s before 3d.

Once the 3d sub-shell starts to fill, the 4s electrons are repelled to a higher energy level and are lost first.

Refer to the electronic configurations of the d-block elements, shown on page 84.

Forming Fe^{2+} and Fe^{3+} ions from iron

Iron has the electronic configuration $[Ar]3d^64s^2$

Iron forms two common ions, Fe^{2+} and Fe^{3+}.

An Fe atom loses **two 4s electrons** $\longrightarrow$ Fe^{2+} ion.

An Fe atom loses **two 4s electrons** and **one 3d electron** $\longrightarrow$ Fe^{3+} ion.

The variety of oxidation states

The table below summarises the stable oxidation states of the elements scandium to zinc. Those in bold type represent the commonest oxidation numbers of the elements in their compounds.

Sc	Ti	V	Cr	Mn	Fe	Co	Ni	Cu	Zn
	+1	+1	+1	+1	+1	+1	+1	+1	
	+2	+2	+2	+2	+2	+2	+2	+2	+2
+3	+3	+3	+3	+3	+3	+3	+3	+3	
	+4	+4	+4	+4	+4	+4	+4		
		+5	+5	+5	+5	+5			
			+6	+6	+6				
				+7					

- The variety of oxidation states results from the close similarity in energy of the 4s and the 3d electrons.
- The number of available oxidation states for the element **increases** from Sc → Mn.
  - All of the available 3d and 4s electrons may be used for bond formation.
- The number of available oxidation states for the element **decreases** from Mn → Zn.
  - There is a decreasing number of unpaired d-electrons available for bond formation.
- Higher oxidation states involve covalency because of the high charge densities involved.
  - In practice, for oxidation states of +4 and above, the bonding electrons are involved in covalent bond formation.

Importance of transition elements (WJEC only)

The transition elements have great economic importance, e.g. iron and other transition elements in the various types of steel.

In living systems, many transition elements are essential as ions in small quantities as 'trace elements', e.g. Cu^{2+} is an essential part of many enzymes, such as *cytochrome c oxidase*.

Progress check

1 Write down the electronic configuration of the following ions:
   (a) Ni^{2+} (b) Cr^{3+} (c) Cu^+ (d) V^{3+}.

2 What is the oxidation state of the metal in each of the following?
   (a) MnO_2 (b) CrO_3 (c) MnO_4^- (d) VO_2^+

1 (a) $[Ar]3d^8$ (b) $[Ar]3d^3$ (c) $[Ar]3d^{10}$ (d) $[Ar]3d^2$.
2 (a) +4 (b) +6 (c) +7 (d) +5.

3.4 Transition element complexes

After studying this section you should be able to:

- *explain what is meant by the terms 'complex ion' and 'ligand'*
- *predict the formula and possible shape of a complex ion*
- *know that ligands can be unidentate, bidentate and multidentate*
- *understand that some complexes have stereoisomers*
- *know that transition metal ions can be identified by their colour*
- *know that electronic transitions are responsible for colour*
- *explain how colorimetry can be used to determine the concentration and formula of a complex ion*

LEARNING SUMMARY

Ligands and complex ions

AQA	M5	SALTERS	M4
EDEXCEL	M5	WJEC	M5
OCR	M5	CCEA	M5

Transition metal ions are small and densely charged. They strongly attract electron-rich species called **ligands** forming **complex ions**.

A ligand has a lone pair of electrons.

> **KEY POINT**
>
> A ligand is a molecule or ion that bonds to a metal ion by:
> - forming a coordinate (dative covalent) bond
> - donating a lone pair of electrons into a vacant orbital.
>
> Common ligands include: H_2O:, :Cl^-, :NH_3, :CN^-

> **KEY POINT**
>
> A **complex ion** is a central metal ion surrounded by ligands.
> The **coordination number** is the total **number** of **coordinate bonds** from ligands to the central transition metal ion of a complex ion.

Key points from AS

- **Electron pair repulsion theory**
  Revise AS pages 60–61

For AS, you learnt that electron pair repulsion is responsible for the shape of a molecule or ion.

Shapes of complex ions

The size of a ligand helps to decide the shape of the complex ion.

Six coordinate complexes

Six water molecules are able to fit around a Cu^{2+} ion to form:

- the complex ion $[Cu(H_2O)_6]^{2+}$
- with a coordination number of **6**.

The **six** electron pairs surrounding the central Cu^{2+} ion in $[Cu(H_2O)_6]^{2+}$ repel one another as far apart as possible to form a complex ion with an **octahedral** shape.

The hexaaquacopper(II) complex ion, $[Cu(H_2O)_6]^{2+}$

Four coordinate complexes

Chloride ions are larger than water molecules and it is only possible for **four** chloride ions to fit around the central copper(II) ion to form:

- the complex ion $[CuCl_4]^{2-}$
- with a coordination number of **4**.

The **four** electron pairs surrounding the central Cu^{2+} ion in $[CuCl_4]^{2-}$ repel one another as far apart as possible to form a complex ion with a **tetrahedral** shape.

The tetrachlorocuprate(II) complex ion [CuCl₄]²⁻

Notice the overall charge on the complex ion $[CuCl_4]^{2-}$ is 2–.

Cu^{2+} and 4 $Cl^- \rightarrow$ 2– ions.

$[CuCl_4]^{2-}$
tetrahedral
4 coordinate

$$\left[\begin{array}{c} Cl \\ | \quad 109.5° \\ Cu \\ Cl \quad \bigg| \quad Cl \\ Cl \end{array} \right]^{2-}$$

General rules for deciding the shape of a complex ion

Although there are exceptions, the following general rules are useful.

> **KEY POINT**
>
> Complex ions with small ligands such as H_2O and NH_3 are usually **6-coordinate** and **octahedral**.
>
> Complex ions with large Cl^- ligands are usually **4-coordinate** and **tetrahedral**.

The table below compares complex ions formed from cobalt(II) and chromium(III) ions.

ligand	complex ions from Co²⁺	complex ions from Cr³⁺	shape
H_2O	$[Co(H_2O)_6]^{2+}$	$[Cr(H_2O)_6]^{3+}$	octahedral
NH_3	$[Co(NH_3)_6]^{2+}$	$[Cr(NH_3)_6]^{3+}$	octahedral
Cl^-	$[CoCl_4]^{2-}$	$[CrCl_4]^-$	tetrahedral

The $[Ag(NH_3)_2]^+$ complex ion present in Tollens' reagent (ammonical silver nitrate) is used as a test for the aldehyde functional group (see pages 116–117).

- Some complex ions contain more than one type of ligand. For example, copper(II) forms a complex ion with a mixture of water and ammonia ligands, $[Cu(NH_3)_4(H_2O)_2]^{2+}$.
- Ag^+ and Cu^+ forms linear complexes that are 2-coordinate: e.g. $[Ag(H_2O)_2]^+$, $[Ag(NH_3)_2]^+$, $[AgCl_2]^-$ and $[CuCl_2]^-$.

Ligands have teeth

AQA   M5   SALTERS
EDEXCEL   CCEA   M5
OCR

Ligands such as H_2O, NH_3 and Cl^- are called **monodentate** ligands. They form only **one** coordinate bond to the central metal ion.

A molecule or ion with more than one oxygen or nitrogen atom may form more than one coordinate bond to the central metal ion.

Ligands that can form **two** coordinate bonds to the central metal ion are called **bidentate** ligands. Examples of bidentate ligands are ethane-1,2-diamine, $NH_2CH_2CH_2NH_2$ and the ethanedioate ion, $(COO^-)_2$.

Monodentate means **one** tooth.

Bidentate ligands have **two** teeth.

Multidentate ligands have **many** teeth.

Each 'tooth' is a coordinate bond.

Multidentate ligands can form many coordinate bonds to the central metal ion. For example, the hexadentate ligand, edta⁴⁻, is able to form **six** coordinate bonds to the central metal ion. The diagram of edta⁴⁻ shows four oxygen atoms and two nitrogen atoms able to form coordinate bonds.

edta⁴⁻

Stereoisomerism in complexes

AQA   M5   EDEXCEL   M5
OCR   CCEA   M5

Stereoisomers have the same structural isomers but different arrangements in space. Transition metal complexes show two types of stereoisomerism:

- *cis-trans* isomerism
- optical isomerism.

Cis-trans isomerism

Cis-trans isomerism occurs in square planar complexes with two ligands of one kind and two of another kind, e.g. $Ni(NH_3)_2Cl_2$.

cis-isomer
groups on same side

trans-isomer
groups on opposite side

Platin

The structures of *cis*-platin and *trans*-platin are shown below. Both structures have a square planar shape with a 90° bond angle.

cis-platin

trans-platin

Cis-platin is used as a chemotherapy drug in the treatment of some cancers. *Cis*-platin forms coordinate bonds to a basic site in the DNA of cancer cells. This process prevents the cancerous DNA from replicating, effectively killing the cancerous cells. Careful treatment is required to avoid harmful side-effects of the drug.

Optical isomerism (OCR only)

Optical isomers are non-superimposable mirror images of one another. Some octahedral complexes with multidentate ligands form optical isomers, e.g. $[Ni(NH_2CH_2CH_2NH_2)_3]^{2+}$.

The two optical isomers rotate plane-polarised light in opposite directions.

Colour and complex ions

AQA	M5	SALTERS	
EDEXCEL		WJEC	M5
OCR		CCEA	M5

Complex ions have different colours

Many transition metal ions have different colours. These colours indicate the possible identity of a transition metal ion. The table on page 90 shows the colours of some **aqua** complex ions of d-block elements. Notice that different oxidation states of a d-block element can have different colours.

	Sc	Ti	V	Cr	Mn	Fe	Co	Ni	Cu	Zn
+1									colourless	
+2					pale pink	pale green	pink	green	blue	colourless
+3	colourless	violet	green	*ruby / green		red-brown				
+5			yellow							
+6				yellow or orange						
+7					purple					

*Cr^{3+} surrounded by 6 water ligands looks violet, but some water is often replaced by other ligands so Cr^{3+} usually looks green.

Why are transition element ions coloured?

AQA ▸ M5   SALTERS ▸ M5
EDEXCEL ▸ M5   WJEC ▸ CH5

Splitting the 3d energy level

All five d-orbitals in the outer d sub-shell of an **isolated** transition element ion have the same energy. However, when ligands bond to the central ion, the d-orbitals move to two different energy levels. The gap between the energy levels depends upon:

- the ligand
- the coordination number
- the transition metal ion.

This diagram shows how water ligands split the energy levels of the five 3d orbitals in a copper(II) ion.

3d orbitals in an isolated copper (II) ion

Ligands split the 3d energy level

$[Cu(H_2O)_6]^{2+}$

energy gap, ΔE

Absorption of light energy

An electron can be promoted from the lower 3d energy level by absorbing energy **exactly** equal to the energy gap ΔE. This energy is provided by radiation from the visible and ultraviolet regions of the spectrum.

Radiation with a frequency f has an energy hf, where h = Planck's constant, 6.63×10^{-34} J s.

So, to promote an electron through as energy gap ΔE, a quantity of light energy is needed given by:

$$\Delta E = hf$$

The diagram below shows what happens when a complex ion absorbs light energy.

Energy and frequency

The difference in energy between the d-orbitals, ΔE, and the frequency of the absorbed radiation, f, are linked by the following relationship:

$\Delta E = hf$

h = Planck's constant, 6.63×10^{-34} J s.

absorption of light energy, hf

promotion of electron to higher energy level

$[Cu(H_2O)_6]^{2+}$ ion in **ground** state

$[Cu(H_2O)_6]^{2+}$ ion in **excited** state

Only those ions that have a partially filled d sub-shell are coloured.
Cu^{2+} is [Ar]3d^{9}4s^2

Ag$^+$ complexes are also colourless because the 3d sub-shell is full.

The blue colour of [Cu(H$_2$O)$_6$]$^{2+}$ results from:

- **absorption** of light energy in the red, yellow and green regions of the spectrum. This absorbed radiation provides the energy for an electron to be excited to a higher energy level
- **reflection** of blue light only, giving the copper(II) hexaaqua ion its characteristic blue colour. The blue region of the spectrum is **not** absorbed.

Copper has another ion, Cu$^+$, with the electronic configuration: [Ar]3d^{10}.

The Cu$^+$ ion is colourless because:

- the 3d sub-shell is full, [Ar]3d^{10}, preventing electron transfer.

> The colour of a transition metal complex ion results from the **transfer** of an **electron** between the orbitals of an **unfilled** d sub-shell.

Absorbance and reflectance spectra (Salters only)

The absorbance and reflectance spectra of a pigment are shown below. Note that the spectra show peaks the opposite way around.

Key points from AS

- **Atomic spectra of hydrogen**
  Revise AS pages 28–29

Different techniques can be used to piece together analytical information about substances.

For example, analysis of paintings may include:

- gas chromatography, GC (see pages 163–164)
- atomic emission and absorption spectroscopy (see AS pages 28–29)
- visible spectroscopy (reflection and absorption).

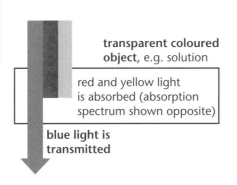

transparent coloured object, e.g. solution

red and yellow light is absorbed (absorption spectrum shown opposite)

blue light is transmitted

absorption spectrum

blue light is reflected (reflectance spectrum shown opposite)

paint layer absorbs red and yellow light

opaque coloured object, e.g. painting

reflectance spectrum

Using colorimetry

AQA ▸ MS   SALTERS ▸ M4
CCEA ▸ MS

A colorimeter measures the amount of light absorbed by a coloured solution. The concentration and formula of a complex ion can be determined from the intensity of light absorbed by the colorimeter.

Finding an unknown concentration of a coloured solution (AQA and Salters only)

- A suitable filter and wavelength is chosen for the readings.
- Solutions of a coloured complex ion of known concentration are prepared.

- The absorbances of the solutions of known concentration are measured.
- A calibration curve is plotted.
- The solution of the complex ion with an unknown concentration is then placed in the colorimeter and the absorbance measured.
- The unknown concentration can be determined from the calibration curve.

Colorimetry can also be used as a method for monitoring reaction rates (see pages 18–27).

Finding the formula of a coloured complex (CCEA only)

Colorimetry can be used to determine the formula of a coloured complexion.

The formula of a complex ion by colorimetry

The CCEA specification lists $[Cu(NH_3)_4(H_2O)_2]^{2+}$ and $[Fe(SCN)(H_2O)_5]^{2+}$ as specific examples.

- A suitable filter and wavelength is chosen for the readings.
- Aqueous solutions of $Cu^{2+}(aq)$ and $NH_3(aq)$ are prepared.
- Mixtures are now prepared with different relative proportions of the $Cu^{2+}(aq)$ and $NH_3(aq)$.
- The absorbance of each solution is measured in a colorimeter.
- A graph is plotted to find the maximum absorbance – this shows the amounts of the Cu^{2+} and NH_3 present on the complex ion.

The maximum absorption of light occurs when the Cu^{2+}/NH_3 ratio is closest to the actual formula of the complex ion.

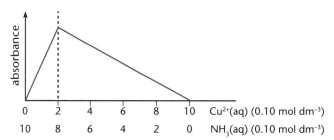

0	2	4	6	8	10	$Cu^{2+}(aq)$ (0.10 mol dm^{-3})
10	8	6	4	2	0	$NH_3(aq)$ (0.10 mol dm^{-3})

From the graph, the maximum absorbance is when $Cu^{2+} : NH_3 = 1 : 4$

In $Cu^{2+}(aq)$, the complex ion is $[Cu(H_2O)_6]^{2+}$

$\therefore$ 4 NH_3 molecules have replaced 4 H_2O ligands in $[Cu(H_2O)_6]^{2+}$ and the formula of the complex ion is $[Cu(NH_3)_4(H_2O)_2]^{2+}$

The same method could be used to find the formula of any other complex ion.

Progress check

1 Predict the formula of the following complex ions:
   (a) iron(III) with water ligands
   (b) vanadium(III) with chloride ligands
   (c) nickel(II) with ammonia ligands.

2 Explain the origin of colour in a complex ion that is yellow.

2 The 3d orbitals are split between two energy levels. A 3d electron from the lower 3d energy level absorbs visible light in the blue and red regions of the spectrum. The absorbed energy allows the d-electron to be promoted to a vacancy in the orbitals of the higher 3d energy level. With the red and blue regions of the spectrum absorbed, only yellow light is reflected, giving the complex ion its colour.

1 (a) $[Fe(H_2O)_6]^{3+}$ (b) $[VCl_6]^-$ (c) $[Ni(NH_3)_6]^{2+}$.

3.5 Ligand substitution of complex ions

After studying this section you should be able to:

- describe what is meant by ligand substitution
- understand that ligand exchange may produce changes in colour and coordination number
- understand stability of complex ions in terms of stability constants
- understand edta titrations of complex ions

Exchange between ligands

Ligand substitution usually produces a change in colour.

A **ligand substitution** reaction takes place when a ligand in a complex ion exchanges for another ligand. A change in ligand usually changes the energy gap between the 3d energy levels. With a different ΔE value, light with a different frequency is absorbed, producing a colour change.

Exchange between H_2O and NH_3 ligands

The **similar sizes** of water and ammonia molecules ensure that ligand exchange takes place with **no change in coordination number**.

For example, addition of an excess of concentrated aqueous ammonia to aqueous nickel(II) ions results in ligand substitution of ammonia ligands for water ligands. Both complex ions have an octahedral shape with a coordination number of six.

*Remember that the **size** of the ligand is a major factor in deciding the coordination number (see page 87):*
- H_2O *and* NH_3 *have similar sizes – same coordination number*
- H_2O *and* Cl^- *have different sizes – different coordination numbers.*

octahedral 6 coordinate → **same coordinate number** → octahedral 6 coordinate

green solution → blue solution

$$[Ni(H_2O)_6]^{2+} + 6\overset{\bullet\bullet}{N}H_3 \longrightarrow [Ni(NH_3)_6]^{2+} + 6H_2\overset{\bullet\bullet}{O}$$

Exchange between H_2O and Cl^- ligands

You should also be able to construct a balanced equation for any ligand exchange reaction.

The **different sizes** of water molecules and chloride ions ensure that ligand exchange takes place with a **change in coordination number**.

For example, addition of an excess of concentrated hydrochloric acid (as a source of Cl^- ions) to aqueous cobalt(II) ions results in ligand substitution of chloride ligands for water ligands. The complex ions have different shapes with different coordination numbers.

octahedral 6 coordinate → **change in coordinate number** → tetrahedral 4 coordinate

blue solution → yellow solution

$$[Co(H_2O)_6]^{2+} + 4\overset{\bullet\bullet}{C}l^- \longrightarrow [CoCl_4]^{2-} + 6H_2\overset{\bullet\bullet}{O}$$

Ligand substitutions of copper(II) and cobalt(II)

Ligand substitutions of copper(II) and cobalt(II) complexes are shown below.

Concentrated hydrochloric acid is used as a source of Cl⁻ ligands.

Concentrated aqueous ammonia is used as a source of NH₃ ligands.

Copper

$[Cu(H_2O)_6]^{2+}$
blue
octahedral

HCl(aq)   NH₃(aq)

$[CuCl_4]^{2-}$ $[Cu(NH_3)_4(H_2O)_2]^{2+}$
yellow   deep blue
tetrahedral   octahedral

Cobalt

$[Co(H_2O)_6]^{2+}$
pink
octahedral

HCl(aq)   NH₃(aq)

$[CoCl_4]^{2-}$ $[Co(NH_3)_6]^{2+}$
blue   pale straw
tetrahedral   octahedral

Incomplete substitution

Substitution may be incomplete.

- Aqueous ammonia only exchanges **four** from the six water ligands in $[Cu(H_2O)_6]^{2+}$.

$$[Cu(H_2O)_6]^{2+} + 4NH_3 \longrightarrow [Cu(NH_3)_4(H_2O)_2]^{2+} + 4H_2O$$
blue solution   deep blue solution

This reaction is used as a test for the Fe³⁺ ion – the deep red-brown colour of $[Fe(H_2O)_5(SCN)]^{2+}$ is intense and extremely small traces of Fe³⁺ ions can be detected using this ligand substitution reaction.

- Aqueous thiocyanate ions, SCN⁻, only exchanges **one** from the six water ligands in $[Fe(H_2O)_6]^{3+}$.

$$[Fe(H_2O)_6]^{3+} + SCN^-(aq) \longrightarrow [Fe(H_2O)_5(SCN)]^{2+} + H_2O$$
pale brown solution   deep red solution

Haemoglobin

AQA   M5   CCEA   M5
OCR   M5

Haemoglobin is a complex ion of Fe²⁺. A haem group consists of an Fe²⁺ ion, held in a ring, known as a porphyrin. The Fe²⁺ ion, which is the site of oxygen binding, bonds with four nitrogen atoms in the centre of the ring in a square planar geometry. The protein globin bonds below and O₂ is able to bond weakly above.

The porphyrin ring in haemoglobin

O₂ can be easily lost when required and is replaced by a weakly bonded water ligand. When haemoglobin reaches the lungs again, fresh O₂ is able to bond to Fe²⁺, displacing the water.

Carbon monoxide, CO, is a serious poison for the blood as the polar molecule binds much more strongly to the Fe²⁺ ion than O₂ and cannot be displaced. This prevents oxygen from being carried by the blood.

Stability of complex ions

AQA M5 CCEA M5
OCR M5

The stability of a complex ion depends upon the ligands. For example, a complex ion of copper(II) is more stable with ligands of ammonia than with water ligands.

Bidentate or multidentate ligands (such as edta) lead to a particularly stable complex. This is due to an increase in entropy:

e.g. $\underbrace{[Cu(H_2O)_6]^{2+} + edta^{4-}}_{2\ mol} \xrightarrow{\text{INCREASE IN ENTROPY}} \underbrace{[Cu(edta)]^{2-} + 6H_2O}_{7\ mol}$

The concentration of water is virtually constant. Constant terms are not included in expressions for equilibrium constants.

The stability of a complex ion is measured as a **stability constant**, K_{stab}:

$$[Cu(H_2O)_6(aq)]^{2+} + 4NH_3(aq) \rightleftharpoons [Cu(NH_3)_4(H_2O)_2(aq)]^{2+} + 4H_2O(l)$$

$$K_{stab} = \frac{[Cu(NH_3)_4(H_2O)_2]^{2+}}{[Cu(H_2O)_6]^{2+}\,[NH_3(aq)]^4} = 1 \times 10^{12}\ dm^{12}\ mol^{-4}\ at\ 25°C$$

The data is expressed on a logarithmic scale because of the very large range of values.

Stability constants usually measure the stability of a complex compared with the metal aqua-ion. The greater the value of K_{stab}, the more stable the complex ion.

complex ion	log K_{stab}	
$[Fe(H_2O)_5(SCN)]^{2+}$	3.9	
$[CuCl_4]^{2-}$	5.6	increased
$[Ni(NH_3)_6]^{2+}$	8.6	stability
$[Ni(edta)]^{2-}$	19.3	

Edta titrations

CCEA M5

The structure of edta is discussed on page 88.

Edta can be used for complexometric titrations in which the formation of a coloured complex is used to indicate the end-point of a titration.

Disodium edta, $Na_2H_2(edta)$, is typically used to prepare solutions for titration. In a weak alkaline aqueous solution, this dissociates forming $edta^{4-}(aq)$ ions.

Most 2+ metal cations form a complex with $edta^{4-}$:

$$M^{2+}(aq) + edta^{4-}(aq) \longrightarrow M(edta)^{2-}(aq)$$

Note that the M^{2+} and $edta^{4-}$ reacts in a **1 : 1** molar ratio.

Edta is used as a 'sequestering agent' in detergents to soften hard water. Edta acts by complexing to Ca^{2+} ions that cause water hardness.

Edta is used in medicine to prevent clotting of blood. Edta acts by complexing to Ca^{2+} ions, required for the clotting action.

Determination of Ca^{2+}(aq) or Mg^{2+}(aq) ions

The titration is carried out in the presence of a weakly alkaline buffer.

Calculation from titration results follows common principles. See pages 43–44 (acid-base) and pages 96–98 (redox) for examples and hints.

A complexometric indicator, such as Eriochrome Black T, is used.

- The indicator binds to the metal cations in solution to form a red-coloured complex.
- Edta is added from the burette.
- Edta binds to metal cations much more strongly than the indicator and displaces the indicator from the metal cations as it is added.
- The colour change at the end point is from **red to blue**.

Progress check

1 Write equations for the following ligand substitutions:
(a) $[Ni(H_2O)_6]^{2+}$ with six ammonia ligands
(b) $[Fe(H_2O)_6]^{3+}$ with four chloride ligands
(c) $[Cr(H_2O)_6]^{3+}$ with six cyanide ligands, CN^-.

(c) $[Cr(H_2O)_6]^{3+} + 6CN^- \longrightarrow [Cr(CN)_6]^{3-} + 6H_2O$.
(b) $[Fe(H_2O)_6]^{3+} + 4Cl^- \longrightarrow [FeCl_4]^- + 6H_2O$
1 (a) $[Ni(H_2O)_6]^{2+} + 6NH_3 \longrightarrow [Ni(NH_3)_6]^{2+} + 6H_2O$

3.6 Redox reactions of transition metal ions

After studying this section you should be able to:

- describe redox behaviour in transition elements
- construct redox equations, using relevant half-equations
- perform calculations involving simple redox titrations

Redox reactions

Transition elements have variable oxidation states with characteristic colours (see page 90). Many **redox** reactions take place in which transition metals ions change their oxidation state by **gaining** or **losing** electrons.

The iron(II) – manganate(VII) reaction

Key points from AS

- **Redox reactions**
  Revise AS pages 42–44

Iron(II) ions are oxidised by acidified manganate(VII) ions.

- Acidified manganate(VII) ions are reduced to manganate(II) ions.

reduction:

$$MnO_4^-(aq) + 8H^+(aq) + 5e^- \longrightarrow Mn^{2+}(aq) + 4H_2O(l)$$
$$+7 \qquad\qquad\qquad\qquad\qquad +2$$

purple ⟶ very pale pink

The reaction is easy to see.

The deep purple MnO_4^- ions are reduced to the very pale pink Mn^{2+} ions, which are virtually colourless in solution.

- Iron(II) ions are oxidised to iron(III) ions.

oxidation:

$$Fe^{2+}(aq) \longrightarrow Fe^{3+}(aq) + e^-$$
$$+2 \qquad\qquad +3$$

To give the overall equation:

- the electrons are balanced

$$MnO_4^-(aq) + 8H^+(aq) + 5e^- \longrightarrow Mn^{2+}(aq) + 4H_2O(l)$$
$$5Fe^{2+}(aq) \longrightarrow 5Fe^{3+}(aq) + 5e^-$$

- the half-equations are added

$$5Fe^{2+}(aq) + MnO_4^-(aq) + 8H^+(aq) \longrightarrow 5Fe^{3+}(aq) + Mn^{2+}(aq) + 4H_2O(l)$$

Redox titrations

Redox titrations can be used in analysis.

Essential requirements are:

- an oxidising agent
- a reducing agent
- an easily-seen colour change (with or without an indicator).

It is important to use an acid that does **not** react with either the oxidising or reducing agent – dilute sulphuric acid is usually used.

For example, hydrochloric acid cannot be used because HC is oxidised to Cl_2 by MnO_4^-.

The most common redox titration encountered at A Level is that between manganate(VII) and acidified iron(II) ions.

Redox titrations using the acidified manganate(VII)

- $KMnO_4$ is reacted with a reducing agent such as Fe^{2+} under acidic conditions (see above).
- Purple $KMnO_4$ is added from the burette to a measured amount of the acidified reducing agent.
- At the end-point, the colour changes from colourless to pale pink, showing that all the reducing agent has exactly reacted. The pale-pink colour indicates the first trace of an excess of purple manganate(VII) ions.
- This titration is **self-indicating** – no indicator is required, the colour change from the reduction of MnO_4^- ions being sufficient to indicate that the reaction is complete.

Key points from AS

- Calculations in acid–base titrations
  Revise AS page 40

The estimation of iron in iron tablets

Five iron tablets with a combined mass of 0.900 g were dissolved in acid and made up to 100 cm³ of solution. In a titration, 10.0 cm³ of this solution reacted exactly with 10.4 cm³ of 0.0100 mol dm⁻³ potassium manganate(VII). What is the percentage by mass of iron in the tablets?

As with all titrations, we must consider **five** pieces of information:

Calculations for redox titrations follow the same principles as those used for acid–base titrations, studied during AS Chemistry.

- the balanced equation
  $$5Fe^{2+}(aq) + MnO_4^-(aq) + 8H^+(aq) \rightarrow 5Fe^{3+}(aq) + Mn^{2+}(aq) + 4H_2O(l)$$
- the concentration c_1 and reacting volume V_1 of $KMnO_4(aq)$
- the concentration c_2 and reacting volume V_2 of $Fe^{2+}(aq)$.

From the titration results, the amount of $KMnO_4$ can be calculated:

The concentration and volume of $KMnO_4$ are known so the amount of MnO_4^- ions can be found.

$$\text{amount of } KMnO_4 = c \times \frac{V}{1000} = 0.0100 \times \frac{10.4}{1000} = 1.04 \times 10^{-4} \text{ mol}$$

From the equation, the amount of Fe^{2+} can be determined:

$$5Fe^{2+}(aq) + MnO_4^-(aq) + 8H^+(aq) \rightarrow 5Fe^{3+}(aq) + Mn^{2+}(aq) + 4H_2O(l)$$

Using the equation determine the number of moles of the second reagent.

5 mol   1 mol *(balancing numbers)*

$\therefore$ **5** x 1.04×10^{-4} mol Fe^{2+} reacts with 1.04×10^{-4} mol MnO_4^-

$\therefore$ amount of Fe^{2+} that reacted = 5.20×10^{-4} mol

Find the amount of Fe^{2+} in the solution prepared from the tablets:

If there is sampling of the original solution, you will need to scale.

10.0 cm³ of $Fe^{2+}(aq)$ contains 5.20×10^{-4} mol Fe^{2+} (aq)

The 100 cm³ solution of iron tablets contains $10 \times (5.20 \times 10^{-4})$

$$= 5.20 \times 10^{-3} \text{ mol } Fe^{2+}$$

Find the percentage of Fe^{2+} in the tablets (A_r: Fe, 55.8)

5.20×10^{-3} mol Fe^{2+} has a mass of $5.20 \times 10^{-3} \times 55.8 = 0.290$ g

$$\therefore \% \text{ of } Fe^{2+} \text{ in tablets} = \frac{\text{mass of } Fe^{2+}}{\text{mass of tablets}} \times 100 = \frac{0.290}{0.900} \times 100 = 32.2\%$$

For A2, you are expected to tackle unstructured titration calculations. The principles are the same, irrespective of the type of titration.

Further redox titrations

Providing there is a visible colour change, many other redox reactions can be used for redox titrations.

For example, acidified dichromate(VI) ions is an oxidising agent which can also oxidise Fe^{2+} ions to Fe^{3+}.

reduction: $Cr_2O_7^{2-}(aq) + 14H^+(aq) + 6e^- \longrightarrow 2Cr^{3+}(aq) + 7H_2O(l)$
oxidation: $Fe^{2+}(aq) \longrightarrow Fe^{3+}(aq) + e^-$
overall: $6Fe^{2+}(aq) + Cr_2O_7^{2-}(aq) + 14H^+(aq) \rightarrow 6Fe^{3+}(aq) + 2Cr^{3+}(aq) + 7H_2O(l)$

Iodine-thiosulfate titrations

EDEXCEL   MS   WJEC   MS
OCR   CCEA   MS

Iodine-thiosulfate titrations can be used to determine iodine concentrations, either directly or by liberating iodine using a more powerful oxidising agent.

Iodine oxidises thiosulfate ions:

$$I_2(aq) + 2S_2O_3^{2-}(aq) \longrightarrow 2I^-(aq) + S_4O_6^{2-}(aq)$$

Starch is a sensitive test for the presence of iodine. If added too early in this titration, the deep-blue colour perseveres and the end-point cannot be detected with accuracy.

- Thiosulfate is added from the burette to the iodine solution until the deep-brown colour of iodine becomes a light-straw colour.
- Starch is now added, forming a deep-blue colour.
- Thiosulfate is added dropwise until the deep-blue colour becomes colourless. This is the end-point and shows that all the iodine has just been reacted.

The estimation of copper in a brass

1.65 g of a sample of brass was reacted with concentrated nitric acid. The solution was neutralised and diluted to 250 cm³ with water.

25.0 cm³ of this solution was pipetted into a flask. An excess of aqueous potassium iodide was then added to liberate iodine:

equation 1 $2Cu^{2+}(aq) + 4I^-(aq) \longrightarrow 2CuI(s) + I_2(s)$

Titration of this solution required 21.20 cm³ of 0.100 mol dm⁻³ sodium thiosulfate.

equation 2 $2S_2O_3^{2-}(aq) + I_2(aq) \longrightarrow 2I^-(aq) + S_4O_6^{2-}(aq)$

Calculate the percentage, by mass, of copper in the brass sample.

From the titration results, the amount (in moles) of $Na_2S_2O_3$ can be calculated:

$$\text{amount of } Na_2S_2O_3 = c \times \frac{V}{1000} = 0.100 \times \frac{21.20}{1000} = 0.00212 \text{ mol}$$

From the equations, the amount (in moles) of Cu^{2+} can be determined:

From *equation 2*, 2 mol $S_2O_3^{2-}$ reacts with 1 mol I_2

From *equation 1*, 1 mol I_2 produced from 2 mol Cu^{2+}

∴ 2 mol $S_2O_3^{2-}$ ≡ 1 mol I_2 ≡ 2 mol Cu^{2+}

∴ 0.00212 mol $S_2O_3^{2-}$ ≡ 0.00212/2 mol I_2 ≡ 0.00212 mol Cu^{2+}

amount of Cu^{2+} = 0.00212 mol

The amount of Cu^{2+} of the 250 cm³ solution prepared from the brass can be deduced:

25.0 cm³ of $Cu^{2+}(aq)$ in the titration contains 0.00212 mol Cu^{2+}.

250 cm³ of the $Cu^{2+}(aq)$ solution from the brass contains 10 × 0.00212 = 0.0212 mol Cu^{2+}.

Finally, the percentage, by mass, of copper in brass can be calculated:

0.0212 mol Cu produced 0.0212 mol Cu^{2+}

mass of copper in brass = 63.5 × 0.0212 = 1.35 g

∴ percentage of copper in brass = $\frac{1.35}{1.65}$ × 100 = 81.8 %.

Left margin notes:

This is an unstructured titration calculation. See also pages 43–44 for general hints on solving this type of problem.

The concentration and volume of $Na_2S_2O_3$ are known.

Using the equations, determine the number of moles of Cu^{2+}.

Don't forget to take into account any need to scale up.

The final step is now easy provided that you have remembered:

$n = \dfrac{\text{mass}}{\text{molar mass}}$

Further examples of redox reactions

| AQA | M5 | WJEC | M5 |
| EDEXCEL | M5 | CCEA | M5 |

Vanadium

Vanadium(V) can be reduced to vanadium(II) using zinc in acid solution as the reducing agent. The formation of different oxidation states of vanadium can be followed by colour changes.

oxidation state	+2	+3	+4	+5
species	V^{2+} (violet)	V^{3+} (green)	VO^{2+} (blue)	VO_2^+ (yellow)

Zn/HCl ←

Chromium

The oxidation states of chromium can be interchanged using:

• hydrogen peroxide in alkaline solution as the oxidising agent
• zinc in acid solution as the reducing agent.

Left margin notes:

For moving **up** oxidation states H_2O_2/OH^- is a good general oxidising agent.

For moving **down** oxidation states Zn/HCl is a good general reducing agent.

oxidation state	+2	+3		+6
			$H_2O_2/OH^- \longrightarrow$	CrO_4^{2-} (yellow)
species	Cr^{2+}	Cr^{3+}		
	← Zn/HCl →			$Cr_2O_7^{2-}$ (orange)
	(blue)	(green)		

This conversion can be explained by applying le Chatelier's principle.

Addition of acid increases [H⁺(aq) – the equilibrium moves to the right, countering this change.

Added alkali reacts with H⁺(aq) ions present in the equilibrium system. [H⁺(aq) decreases – the equilibrium moves to the left, countering this change.

The dichromate(VI) – chromate(VI) conversion

The table on the previous page shows that the dichromate(VI) ion $Cr_2O_7^{2-}$ and the chromate(VI) ion CrO_4^{2-} have the **same** oxidation state. These ions can be interconverted at different pH values.

$$2H^+(aq) + 2CrO_4^{2-}(aq) \underset{\text{alkali}}{\overset{\text{acid}}{\rightleftharpoons}} Cr_2O_7^{2-}(aq) + H_2O(l)$$

yellow   orange

Cobalt

Cobalt(II) can be oxidised to Co(III) using:

- hydrogen peroxide in alkaline solution or
- air in the presence of ammonia as an oxidising agent.

oxidation state	+2		+3
species	$[Co(NH_3)_6]^{2+}$ (light straw)	H_2O_2/OH^- *or* NH_3/air $\longrightarrow$	$[Co(NH_3)_6]^{3+}$ (brown)

Disproportionation in transition metal chemistry (Edexcel only)

Copper metal has an oxidation state of 0. Copper compounds exist in the oxidation states +1 and +2. We can predict redox reactions that may take place involving these three species by treating redox systems as if they are half-cells.

$$Cu^{2+}(aq) + e^- \rightleftharpoons Cu^+(aq) \qquad E^\ominus = +0.16 \text{ V}$$
$$Cu^+(aq) + e^- \rightleftharpoons Cu(s) \qquad E^\ominus = +0.52 \text{ V}$$

- $E_{cell} = (+0.52) - (+0.16) = +0.36$ V.

The Cu^{2+}/Cu^+ system has the more negative $E^\ominus$ value and its half-reaction will proceed to the left, donating electrons:

$$Cu^{2+}(aq) + e^- \longleftarrow \mathbf{Cu^+}(aq) \qquad E^\ominus = +0.16 \text{ V}$$
$$\mathbf{Cu^+}(aq) + e^- \longrightarrow Cu(s) \qquad E^\ominus = +0.52 \text{ V}$$

Notice that a reaction takes place between reactants from different sides of each half-equation.

We would therefore predict that **Cu⁺** would react with **Cu⁺** and that this **disproportionation** will take place:

$$2Cu^+ \longrightarrow Cu^{2+} + Cu$$

+1 $\longrightarrow$ 0	*reduction*	$Cu^+ + e^- \longrightarrow Cu$
+1 $\longrightarrow$ +2	*oxidation*	$Cu^+ \longrightarrow Cu^{2+} + e^-$

This disproportionation reaction can be predicted from electrode potentials.

Progress check

1. Write a full equation from the following pairs of half-equations. For each pair, identify the changes in oxidation number, what has been oxidised and what has been reduced.
   (a) $Zn \longrightarrow Zn^{2+} + 2e^-$
   $VO_2^+ + 4H^+ + 3e^- \longrightarrow V^{2+} + 2H_2O$
   (b) $Cr^{3+} + 8OH^- \longrightarrow CrO_4^{2-} + 4H_2O + 3e^-$
   $H_2O_2 + 2e^- \longrightarrow 2OH^-$

2. In a redox titration, 25.0 cm³ of an acidified solution containing $Fe^{2+}(aq)$ ions reacted exactly with 21.8 cm³ of 0.0200 mol dm⁻³ potassium manganate(VII). Calculate the concentration of $Fe^{2+}(aq)$ ions.

2 0.0872 mol dm⁻³

Cr: +3 → +6; O: –1 → –2; Cr^{3+} oxidised; H_2O_2 reduced.
(b) $2Cr^{3+} + 3H_2O_2 + 10OH^- \longrightarrow 2CrO_4^{2-} + 8H_2O$
V: +5 → +2; Zn: 0 → +2; VO_2^+ reduced; Zn oxidised.
1 (a) $2VO_2^+ + 3Zn + 8H^+ \longrightarrow 2V^{2+} + 3Zn^{2+} + 4H_2O$

3.7 Catalysis

Key points from AS

- **How do catalysts work?**
  Revise AS page 108

- **Heterogeneous and homogeneous catalysis**
  Revise AS pages 108–109

The two different classes of catalyst, homogeneous and heterogeneous, were discussed in detail during AS Chemistry. Both types are used industrially but heterogeneous catalysis is much more common. The AS coverage is summarised below and expanded to include extra examples of transition element catalysis.

Transition elements as catalysts

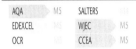

AQA	▸ M5	SALTERS	
EDEXCEL		WJEC	▸ M5
OCR		CCEA	▸ M5

An important use of transition metals and their compounds is as catalysts for many industrial processes. Nickel and platinum are extensively used in the petroleum and polymer industries.

> Transition metal ions are able to act as catalysts by changing their oxidation states. This is made possible by using the partially full d-orbitals for gaining or losing electrons.
>
> KEY POINT

Heterogeneous catalysis

AQA	▸ M5	SALTERS	
EDEXCEL		WJEC	▸ M5
		CCEA	▸ M5

Key points from AS

- **Heterogeneous and homogeneous catalysis**
  Revise AS pages 108–109

Transition metals use d and s-electrons to form weak bonds with reactant molecules at the catalyst surface.

A **heterog**eneous catalyst has a **different** phase from the reactants.

- Many examples of heterogeneous catalysts involve reactions between **gases**, catalysed by a **solid** catalyst which is often a transition metal or one of its compounds.

The process involves:

- **diffusion** of gas molecules onto the surface of the iron catalyst
- **adsorption** of the gases to the surface of the catalyst
- **weakening of bonds**, allowing a chemical reaction to take place
- **diffusion** of the product molecules from the surface of the catalyst, allowing more gas molecules to diffuse onto its surface.

Transition metal ions as heterogeneous catalysts

The catalytic converter

- Rh/Pt/Pd catalysts are used in catalytic converters for the removal of polluting gases, such as CO and NO, produced in a car engine.
- The **solid** Rh/Pt/Pd catalyst is held on a ceramic honeycomb support, giving a large surface area for the catalyst which can be spread extremely thinly on the support. The high surface area of the catalyst increases the chances of a reaction and less of the expensive catalysts are needed, keeping down costs.

Iron

Iron-containing catalysts are used in many industrial processes, the most important being to form ammonia from nitrogen and hydrogen in the Haber process.

$$N_2(g) + 3H_2(g) \rightleftharpoons 2NH_3(g)$$

Chromium(III) oxide

Chromium(III) oxide, Cr_2O_3, is used to catalyse the manufacture of methanol from carbon monoxide and hydrogen.

$$CO(g) + 2H_2(g) \rightleftharpoons CH_3OH(g)$$

Vanadium

A vanadium(V) oxide, V_2O_5, catalyst is used in the contact process for sulfur trioxide production from which sulfuric acid is manufactured.

This catalysis proceeds via an **intermediate state**.

- The vanadium(V) oxide catalyst first oxidises sulfur dioxide to sulfur trioxide. The vanadium is **reduced** from the +5 to +4 oxidation state, forming vanadium(IV) oxide $V_2O_4(s)$ as the **intermediate state**.

$$SO_2(g) + V_2O_5(s) \longrightarrow SO_3(g) + V_2O_4(s) \qquad \textit{vanadium reduced}$$

+4 $\longrightarrow$ +6

+5 $\longrightarrow$ +4
+5 $\longrightarrow$ +4

Notice that the oxidation number of each atom has been shown. Two V atoms have been reduced from +5 to +4.

- The intermediate V_2O_4 is then oxidised back to the +5 oxidation state by oxygen, forming vanadium(V) oxide.

$$\tfrac{1}{2}O_2(g) + V_2O_4(s) \longrightarrow V_2O_5(s) \qquad \textit{vanadium oxidised}$$

0 $\longrightarrow$ −2

+4 $\longrightarrow$ +5
+4 $\longrightarrow$ +5

This is a chain reaction because the vanadium(V) oxide is then able to oxidise further sulfur dioxide.

Note:

- Although V_2O_5 has been involved in the mechanism, the overall reaction does not include V_2O_5. V_2O_5 is unchanged at the end of the reaction.
- The equation for the overall reaction taking place is:
  $$SO_2(g) + \tfrac{1}{2}O_2(g) \rightleftharpoons SO_3(g) \qquad \textit{The contact process}$$
- The vanadium is able to catalyse the reaction by interchanging its +5 and +4 oxidation states.

Poisoning of catalysts (AQA only)

A key requirement of a heterogeneous catalyst is to bind reactant molecules to the surface of the catalyst by adsorption.

The strength of this adsorption is critical.

Catalysts can become coated by impurities and can become 'poisoned' – efficiency is reduced as active sites are blocked.

- Lead bonds very strongly to the catalyst. The adsorption is so strong that molecules are unable to leave the surface and the catalyst is poisoned and rendered useless. Before catalytic converters could be introduced, unleaded petrol was developed. Otherwise, the lead in the fuel would poison the catalyst.

Nickel and platinum are commonly used as heterogeneous catalysts.

- The bonding of reactant molecules is sufficient to allow a reaction to take place but weak enough to allow molecules to escape following reaction.

Homogeneous catalysis

AQA   MS   SALTERS

Key points from AS

- **How catalysts work**
  Revise AS page 108
- **Homogeneous catalysis**
  Revise AS page 109

A **homo**geneous catalyst has the **same** phase as the reactants.

- During homogeneous catalysis, the catalyst initially reacts with the reactants forming an **intermediate state** with **lower activation energy**.

- The catalyst is then regenerated as the reaction completes, allowing the catalyst to react with further reactants.

- Overall the catalyst is **not** used up, but it is actively involved in the process by a chain reaction.

Transition metal ions as homogeneous catalysts

The reaction between I^- and $S_2O_8^{2-}$, catalysed by $Fe^{2+}(aq)$ ions

The reaction between I^- and $S_2O_8^{2-}$ is very slow:

- it can be catalysed by Fe^{2+} ions
- the Fe^{2+} ions are recycled in a chain reaction.

Transition metal ions often change their oxidation states during their catalytic action.

Here: $Fe^{2+} \longrightarrow Fe^{3+}$

chain reaction: $\quad S_2O_8^{2-}(aq) + 2Fe^{2+}(aq) \longrightarrow 2SO_4^{2-}(aq) + 2Fe^{3+}(aq)$

$\quad\quad\quad\quad\quad\quad\quad\quad 2Fe^{3+}(aq) + 2I^-(aq) \longrightarrow I_2(aq) + 2Fe^{2+}(aq)$

overall: $\quad\quad\quad\quad\quad\quad S_2O_8^{2-}(aq) + 2I^-(aq) \longrightarrow I_2(aq) + 2SO_4^{2-}(aq)$

The overall reaction between I^- and $S_2O_8^{2-}$ involves two negative ions. These negative ions will repel one another, making it difficult for a reaction to take place. Notice that, in each stage of the mechanism, the catalyst provides positive ions which attract, and react with, each negative ion in turn.

Autocatalysis by $Mn^{2+}(aq)$ ions

Notice that, as in the first example, this reaction is between two negative ions.

The mechanism of this reaction proceeds in a similar manner – the catalyst provides positive ions which react with each negative ion.

In warm acidic conditions, manganate(VII) ions, MnO_4^-, oxidise ethanedioate ions, $C_2O_4^{2-}$, to carbon dioxide.

$$2MnO_4^-(aq) + 5C_2O_4^{2-}(aq) + 8H^+(aq) \rightarrow 2Mn^{2+}(aq) + 10CO_2(g) + H_2O(l)$$

The purple manganate(VII) colour disappears very slowly at first, indicating a slow reaction. As soon as $Mn^{2+}(aq)$ ions start to form, the colour disappears immediately because the $Mn^{2+}(aq)$ ions catalyse the reaction. This is called **autocatalysis**.

The development of new catalysts

EDEXCEL   M5

The development of new catalysts is a priority area for chemical research today.

- Ethanoic acid, CH_3COOH, used to be prepared by oxidation of butane. The preparation had a poor percentage yield and was wasteful in terms of energy and atom economy.

- Research into a new synthesis of ethanoic acid, using new catalysts, has resulted in the development of a new synthesis, using methanol and carbon monoxide.

New catalysts provide bigger profits for companies, better use of available raw materials and less pollution to the environment.

$$CH_3OH(l) + CO(g) \longrightarrow CH_3COOH$$

- An iridium-based catalyst is now used that produces near to 100% atom economy.

New catalysts lead to more efficient processes in which products can be made quicker, with less energy, better yields and improved atom economies.

Progress check

1 (a) State what is meant by homogeneous and heterogeneous catalysis.
  (b) State an example of a homogeneous and heterogeneous catalyst.

2 How does a transition metal act as a catalyst?

2 Transition metal changes oxidation states by gaining or losing electrons from partially filled d-orbitals.
Heterogeneous: iron catalysing the reaction between N_2 and H_2 for ammonia production (Haber process).
(b) Homogeneous: Fe^{2+} catalysing the reaction between I^- and $S_2O_8^{2-}$.
Heterogeneous: catalyst and reactants have different phases.
1 (a) Homogeneous: catalyst and reactants have the same phase.

3.8 Reactions of metal aqua-ions

After studying this section you should be able to:

- *describe the precipitation reactions of metal aqua-ions with bases*
- *describe the acidity of transition metal aqua-ions*
- *describe the acid–base behaviour of metal hydroxides*

Simple precipitation reactions of metal aqua-ions

AQA	M5	SALTERS	M4
EDEXCEL	M5	WJEC	M5
OCR	M5	CCEA	M5

Notice that NaOH(aq) and NH_3(aq) **both** act as bases.

A precipitation reaction takes place between **aqueous alkali** and an aqueous solution of a **metal(II)** or **metal(III) cation**.

This results in the formation of a **precipitate** of the **metal hydroxide**, often with a characteristic colour.

Suitable aqueous alkalis include aqueous sodium hydroxide, NaOH(aq), and aqueous ammonia, NH_3(aq).

Aqueous ammonia provides OH^-(aq) ions.
$$NH_3(aq) + H_2O(l) \rightleftharpoons NH_4^+(aq) + OH^-(aq)$$

These precipitation reactions can be represented simply as follows.

$$Cu^{2+}(aq) + 2OH^-(aq) \longrightarrow Cu(OH)_2(s)$$
$$Cr^{3+}(aq) + 3OH^-(aq) \longrightarrow Cr(OH)_3(s)$$

The hydroxides of all transition metals are insoluble in water.

The characteristic colour of the precipitate can help to identify the metal ion. The colours of some hydroxide precipitates are shown below.

hydroxide	$Fe(OH)_2(s)$	$Fe(OH)_3(s)$	$Co(OH)_2(s)$	$Cu(OH)_2(s)$	$Cr(OH)_3(s)$
colour	green	brown	blue–green	blue	green

See also:
'Reactions of metal aqua-ions with aqueous bases' page 105.

The precipitate has no charge.

AQA and Edexcel prefer to show hydroxide precipitates based on six ligands, e.g. $Cu(OH)_2(H_2O)_4(s)$ – but most Boards will credit the simplified formulae such as $Cu(OH)_2(s)$.

Species in solution are charged.

Reaction of complex metal aqua-ions with aqueous alkali

Equations can be written using complex aqua-ions for the reactions above, each producing a **precipitate** of the **hydrated hydroxide**.

$$[Cu(H_2O)_6]^{2+}(aq) + 2OH^-(aq) \longrightarrow Cu(OH)_2(H_2O)_4(s) + 2H_2O(l)$$
blue precipitate

$$[Cr(H_2O)_6]^{3+}(aq) + 3OH^-(aq) \longrightarrow Cr(OH)_3(H_2O)_3(s) + 3H_2O(l)$$
green precipitate

Excess aqueous sodium hydroxide

Metal(III) hydroxides **dissolve** in an **excess** of dilute aqueous sodium hydroxide forming charged **anions**.

$$Al(OH)_3(H_2O)_3(s) + OH^-(aq) \rightleftharpoons [Al(OH)_4]^-(aq) + 3H_2O(l)$$
aluminate ions

$$Cr(OH)_3(H_2O)_3(s) + 3OH^-(aq) \rightleftharpoons [Cr(OH)_6]^{3-}(aq) + 3H_2O(l)$$
chromate(III) ions

Excess aqueous ammonia

Excess ammonia usually results in **ligand exchange** forming a soluble ammine complex:

$$Cr(H_2O)_3(OH)_3(s) + 6NH_3(aq) \longrightarrow [Cr(NH_3)_6]^{3+}(aq) + 3H_2O(l) + 3OH^-(aq)$$
purple solution

Salters prefer to use $[Cu(NH_3)_4]^{2+}$ for the complex formed between Cu^{2+} and NH_3.

If aqueous ammonia is slowly added to an aqueous solution of a **metal(II)** or **metal(III) cation**:

- a precipitate of the metal hydroxide first forms: **acid–base reaction**
- the precipitate then dissolves in excess ammonia: **ligand substitution**.

Precipitation as acid–base equilibria

AQA   MS   CCEA   MS
EDEXCEL   MS

Fission of O–H bond

Metal(II) aqua-ions are weaker acids than metal(III) aqua-ions.

Examples of metal aqua-ions with 3+ cations include $[Al(H_2O)_6]^{3+}$, $[V(H_2O)_6]^{3+}$, $[Cr(H_2O)_6]^{3+}$ and $[Fe(H_2O)_6]^{3+}$.

Acidity of metal aqua-ions

Metal(II) and metal(III) aqua-ions behave as weak acids. This is achieved by:

- fission of the O–H bond in one of the water ligands of the complex ion
- donation of a proton to a water molecule.

Metal(II) cations

- With a metal(II) aqua-ion, the equilibrium lies well to the left-hand side: only a very weak acid is formed.

$$[Cu(H_2O)_6]^{2+}(aq) + H_2O(l) \rightleftharpoons [Cu(H_2O)_5(OH)]^+(aq) + H_3O^+(l)$$
very weakly acidic

Metal(III) cations

- Metal(III) aqua-ions are slightly more acidic than metal(II) aqua-ions: the equilibrium position has moved slightly to the right.

$$[Fe(H_2O)_6]^{3+}(aq) + H_2O(l) \rightleftharpoons [Fe(H_2O)_5(OH)]^{2+}(aq) + H_3O^+(aq)$$
weakly acidic

> The aqua-ions of metal(III) cations have a greater charge/size ratio and greater acidity than aqua-ions of metal(II) cations. **KEY POINT**

Precipitation

The precipitation reaction of a transition metal with hydroxide ions can be expressed as a series of acid–base equilibria.

The equilibrium set up in water by the hexaaqua ion of iron(III) is:

$$[Fe(H_2O)_6]^{3+}(aq) + H_2O(l) \rightleftharpoons [Fe(H_2O)_5(OH)]^{2+}(aq) + H_3O^+(aq)$$

OH^- reacts with H_3O^+:

$OH^- + H_3O^+ \longrightarrow 2H_2O$

With the weak alkali $Na_2CO_3(aq)$, metal(II) aqua-ions precipitate metal **carbonates**.

The more acidic metal(III) aqua-ions precipitate metal **hydroxides**.

Metal(III) carbonates do not exist.

On addition of aqueous hydroxide ions:
- H_3O^+ ions are removed from the equilibrium, which shifts to the right.

The $[Fe(H_2O)_5(OH)]^{2+}$ ion now sets up a second equilibrium:

$$[Fe(H_2O)_5(OH)]^{2+}(aq) + H_2O(l) \rightleftharpoons [Fe(H_2O)_4(OH)_2]^+(aq) + H_3O^+(aq)$$

- Again, the hydroxide ion removes H_3O^+ ions from the aqueous equilibrium, which shifts to the right.
- This process continues. Eventually **all charge** will be **removed** from the iron(III) complex. Then the uncharged **hydrated metal hydroxide** precipitates.

$$[Fe(H_2O)_4(OH)_2]^+(aq) + H_2O(l) \rightleftharpoons [Fe(H_2O)_3(OH)_3](s) + H_3O^+(aq)$$

Acid–base properties of metal hydroxides

AQA   MS   WJEC   MS
EDEXCEL   MS   CCEA   MS

Many insoluble metal hydroxides are **amphoteric** – they can act as both an acid and a base.

Metal hydroxides as acids

See also:
'Reaction of complex metal aqua-ions with aqueous alkali' page 103.

Some insoluble metal hydroxides can act as acids, donating protons to strong alkalis and forming soluble anions.

- Metal(III) hydroxides react with dilute aqueous sodium hydroxide.

$$Al(OH)_3(H_2O)_3(s) + OH^-(aq) \rightleftharpoons [Al(OH)_4(H_2O)_2]^-(aq) + H_2O(l)$$
aluminate ions

$$Cr(OH)_3(H_2O)_3(s) + 3OH^-(aq) \rightleftharpoons [Cr(OH)_6]^{3-}(aq) + 3H_2O(l)$$
chromate(III) ions

In general, metal(III) hydroxides form anions with the general formula $[M(OH)_6]^{3-}$.

• Metal(II) hydroxides are weaker acids and do not usually react with aqueous sodium hydroxide.

Metal hydroxides as bases (AQA and Edexcel only)

All metal hydroxides can act as bases, removing protons from strong acids.

This reaction reverses the direction of the acid-base equilibria that produce hydrated hydroxide precipitates (see page 104).

$$Cu(OH)_2(H_2O)_4(s) + 2H^+(aq) \rightleftharpoons [Cu(H_2O)_6]^{2+}(aq)$$
blue precipitate

$$Cr(OH)_3(H_2O)_3(s) + 3H^+(aq) \rightleftharpoons [Cr(H_2O)_6]^{3+}(aq)$$
green precipitate

Reactions of metal aqua-ions with aqueous bases

AQA	M5	SALTERS	
EDEXCEL	M5	WJEC	M5
OCR		CCEA	M5

Make sure that you check carefully with your specification here. The Boards themselves do not always agree. The reactions are complex and different simplifications have been made by different Boards.

In exams, a great emphasis is placed on these reactions. You will need to learn the colours of all solutions and precipitates in your course.

Reactions with OH⁻(aq) and NH₃(aq)

You are expected to know reactions of metal aqua ions with aqueous OH⁻(aq) ions (probably as NaOH(aq)) and with aqueous ammonia, NH₃(aq).

The tables below summarise what you are expected to know for each Board so make sure that you do not learn more than you have to!

	AQ	ED	OC	SA	WJ	CC
Fe^{2+}	✓	✓	✓	✓	✓	✓
Co^{2+}	✓		✓			✓
Cu^{2+}	✓	✓	✓	✓	✓	✓
Mn^{2+}	✓					✓
Ni^{2+}	✓					✓
Zn^{2+}	✓					
Cr^{3+}	✓	✓			✓	✓
Fe^{3+}	✓	✓	✓	✓	✓	✓
Al^{3+}	✓					

	OH⁻	OH⁻ excess	NH₃	NH₃ excess
AQA	✓	✓	✓	✓
EDEXCEL	✓	✓	✓	✓
OCR	✓		✓	✓
SALTERS	✓		✓	✓
WJEC	✓		✓	✓
CCEA	✓	✓	✓	✓

aqua ion	OH⁻(aq) or NH₃(aq) as a base		NH₃(aq) as a ligand
	NaOH(aq) or dilute NH₃(aq) (small amount)	excess NaOH(aq)	excess/conc. NH₃(aq)
$[Fe(H_2O)_6]^{2+}$ green	$Fe(OH)_2(s)$ / $Fe(OH)_2(H_2O)_4(s)$ green	no change	no change
$[Co(H_2O)_6]^{2+}$ pink	$Co(OH)_2(s)$ / $Co(OH)_2(H_2O)_4(s)$ blue–green	no change	$[Co(NH_3)_6]^{2+}$ pale straw
$[Cu(H_2O)_6]^{2+}$ blue	$Cu(OH)_2(s)$ / $Cu(OH)_2(H_2O)_4(s)$ pale blue	no change	$[Cu(NH_3)_4(H_2O)_2]^{2+}$ deep blue
$[Mn(H_2O)_6]^{2+}$ pale pink	$Mn(OH)_2(s)$ / $Mn(OH)_2(H_2O)_4(s)$ pale brown	no change	no change
$[Ni(H_2O)_6]^{2+}$ green	$Ni(OH)_2(s)$ / $Ni(OH)_2(H_2O)_4(s)$ pale green	no change	$[Ni(NH_3)_6]^{2+}$ blue
$[Zn(H_2O)_6]^{2+}$ colourless	$Zn(OH)_2(s)$ / $Zn(OH)_2(H_2O)_4(s)$ white	$[Zn(OH)_4]^{2-}(aq)$ colourless	$[Zn(NH_3)_4]^{2+}$ colourless
$[Cr(H_2O)_6]^{3+}$ ruby	$Cr(OH)_3(s)$ / $Cr(OH)_3(H_2O)_3(s)$ green	$[Cr(OH)_6]^{3-}(aq)$ green	$[Cr(NH_3)_6]^{3+}(aq)$ purple
$[Fe(H_2O)_6]^{3+}$ red-brown	$Fe(OH)_3(s)$ / $Fe(OH)_3(H_2O)_3(s)$ brown	no change	no change
$[Al(H_2O)_6]^{3+}$ colourless	$Al(OH)_3(s)$ / $Al(OH)_3(H_2O)_3(s)$ white	$[Al(OH)_4(H_2O)_2]^-(aq)$ colourless	no change

- Of the 2+ ions, only $Zn(OH)_2(s)$ reacts with excess $OH^-(aq)$.
- $Fe(OH)_2(s)$ is oxidised in air forming a brown precipitate of $Fe(OH)_3(s)$.
- $[Co(NH_3)_6]^{2+}$ darkens in air as it is oxidised to $[Co(NH_3)_6]^{3+}$ (see page 99).
- $Mn(OH_2)(s)$ forms a dark-brown precipitate of manganese(III) oxide, $Mn_2O_3(s)$.

Reactions of metal aqua-ions with aqueous carbonate ions (AQA only)

Metal(II) ions

Metal(II) ions form a precipitate of the metal carbonate. The chemistry is complex and the equations here are acceptable simplifications.

$$Fe^{2+}(aq) + CO_3^{2-}(aq) \longrightarrow FeCO_3(s)$$
$$green$$
$$Co^{2+}(aq) + CO_3^{2-}(aq) \longrightarrow CoCO_3(s)$$
$$pink$$
$$Cu^{2+}(aq) + CO_3^{2-}(aq) \longrightarrow CuCO_3(s)$$
$$green$$

Metal(III) ions

Metal(III) aqua-ions behave as weak acids (see page 104).

A hydroxide precipitate is formed together with CO_2 gas.

$$[Cr(H_2O)_6]^{3+} + 1\tfrac{1}{2}CO_3^{2-}(aq) \longrightarrow Cr(OH)_3(H_2O)_3(s) + 1\tfrac{1}{2}CO_2(g) + 1\tfrac{1}{2}H_2O(l)$$
$$green$$
$$[Fe(H_2O)_6]^{3+} + 1\tfrac{1}{2}CO_3^{2-}(aq) \longrightarrow Fe(OH)_3(H_2O)_3(s) + 1\tfrac{1}{2}CO_2(g) + 1\tfrac{1}{2}H_2O(l)$$
$$brown$$
$$[Al(H_2O)_6]^{3+} + 1\tfrac{1}{2}CO_3^{2-}(aq) \longrightarrow Al(OH)_3(H_2O)_3(s) + 1\tfrac{1}{2}CO_2(g) + 1\tfrac{1}{2}H_2O(l)$$
$$white$$

- The acidity of the metal(III) ion means that this reaction is essentially the same as that between an acid such as hydrochloric acid and a carbonate.
- Consequently, metal(III) carbonates do not exist.

Progress check

1 Write equations for the following reactions of metal aqua-ions with aqueous alkali. What is the colour of each precipitate?

(a) $Co^{2+}(aq)$ (b) $Fe^{2+}(aq)$ (c) $[Ni(H_2O)_6]^{2+}$ (d) $[Fe(H_2O)_6]^{3+}$.

2 Write equations to show what happens when excess $NH_3(aq)$ is added to the following hydrated metal hydroxides.

(a) $Co(OH)_2(H_2O)_4$ (b) $Cu(OH)_2(H_2O)_4$ (c) $Cr(OH)_3(H_2O)_3$.

2 (a) $Co(OH)_2(H_2O)_4(s) + 6NH_3(aq) \longrightarrow [Co(NH_3)_6]^{2+}(aq) + 4H_2O(l) + 2OH^-(aq)$ (pale straw)
(b) $Cu(OH)_2(H_2O)_4(s) + 4NH_3(aq) \longrightarrow [Cu(NH_3)_4(H_2O)_2]^{2+}(aq) + 2H_2O(l) + 2OH^-(aq)$ (deep-blue)
(c) $Cr(OH)_3(H_2O)_3(s) + 6NH_3(aq) \longrightarrow [Cr(NH_3)_6]^{3+}(aq) + 3H_2O(l) + 3OH^-(aq)$ (purple)

1 (a) $Co^{2+}(aq) + 2OH^-(aq) \longrightarrow Co(OH)_2(s)$ (blue–green)
(b) $Fe^{2+}(aq) + 2OH^-(aq) \longrightarrow Fe(OH)_2(s)$ (green)
(c) $[Ni(H_2O)_6]^{2+}(aq) + 2OH^-(aq) \longrightarrow Ni(OH)_2(H_2O)_4(s) + 2H_2O(l)$ (pale green)
(d) $[Fe(H_2O)_6]^{3+}(aq) + 3OH^-(aq) \longrightarrow Fe(OH)_3(H_2O)_3(s) + 3H_2O(l)$ (brown)

Sample question and model answer

This question looks at different transition elements.

(a) Write the electron configuration of a vanadium atom and a cobalt(III) ion.

V: $1s^2 2s^2 2p^6 3s^2 3p^6 3d^3 4s^2$ ✓
Co^{3+}: $1s^2 2s^2 2p^6 3s^2 3p^6 3d^6$ ✓ [2]

(b) When concentrated hydrochloric acid is added to an aqueous solution containing copper(II) ions, the colour changes. Write the formulae of the two copper complexes involved, their colours, their shapes and the co-ordination numbers of the copper. Write an equation for the reaction and explain why the addition of dilute hydrochloric acid does not produce the same colour change.

This is standard bookwork but it also reinforces important principles:

H_2O forms octahedral complexes with a coordination number of 6.

Cl^- forms tetrahedral complexes with a coordination number of 4.

Aqueous copper(II) contains the complex $[Cu(H_2O)_6]^{2+}$ ✓ which is pale blue. ✓ It has an octahedral shape ✓ and a coordination number of 6. ✓

After addition of concentrated hydrochloric acid, the complex ion $[CuCl_4]^{2-}$ ✓ is formed, which is yellow. ✓ It has a tetrahedral shape ✓ and a coordination number of 4. ✓

Equation: $[Cu(H_2O)_6]^{2+} + 4Cl^- \longrightarrow CuCl_4^{2-} + 6H_2O$ ✓✓

The concentration of chloride ligands in **dilute** hydrochloric acid is much smaller than the concentration of water ligands. ✓ [11]

(c) A 0.626 g sample of hydrated iron(II) sulfate was dissolved in water and the solution made up to 250 cm³. A 25.0 cm³ sample of this solution was acidified with dilute sulfuric acid and titrated with potassium dichromate(VI). 25.85 cm³ of 0.00145 mol dm⁻³ potassium dichromate(VI) were required for exact reaction.

The half-equation for the reduction of acidified dichromate ions is shown below.
$$Cr_2O_7^{2-}(aq) + 14H^+(aq) + 6e^- \longrightarrow 2Cr^{3+}(aq) + 7H_2O(l)$$

Learn how to combine redox half-equations. Although this titration is different from the permanganate titrations that you have carried out, the same principles are used.

Even if your equation is wrong, you can still score in the next part provided that your method is built upon sound chemical principles.

Show your working!

(i) Write the half-equation for the oxidation of Fe^{2+} and construct the equation for the overall reaction with acidified dichromate (VI) ions.

$Fe^{2+}(aq) \longrightarrow Fe^{3+}(aq) + e^-$ ✓

$6Fe^{2+}(aq) + Cr_2O_7^{2-}(aq) + 14H^+(aq) \longrightarrow 6Fe^{3+}(aq) + 2Cr^{3+}(aq) + 7H_2O(l)$ ✓

(ii) Using the information above, calculate the molar mass of the hydrated iron(II) sulfate and determine the number of moles of water of crystallisation present in one mole of hydrated iron(II) sulfate.

amount of $Cr_2O_7^{2-} = c \times \dfrac{V}{1000} = 0.00145 \times \dfrac{25.85}{1000} = 3.75 \times 10^{-5}$ mol ✓

From the equation, the amount of $Fe^{2+} = 6 \times$ amount of $Cr_2O_7^{2-}$. ✓

∴ amount of Fe^{2+} that reacted $= 6 \times 3.75 \times 10^{-5}$ mol $= 2.25 \times 10^{-4}$ mol ✓

In 250 cm³ solution, amount of $Fe^{2+} = 10 \times 2.25 \times 10^{-4} = 2.25 \times 10^{-3}$ mol ✓

Don't forget the scaling stage. The original solution was 250 cm³ but you have only taken 25 cm³ for the titration.

Show your working!

∴ $\dfrac{0.626}{2.25 \times 10^{-3}} = 278$ g of hydrated iron(II) sulfate contains 1 mol Fe^{2+} ✓

M of hydrated iron(II) sulfate $= 278$ g mol⁻¹ ✓

M of $FeSO_4 = 55.8 + 32.1 + 16.0 \times 4 = 151.9$ g mol⁻¹ ✓

Number of waters of crystallisation $= \dfrac{278.0 - 151.9}{18.0} = \dfrac{126.1}{18.0} = 7$ ✓

(iii) Explain how the volume of potassium dichromate(VII) solution needed in the titration would be different if the solution of iron(II) sulfate had been left for some time before being titrated.

Less dichromate(VI) solution would have been needed ✓ because some of the Fe^{2+} would have been oxidised by air to Fe^{3+} ✓ [12]

[Total: 25]

Practice examination questions

1 (a) Complete the following table for the Period 3 oxides.

	Mg	Al	Si	P	S
Oxidation number	+2	+3	+4	+5	+6
Formula of oxide					

[2]

(b) Explain the trend in oxidation number. [1]

(c) Write equations, if any, to show the effect of water on these oxides. [4]

(d) What is the relationship between the structure and bonding of the Period 3 oxides and the pH of any resulting solution? [3]

[Total: 10]

2 (a) Write the electron configuration of a nickel atom and a Ni^{2+} ion. [2]

(b) Nickel is both a d-block element and a transition element. State what is meant by each of the terms. [2]

(c) In aqueous solution, nickel sulfate, forms a six-coordinate complex ion which gives a green colour to the solution.

(i) What feature of the water molecule allows it to form a complex ion with Ni^{2+}?

(ii) Write the formula of the nickel complex ion formed in aqueous solution.

(iii) What types of bond are present in this nickel complex ion?

(iv) Suggest the shape of, and bond angles in, this complex ion. [6]

(d) Consider the following reactions.

(i) Write the formula of the green precipitate, **A**, and write an ionic equation for its formation above.

(ii) Predict the formula of **B**.

(iii) Write an equation for the formation of $[Ni(NH_3)_6]^{2+}$ from the aqueous nickel complex ion and suggest the type of reaction taking place. [5]

[Total: 15]

Practice examination questions *(continued)*

3 (a) Deduce the oxidation state of the transition metal in each of the following species.

(i) Cr in CrO_4^{2-}

(ii) Cu in $[CuCl_2]$

(iii) V in $[VO(H_2O)_5]^{2+}$ [3]

(b) Chromium(III) can be oxidised in alkaline conditions by hydrogen peroxide, H_2O_2. Use the half-equations below to construct the overall equation for this reaction.

$$Cr^{3+} + 8OH^- \longrightarrow CrO_4^{2-} + 4H_2O + 3e^-$$

$$H_2O_2 + 2e^- \longrightarrow 2OH^-$$ [1]

(c) Give two examples of the use of transition elements as catalysts in industrial processes. [2]

[Total: 6]

4 The concentration of an aqueous solution of hydrogen peroxide can be determined by titration. Aqueous potassium manganate(VII), $KMnO_4$, is titrated against a solution of hydrogen peroxide in the presence of acid.

The half-equations are shown below.

$$H_2O_2(aq) \longrightarrow 2H^+(aq) + O_2(g) + 2e^-$$

$$MnO_4^-(aq) + 8H^+(aq) + 5e^- \longrightarrow Mn^{2+}(aq) + 4H_2O(l)$$

(a) (i) Deduce the oxidation state of the manganese in the MnO_4^- ion.

(ii) Construct the equation for the reaction between H_2O_2, MnO_4^- ions and H^+ ions. [2]

(b) State the role of hydrogen peroxide in this reaction. [1]

(c) After acidification, 25.0 cm³ of a dilute aqueous solution of hydrogen peroxide reacted exactly with 23.80 cm³ of 0.0150 mol dm⁻³ $KMnO_4$. Calculate the concentration, in mol dm⁻³, of hydrogen peroxide in the solution. [3]

[Total: 6]

5 Cobalt and nickel both form complex ions with the ligand ethane-1,2-diamine, $NH_2CH_2CH_2NH_2$.

(a) (i) Explain how ligands bond to a metal ion in a complex ion.

(ii) What name is given to this type of ligand? [3]

(b) Cobalt forms the complex compound, $[Co(NH_2CH_2CH_2NH_2)_3]Cl_3$.

(i) What is the oxidation state of cobalt in this compound?

(ii) What is the electron configuration of cobalt in this compound? [2]

(c) In solution, nickel forms the complex ion $[Ni(NH_2CH_2CH_2NH_2)_3]^{2+}$.

(i) Deduce the co-ordination number of nickel in this complex ion.

(ii) State the shape around nickel in this complex ion.

(iii) Outline how you could prepare a solution containing the complex ion $[Ni(NH_2CH_2CH_2NH_2)_3]^{2+}$ from an aqueous solution of nickel(II) chloride. Include an equation in your answer. [4]

[Total: 9]

Chapter 4
Chemistry of organic functional groups

The following topics are covered in this chapter:

- Isomerism and functional groups
- Aldehydes and ketones
- Carboxylic acids
- Esters
- Acylation

- Arenes
- Reactions of arenes
- Phenols and alcohols
- Amines

4.1 Isomerism and functional groups

After studying this section you should be able to:

- understand what is meant by structural isomerism and stereoisomerism
- explain the term chiral centre and identify any chiral centres in a molecule of given structural formula
- understand that many natural compounds are present as one optical isomer only
- understand that many pharmaceuticals are chiral drugs
- recognise common functional groups

LEARNING SUMMARY

Key points from AS

- **Basic concepts**
  Revise AS pages 119–124

During AS Chemistry, you learnt about the basic concepts used in organic chemistry. You should revise these thoroughly before you start the A2 part of this course.

Isomerism

AQA	M4	SALTERS	M4, M5
EDEXCEL	M4	WJEC	M4
OCR	M4	CCEA	M4

Structural and E/Z isomerism were introduced during AS Chemistry. These are reviewed below and E/Z isomerism is discussed in the wider context of **stereoisomerism**.

Key points from AS

- **Structural isomerism**
  Revise AS pages 121–122
- **E/Z isomerism**
  Revise AS pages 129–130

Structural isomerism

Structural isomers are molecules with the same molecular formula but with different structural arrangements of atoms (structural formulae).

Two structural isomers of $C_2H_4O_2$ are shown below:

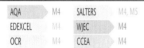

Stereoisomerism

Stereoisomers are molecules with the same structural formula but their atoms have different positions in space.

There are two types of stereoisomerism, each arising from a different structural feature:

- E/Z isomerism about a **C=C** double bond
- **optical** isomerism about a **chiral** carbon centre.

E/Z isomerism

E/Z isomerism occurs in molecules with:

- a C=C double bond and
- two **different** groups attached to **each** carbon in the C=C bond.

The double bond prevents rotation.

E.g. *E/Z* Isomers of 1,2-dichloroethene, ClCH=CHCl.

cis-isomer
(groups on one side)
Z isomer

trans-isomer
(groups on opposite sides)
E isomer

Optical isomers

> Optical isomerism occurs in the molecules of a compound with a **chiral** (or *asymmetric*) carbon atom. A chiral carbon atom has **four** different groups attached to it.
>
> **KEY POINT**

E.g. the optical isomers of an amino acid, $RCHNH_2COOH$ (see pages 148–151).

chiral (or asymmetric)
carbon atom
– 4 different
groups attached

mirror plane

Optical isomers:

- are non-superimposable mirror images of one another
- rotate plane-polarised light in opposite directions
- are chemically identical.

Chirality and drug synthesis

A synthetic amino acid, made in the laboratory, is optically inactive and does not rotate plane-polarised light:

- it contains equal amounts of each optical isomer – a **racemic** mixture or **racemate**.

A natural amino acid, made by living systems, is optically active:

- it contains only one of the optical isomers.

The difference between the optical isomers present in natural and synthetic organic molecules has important consequences for drug design. The synthesis of pharmaceuticals often requires the production of chiral drugs containing a single optical isomer. Although one of the optical isomers may have beneficial effects, the other may be harmful and may lead to undesirable side effects. See also page 173.

Sidebar notes (left column):

In simple cases in which 2 of the groups are the same, *E/Z* isomerism is often referred to as *cis-trans* isomerism.

Each *E/Z* isomer has the same structural formula.

Optical isomers are usually drawn as 3D diagrams. It is then easy to picture the mirror images.

Optical isomers are also called **enantiomers**.

Society discovered the consequences of harmful side-effects from the 'wrong' optical isomer with the use of thalidomide. One optical isomer combated the effects of morning sickness in pregnant women. The other optical isomer was the cause of deformed limbs of unborn babies.

Partly through this lesson, drugs are now often used as the optically pure form, comprising just the required optical isomer.

Progress check

1. Show the alkenes that are structural isomers of C_4H_8.
2. Show the *E/Z* isomers of C_4H_8.
3. Show the optical isomers of C_4H_9OH.

3

2

1 $CH_3CH_2CH=CH_2; CH_3CH=CHCH_3; (CH_3)_2C=CH_2$

111

Chemistry of organic functional groups

Common functional groups

AQA	M4	SALTERS	M4, M5
EDEXCEL	M4	WJEC	M4
OCR	M4	CCEA	M4

It is essential that you can instantly identify a functional group within a molecule so that you can apply the relevant chemistry.

You must learn all of these!

Key points from AS

• **Functional groups**
  Revise AS pages 120–121

The nitrile carbon atom is included in the name.

CH_3CN contains the longest carbon chain with **two** carbon atoms and its name is based upon ethane – hence ethanenitrile.

The functional groups in the table below contain most of the functional groups you will meet in the A Level Chemistry course, including those met in AS Chemistry. It cannot be stressed enough just how important it is that you can instantly recognise a functional group. Without this, your progress in Organic Chemistry will be very limited.

name	functional group	examples structural formula		prefix or suffix (for naming)
alkane	C–H	$CH_3CH_2CH_3$ propane	$CH_3CH_2CH_3$	-ane
alkene	>C=C<	CH_3CHCH_2 propene		-ene
halogenoalkane	—Br	CH_3CH_2Br bromoethane	CH_3CH_2—Br	bromo-
alcohol	—OH	CH_3CH_2OH ethanol	CH_3CH_2—OH	-ol
aldehyde	—CHO	CH_3CHO ethanal	H_3C—CHO	-al
ketone	—CO—	CH_3COCH_3 propanone	H_3C—COCH_3	-one
carboxylic acid	—COOH	CH_3COOH ethanoic acid	H_3C—COOH	-oic acid
ester	—COO—	CH_3COOCH_3 methyl ethanoate	H_3C—COOCH_3	-oate
acyl chloride	—COCl	CH_3COCl ethanoyl chloride	H_3C—COCl	-oyl chloride
amine	—NH_2	$CH_3CH_2NH_2$ ethylamine	CH_3CH_2—NH_2	-amine
amide	—CONH_2	CH_3CONH_2 ethanamide	H_3C—CONH_2	-amide
nitrile	—C≡N	CH_3CN ethanenitrile	H_3C—CN	-nitrile

112

4.2 Aldehydes and ketones

After studying this section you should be able to:

- understand the polarity and physical properties of carbonyl compounds
- describe the oxidation of alcohols
- describe the reduction of carbonyl compounds to form alcohols
- describe nucleophilic addition to aldehydes and ketones
- describe a test to detect the presence of the carbonyl group
- describe tests to detect the presence of an aldehyde group

LEARNING SUMMARY

Carbonyl compounds General formula: $C_nH_{2n}O$

AQA	M4	SALTERS	M1
EDEXCEL	M4	WJEC	M4
OCR	M4	CCEA	M4

During AS Chemistry, you learnt about how alcohols can be oxidised to carbonyl compounds: aldehydes and ketones. For A2 Chemistry, you will learn about the reactions of aldehydes and ketones.

Key points from AS

- **Oxidation of alcohols**
  Revise AS pages 135–136

Types and naming of carbonyl compounds

The carbonyl group, $C=O$, is the functional group in aldehydes and ketones.

aldehyde, RCHO

ketone, RCOR'

ethanal
CH_3CHO

butanal
$CH_3CH_2CH_2CHO$

propanone
CH_3COCH_3

pentan-2-one
$CH_3COCH_2CH_2CH_3$

Polarity of carbonyl compounds

Carbon and oxygen have different electronegativities, resulting in a polar $C=O$ bond: carbonyl compounds have polar molecules.

oxygen is more electronegative
than carbon producing a dipole

The properties of aldehydes and ketones are dominated by the polar carbonyl group, $C=O$.

Physical properties of carbonyl compounds

The polarity of the carbonyl group is less than that of the hydroxyl group in alcohols. Thus, aldehydes and ketones have weaker dipole–dipole interactions and lower boiling points than alcohols of comparable molecular mass.

stronger dipole

carbonyl group in
aldehyde or ketone

hydroxyl group in
alcohols

The polarity in propanone is such that it mixes with polar solvents such as water and also dissolves many organic compounds.

The low boiling point also makes it easy to remove by evaporation, a property exploited by its use in paints and varnishes.

Oxidation of alcohols

EDEXCEL M4   SALTERS M4
OCR M4   WJEC ▶ M4
              CCEA ▶ M4

Key points from AS

• **Oxidation of alcohols**
  Revise AS pages 135–136

You studied the oxidation of different types of alcohols during AS Chemistry. The link between alcohols, carbonyl compounds and carboxylic acids is shown below.

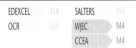

primary alcohol   aldehyde   carboxylic acid

secondary alcohol   ketone

Oxidation of primary alcohols

A primary alcohol can be oxidised to an aldehyde and then to a carboxylic acid.

The orange dichromate ions, $Cr_2O_7^{2-}$, are reduced to green Cr^{3+} ions.

This is carried out using an oxidising agent such as a mixture of concentrated sulfuric acid, H_2SO_4 (source of H^+) and potassium dichromate, $K_2Cr_2O_7$ (source of $Cr_2O_7^{2-}$).

• By heating and distilling the product immediately, oxidation can be stopped at the aldehyde stage.

$$CH_3CH_2OH + [O] \longrightarrow CH_3CHO + H_2O$$

For balanced equations, the oxidising agent can be shown simply as [O].

• By refluxing with an excess of the oxidising agent, further oxidation takes place to form the carboxylic acid.

$$CH_3CH_2OH + 2[O] \longrightarrow CH_3COOH + H_2O$$

Oxidation of secondary alcohols

By heating with $H^+/Cr_2O_7^{2-}$, a secondary alcohol can be oxidised to a ketone. No further oxidation normally takes place.

The equation for the oxidation of propan-2-ol is shown below.

$$CH_3(CHOH)CH_3 + [O] \longrightarrow CH_3COCH_3 + H_2O$$

Reduction of carbonyl compounds

AQA ▶ M4   WJEC ▶ M4
EDEXCEL M4   CCEA ▶ M4
OCR M4

Aldehydes and ketones can be reduced to alcohols using a reducing agent containing the hydride ion, H^-.

Suitable reducing agents are:
• sodium tetrahydridoborate(III) (*sodium borohydride*), $NaBH_4$, in water (heat)
• lithium tetrahydridoaluminate(III) (*lithium aluminium hydride*), $LiAlH_4$, in ether.

Aldehydes are reduced to primary alcohols:

For balanced equations, the reducing agent can be shown simply as [H].

aldehyde   primary alcohol

Ketones are reduced to secondary alcohols:

$NaBH_4$ and $LiAlH_4$ both reduce the C=O double bond in aldehydes and ketones.

They do **not** reduce the C=C bond in alkenes.

ketone   secondary alcohol

Nucleophilic addition to carbonyl compounds

The electron-deficient carbon atom of the polar $C^{\delta+}=O^{\delta-}$ bond attracts **nucleophiles**. This allows an **addition** reaction to take place across the C=O double bond of aldehydes and ketones. This is called **nucleophilic addition**.

$$\overset{\delta+}{C}=\overset{\delta-}{O} \quad \text{electron-rich nucleophile attracted to } \overset{\delta+}{C}$$

Reduction as nucleophilic addition (AQA and OCR only)

The reduction of aldehydes and ketones to alcohols using $NaBH_4$ or $LiAlH_4$ is an example of **nucleophilic addition**. The reducing agent can be considered to release hydride ions, H^-.

$$NaBH_4 \longrightarrow H^- + NaBH_3^+$$

The hydride ion acts as a nucleophile in the first stage of the reaction.

The reaction of carbonyl compounds with hydrogen cyanide is also an example of nucleophilic addition.

Mechanism

$$H_3C-C \begin{matrix} \delta- \\ O \\ H \end{matrix} :H^- \longrightarrow H_3C-\underset{H}{\overset{O^-}{C}}-H \xrightarrow[\text{with } H_2O]{\text{protonation}} H_3C-\underset{H}{\overset{O-H}{C}}-H + OH^-$$

:H⁻ nucleophile donates electron pair

primary alcohol

Nucleophilic addition of hydrogen cyanide, HCN (not OCR)

In the presence of cyanide ions, CN^-, hydrogen cyanide, HCN, is added across the C=O bond in aldehydes and ketones.

HCN is added across the C=O double bond.

aldehyde + HCN ⟶ hydroxynitrile
CH_3CHO + HCN ⟶ $CH_3CH(OH)CN$

$$H_3C-C\begin{matrix}O\\H\end{matrix} \xrightarrow{KCN(aq)\ /\ HCN} H_3C-\underset{H}{\overset{OH}{C}}-C\equiv N$$

This hydroxynitrile has optical isomers.

Hydrogen cyanide is a very poisonous gas and it is usually generated in solution as H^+ and CN^- ions using:
- sodium cyanide, NaCN as a source of CN^-
- dilute sulfuric acid as a source of H^+.

The presence of cyanide ions is essential to provide the nucleophile for the first step of this mechanism.

Mechanism

$$H_3C-C\begin{matrix}\delta-\\O\\H\end{matrix} :C\equiv N^- \longrightarrow H_3C-\underset{H}{\overset{O^-}{C}}-C\equiv N \xrightarrow{H^+ \text{ transfer}} H_3C-\underset{H}{\overset{O-H}{C}}-C\equiv N$$

:CN⁻ nucleophile donates electron pair

'hydroxynitrile'

Increasing the carbon chain length (not OCR)

The nucleophilic addition of hydrogen cyanide is useful for increasing the length of a carbon chain. The nitrile product can then easily be reacted further in organic synthesis.
- Nitriles are easily *hydrolysed* by water in hot dilute acid to form a carboxylic acid.
- Nitriles are easily *reduced* by sodium in ethanol to form an amine.

The diagrams below show how 2-hydroxypropanenitrile, synthesised above, can be converted into a carboxylic acid and an amine.

nitrile   *carboxylic acid*

$$H_3C-\underset{\underset{H}{|}}{\overset{\overset{OH}{|}}{C}}-C\equiv N \ + \ 2H_2O \ + \ H^+ \ \xrightarrow[\text{hydrolysis}]{\substack{\text{aqueous acid} \\ \text{reflux}}} \ H_3C-\underset{\underset{H}{|}}{\overset{\overset{OH}{|}}{C}}-C\overset{\overset{O}{\diagup}}{\underset{OH}{\diagdown}} \ + \ NH_4^+$$

2-hydroxypropanoic acid
(lactic acid)

$$H_3C-\underset{\underset{H}{|}}{\overset{\overset{OH}{|}}{C}}-C\equiv N \ + \ 2[H] \ \xrightarrow[\text{reduction}]{\substack{\text{Na/ethanol } OR \\ \text{LiAlH}_4}} \ H_3C-\underset{\underset{H}{|}}{\overset{\overset{OH}{|}}{C}}-\underset{\underset{H}{|}}{\overset{\overset{H}{|}}{C}}-NH_2$$

nitrile   *amine*

Nitriles from halogenolkanes (WJEC only)

A nitrile can also be prepared by heating under reflux a halogenoalkane with an ionic cyanide, such as NaCN, in ethanol.

$$CH_3CH_2Br + CN^- \longrightarrow CH_3CH_2C\equiv N + Br^-$$

Testing for the carbonyl group

EDEXCEL M4   WJEC M4
OCR M4   CCEA M4

Carbonyl compounds produce an orange–yellow crystalline solid with Brady's reagent.

The carbonyl group can be detected using **Brady's reagent** – a solution of 2,4-dinitrophenylhydrazine (2,4-DNPH) in dilute acid.

- With 2,4-DNPH, both aldehydes and ketones produce bright **orange–yellow crystals** which identify the carbonyl group, C=O.

2,4-DNPH   ethanal 2,4-dinitrophenylhydrazone

orange crystalline precipitate

The carbonyl group can also be detected using IR spectroscopy, see page 165 (see also AS pages 143–144).

This is a **condensation reaction** – water is lost.

- If first recrystallised, the crystals have very sharp melting points which can be compared with known melting points from databases. Thus, the actual carbonyl compound can be identified.

Testing for the aldehyde group

AQA M4   WJEC M4
EDEXCEL M4   CCEA M4
OCR M4

An aldehyde can be distinguished from a ketone by using a combination of these two tests.

Both aldehyde and ketone produce a yellow–orange precipitate with 2,4-DNPH.

Only the aldehydes react in the three tests below.

Ketones cannot normally be oxidised and they **do not** react with Tollens' reagent, Fehling's solution or acidified dichromate(VI).

The oxidation of aldehydes provides the basis of chemical tests used to identify this functional group.

> **KEY POINT**
>
> In each test:
> - the aldehyde **reduces** the reagents used in the test, producing a visible colour change
> - the aldehyde is **oxidised** to a carboxylic acid.
>   RCHO + [O] ⟶ RCOOH

Heat aldehyde with Tollens' reagent

Tollens' reagent is a solution of silver nitrate in aqueous ammonia. The silver ions are reduced by an aldehyde producing a **silver mirror**.

Tollens' reagent contains the $[Ag(NH_3)_2]$ complex ion (see page 88).

> **KEY POINT**
>
> Tollens' reagent $\longrightarrow$ **silver mirror**
>
> $Ag^+(aq) + e^- \longrightarrow Ag(s)$

Heat aldehyde with Fehling's (or Benedict's) solution

Fehling's and Benedict's solutions contain Cu^{2+} ions dissolved in aqueous alkali. The Cu^{2+} ions are reduced to copper(I) ions, Cu^+, producing a **brick-red precipitate** of Cu_2O.

Key points from AS

- **Oxidation of primary alcohols**
  Revise AS pages 135–136

> **KEY POINT**
>
> Benedict's / Fehling's solution $\longrightarrow$ **brick-red precipitate** of Cu_2O
>
> $2Cu^{2+}(aq) + 2e^- + 2OH^-(aq) \longrightarrow Cu_2O(s) + H_2O(l)$

Acidified dichromate(VI) can also be used to oxidise alcohols.

Heat aldehyde with $H^+/Cr_2O_7^{2-}$

Concentrated sulfuric acid, H_2SO_4, is used as a source of H^+ ions and potassium dichromate(VI), $K_2Cr_2O_7$, as a source of $Cr_2O_7^{2-}$ ions. The **orange** $Cr_2O_7^{2-}$ ions are reduced to **green** chromium(III) ions, Cr^{3+}.

Test for the presence of the methyl carbonyl group

EDEXCEL M4   WJEC M4

The presence of a methyl carbonyl group can be detected by heating a compound with iodine, I_2 in aqueous sodium hydroxide, $OH^-(aq)$.

$$\underset{\text{methyl carbonyl}}{H_3C - \overset{\displaystyle O}{\overset{\displaystyle \|}{C}} -}$$

Methyl carbonyl compounds produce pale-yellow crystals with I_2/OH^-.

- Methyl carbonyls produce **pale-yellow crystals** of triiodomethane (*iodoform*), CHI_3, with an antiseptic smell.
- $CH_3CHO + 3I_2 + 4OH^- \longrightarrow CHI_3 + HCOO^- + 3H_2O + 3I^-$

Progress check

1 (a) Draw the structure of:
    (i) 3-methylpentan-2-one
    (ii) 3-ethyl-2-methylpentanal.

2 Three isomers of C_4H_8O are carbonyl compounds.
  (a) (i) Show the formula for each isomer.
     (ii) Classify each isomer as an *aldehyde* or a *ketone*.
     (iii) Name each isomer.
  (b) Write the structural formulae of the **three** alcohols that could be formed by reduction of these isomers.

(answers, printed inverted at foot of page:)

(b) $CH_3CH_2CH_2CH_2OH$; $CH_3CH_2CHOHCH_3$; $(CH_3)_2CHCH_2OH$

2 (a) (i)

(ii) aldehyde — butanal
$CH_3-CH_2-CH_2-C\overset{\displaystyle H}{\underset{\displaystyle O}{}}$

(iii) ketone — butanone
$CH_3-CH_2-\overset{\displaystyle O}{\overset{\displaystyle \|}{C}}-CH_3$

aldehyde — methylpropanal
$CH_3-CH-C\overset{\displaystyle H}{\underset{\displaystyle O}{}}$ with CH_3

1 (a) (i) $CH_3-CH_2-CH-\overset{\displaystyle O}{\overset{\displaystyle \|}{C}}-CH_3$ with CH_3

(ii) $CH_3-CH_2-CH_2-CH-CH-C\overset{\displaystyle H}{\underset{\displaystyle O}{}}$ with CH_2CH_3 and CH_3

4.3 Carboxylic acids

After studying this section you should be able to:

- understand the polarity and physical properties of carboxylic acids
- describe acid reactions of carboxylic acids to form salts
- describe the esterification of carboxylic acids with alcohols

LEARNING SUMMARY

Carboxylic acids   General formula: $C_nH_{2n+1}COOH$ RCOOH

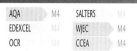

AQA	M4	SALTERS	M4
EDEXCEL	M4	WJEC	M4
OCR	M4	CCEA	M4

The combination of carbonyl and hydroxyl groups in the carboxyl group modifies the chemistry of both groups.

Carboxylic acids have their own set of reactions and react differently from carbonyl compounds and alcohols.

The carboxyl group

The functional group in carboxylic acids is the **carboxyl** group, COOH. Although this combines both the **carbonyl** group and a **hydroxyl** group its properties are very different from either.

carboxyl group

Naming of carboxylic acids

Numbering starts from the carbon atom of the carboxyl group.

4-methylpentanoic acid
$(CH_3)_2CH_2CH_2CH_2COOH$

Natural carboxylic acids

Carboxylic acids are found commonly in nature. Their acidity is comparatively weak and their presence in food often gives a sour taste. Some examples of natural carboxylic acids are shown below.

structure	name	natural source
HCOOH	methanoic acid (*formic acid*)	ants, stinging nettles
CH_3COOH	ethanoic acid (*acetic acid*)	vinegar
COOH | COOH	ethanedioic acid (*oxalic acid*)	rhubarb
OH | $H_3C-CH-COOH$	2-hydroxypropanoic acid (*lactic acid*)	sour milk
CH_2COOH | $HO-C-COOH$ | CH_2COOH	2-hydroxypropane-1,2,3-tricarboxylic acid (*citric acid*)	oranges, lemons

Polarity

The carboxyl group is a combination of two polar groups: the hydroxyl –OH, **and** carbonyl C=O groups. This makes a carboxylic acid molecule more polar than a molecule of an alcohol or a carbonyl compound.

carbonyl group in aldehyde or ketone   →   stronger dipole   →   hydroxyl group in alcohols   →   stronger dipole   →   carboxyl group in carboxylic acids

The properties of carboxylic acids are dominated by the carboxyl group, COOH.

Physical properties of carboxylic acids

The COOH group dominates the physical properties of short-chain carboxylic acids. Hydrogen bonding takes place between carboxylic acid molecules, resulting in:

- higher melting and boiling points than alkanes of comparable M_r
- solubility in water.

The solubility of alcohols in water decreases with increasing carbon chain length as the non-polar contribution to the molecule becomes more important.

Carboxylic acids as 'acids'

AQA	M4	SALTERS	M4
EDEXCEL	M4	WJEC	M4
OCR	M4	CCEA	M4

Carboxylic acids are the 'organic acids'.

Carboxylic acids are only weak acids because they only partially dissociate in water.

$$CH_3COOH \rightleftharpoons CH_3COO^- + H^+$$

- Only 1 molecule in about 100 actually dissociates.
- Only a small proportion of the potential H^+ ions is released.

For more details of the dissociation of weak acids, see pages 35–36.

Carboxylates: salts of carboxylic acids

Carboxylic acid salts, 'carboxylates', are formed by neutralisation of a carboxylic acid by an alkali. In the example below, ethanoic acid produces ethanoate ions.

carboxylic acids exist in acidic conditions

carboxylates exist in alkaline conditions

On evaporation of water, a carboxylate salt crystallises out as an ionic compound. With aqueous sodium hydroxide as the alkali, sodium ethanoate, $CH_3COO^-Na^+$, is produced as the ionic salt.

Acid reactions of carboxylic acids

Key points from AS

- **Typical reactions of an acid**
  Revise AS page 46

Carboxylic acids react by the usual 'acid reactions' producing carboxylate salts.

Carboxylic acids take part in typical acid reactions. Note in the examples below that each salt formed is a carboxylate.

- They are **neutralised by alkalis**, forming a salt and water only.
  $$CH_3COOH + NaOH \longrightarrow CH_3COONa + H_2O$$
- They **react with carbonates**, forming a salt, carbon dioxide and water.
  $$2CH_3COOH + CaCO_3 \longrightarrow (CH_3COO)_2Ca + CO_2 + H_2O$$
- They **react with** reactive **metals**, forming a salt and hydrogen.
  $$2CH_3COOH + Mg \longrightarrow (CH_3COO)_2Mg + H_2$$

> **KEY POINT**
> Carboxylic acids are the only common organic group able to release carbon dioxide gas from carbonates. This provides a useful test to show the presence of a carboxyl group.

Properties of carboxylates

Carboxylates such as sodium ethanoate, $CH_3COO^-Na^+$, are ionic compounds. They have typical properties of an ionic compound.

- They are solids at room temperature with high melting and boiling points.
- They have a giant ionic lattice structure.
- They dissolve in water, totally dissociating into ions.

Reduction of carboxylic acids

Just as primary alcohols can be oxidised to carboxylic acids, the reverse process can take place using a strong reducing agent such as lithium tetrahydridoalumuninate(III) (*lithium aluminium hydride*), $LiAlH_4$ in dry ether. (See also: Reduction of carbonyl compounds, page 114.)

$$RCOOH + 4[H] \longrightarrow RCH_2OH + H_2O$$

Esterification

Esterification is the formation of an ester by reaction of a **carboxylic acid** with an **alcohol** in the presence of an **acid catalyst** (e.g. concentrated sulfuric acid).

$$\text{carboxylic acid} + \text{alcohol} \longrightarrow \text{ester} + \text{water}$$

The esterification of ethanoic acid by methanol is shown below:

$$CH_3COOH + CH_3OH \longrightarrow CH_3COOCH_3 + H_2O$$

Esters are formed by reaction of a carboxylic acid with an alcohol.

Conditions

- An acid catalyst (a few drops of conc. H_2SO_4) and reflux.
- The yield is usually poor due to incomplete reaction.

Purifying solids (Salters only)

Many organic compounds are solids and they will need to be purified. The easiest method of doing this is by **recrystallisation**.

The organic solid must be:

The solvent is critical.

Without a difference in solubility at high and low temperatures, hardly any crystals will appear on cooling.

- very **soluble** in the solvent at **higher temperatures** – the solvent is saturated by the substance at higher temperatures

- **much less soluble**, or nearly so, in the solvent at **lower temperatures** – on cooling, the substance crystallises out, leaving impurities in solution.

Progress check

1 Write down the structural formula of:
   (a) propanoic acid
   (b) the propanoate ion.

2 Explain why a carboxylic acid has a higher boiling point than the corresponding alcohol.

2 Carboxylic acid has both polar carbonyl and hydroxyl groups. Alcohols have hydroxyl group only. Therefore, a carboxylic acid has greater intermolecular forces.

1 (a) CH_3CH_2COOH
(b) $CH_3CH_2COO^-$

4.4 Esters

After studying this section you should be able to:

- understand the physical properties of esters
- describe the acid and base hydrolysis of esters
- understand that fats and oils are saturated and unsaturated esters
- describe the hydrolysis of fats and oils in soap making

LEARNING SUMMARY

Esters

General formula: $C_nH_{2n+1}COOC_mH_{2m+1}$ RCOOR'

AQA	M4	SALTERS	M4
EDEXCEL	M4	WJEC	M4
OCR	M4	CCEA	M4

The functional group in esters is the COOR' group, with an alkyl group in place of the acidic proton of a carboxylic acid.

$$R-C\!\!\begin{array}{c}\nearrow O\\[-2pt]\searrow O-R'\end{array}$$

Naming of esters

The name of an ester is based on the carboxylic acid from which the ester is derived. The ester below, methyl butanoate, is derived from butanoic acid C_3H_7COOH with a methyl group in place of the acidic proton.

butanoate from
butanoic acid
$CH_3CH_2CH_2COOH$

$$CH_3CH_2CH_2-C\!\!\begin{array}{c}\nearrow O\\[-2pt]\searrow O-CH_3\end{array}$$

methyl from methanol
CH_3OH

methyl butanoate

In the name of **methyl butanoate**:

- the alkyl group comes first as the ...*yl*: **methyl**
- the carboxylic acid part comes second as the ...*oate*: **butanoate**

Natural esters

Esters are found commonly in nature as fats and oils. They often have pleasant smells and contribute to the flavouring of many foods as shown below.

Esters are common organic compounds present in fats and oils. They are also used as solvents, plasticisers, in food flavourings and as perfumes.

structure	name	source
$HCOOCH_3$	methyl methanoate	raspberries
$C_3H_7COOC_4H_9$	butyl butanoate	pineapple

Physical properties of esters

Unlike carboxylic acids, esters are neutral. They are less polar than carboxylic acids. The absence of an –OH group means that esters cannot form hydrogen bonds and are generally insoluble in water.

Hydrolysis of esters

AQA	M4	SALTERS	M4
EDEXCEL	M4	WJEC	M4
OCR	M4	CCEA	M4

The hydrolysis of an ester is the reverse reaction to esterification (see page 120):

hydrolysis
ester + water ⇌ carboxylic acid + alcohol
esterification

Note that this reaction is reversible. The direction of reaction can be controlled by the reagents and reaction conditions used.

> Hydrolysis is the breaking down of a compound using **water** as the reagent.
>
> KEY POINT

Hydrolysis takes place by refluxing the ester with dilute aqueous acid or alkali.

- Acid hydrolysis $\longrightarrow$ alcohol + carboxylic acid.
- Alkaline hydrolysis $\longrightarrow$ alcohol + carboxylate.

Esterification **produces** water.

Hydrolysis **reacts** with water.

Fats and oils

AQA	M4	SALTERS	M5
EDEXCEL	M4	CCEA	M4
OCR	M4		

Most fats and oils are mixed esters from different fatty acids. So 'R' in these structures may refer to three different chains.

Natural fats and oils are *triglyceryl esters* of fatty acids (long chain carboxylic acids) and propane-1,2,3-triol (*glycerol*).

$$\begin{array}{lll}
RCOO-CH_2 & R-COOH & HO-CH_2 \\
RCOO-CH & R-COOH & HO-CH \\
RCOO-CH_2 & R-COOH & HO-CH_2 \\
\text{fat or oil} & \text{fatty acids} & \text{glycerol} \\
\text{triglyceryl ester} & &
\end{array}$$

Saturated and unsaturated fats

Saturated fats tend to be solids at room temperature and are mainly from animal products. Unsaturated fats tend to be oils and are mainly from vegetable products.

Saturated and unsaturated fatty acids can be obtained from fats and oils by hydrolysis (see page 123). An unsaturated fatty acid would have at least one double C=C bond in one of the alkyl carbon chains.

Examples of saturated and unsaturated fatty acids

octadecanoic acid, 18,0

octadec-9-enoic acid, 18,1(9);

octadeca-9,12-dienoic acid, 18,2(9,12)

- Note that the unsaturated fatty acids above are *trans* acids. The *cis* isomers are less compact and are far healthier.
- Many triglyceryl esters contain different alkyl carbon chains derived from different fatty acids.

Scientists have discovered that *trans* fatty acids pose health problems, increasing cholesterol with an increased risk of coronary heart disease, strokes and obesity.

In margarine production, vegetable oils are hardened by partial **catalytic hydrogenation**.

- Partial hydrogenation leaves some double bonds intact. The oil is reacted just enough to solidify the oily texture. This prevents the formation of **polysaturates**, with no C=C double bonds, which are linked to unhealthy diets.
- Margarines are often sold as partially hydrogenated **polyunsaturates** to promote the health value of unsaturated fats.

The iodine value of a fat or oil can be determined from the following procedure:

- Add Wij's reagent and keep in the dark for 30 minutes.
- Add excess KI.
- Titrate with sodium thiosulfate. (See pages 97–98).
- Add starch indicator and titrate until the blue–black colour disappears.

Iodine value (CCEA only)

Iodine value is used to measure the degree of unsaturation in a fat or oil.

> **Iodine value** is the mass of iodine in grams that reacts with 100 grams of a fat or oil.

- A solution of iodine is yellow/brown in colour.
- Iodine adds across any C=C bonds in a fat or oil, decolourising the solution.
- The greater the iodine value, the more C=C bonds present and the greater the unsaturation.

Hydrolysis of fats and oils

The hydrolysis of fats and oils is an important reaction used in soap production in which the fat or oil is boiled with aqueous alkali. In alkaline hydrolysis with aqueous NaOH, each mole of the fat or oil produces:

- **three** moles of the fatty acid salt, $3RCOO^-Na^+$ and
- **one** mole of glycerol (a triol), $HOCH_2CHOHCH_2OH$.

$$RCOO-CH_2 \atop RCOO-CH \atop RCOO-CH_2} + 3\ NaOH \xrightarrow{NaOH(aq)} 3\ RCOO^-Na^+ \ {\text{salt of}\atop\text{carboxylic acid}} + {HO-CH_2\atop HO-CH\atop HO-CH_2}$$

Saponification number (CCEA only)

Saponification number is the number of mg of KOH needed to saponify 1g of fat.

- If a fat has long fatty acid chains, a relatively large proportion of the total mass of the fat comes from the carbon chains.

- The greater the saponification number, the smaller the relative molecular mass of the fat and the shorter the fatty acid chain length.

Biodiesel

AQA   M4   OCR   M4
EDEXCEL   M4

Transesterification is also being investigated for the development of low-fat spreads as an alternative to hydrogenation of vegetable oils to produce margarine.

Esters of fatty acids are being developed for use as fuels such as biodiesel. Use of biodiesel as a fuel increases the contribution to energy requirements from renewable fuels. Biodiesel is a mixture of esters of long chain carboxylic acids.

Vegetable oils can be converted into biodiesel by **transesterification** using methanol in the presence of an alkaline catalyst, e.g.

$$RCOO-CH_2 \atop RCOO-CH \atop RCOO-CH_2} + 3\ CH_3OH \xrightarrow[\text{catalyst}]{NaOH} 3\ RCOO-CH_3 \ {\text{methyl ester}} + {HO-CH_2\atop HO-CH\atop HO-CH_2}$$

The methyl ester is the easiest to prepare but methanol has the disadvantage that it is toxic. Research is being done to develop the ethyl ester made from ethanol. Ethanol is easy to produce from fermentation of sugars.

Progress check

1 How is methyl propanoate made from a named carboxylic acid and alcohol?
2 (a) Explain what is meant by the term *hydrolysis*.
  (b) Name the products of the acid and alkaline hydrolysis of ethyl butanoate.

Alkaline hydrolysis: ethanol and butanoate ions.
(b) Acid hydrolysis: ethanol and butanoic acid.
2 (a) Breaking down a molecule using water.
1 Reflux propanoic acid with methanol in the presence of a few drops of concentrated sulfuric acid as catalyst.

4.5 Acylation

LEARNING SUMMARY

After studying this section you should be able to:

- *describe the formation of acyl chlorides from carboxylic acids*
- *describe nucleophilic addition-elimination reactions of acyl chlorides*
- *know the advantages of ethanoic anhydride for industrial acylations*

Acyl chlorides

General formula: $C_nH_{2n+1}COCl$ $RCOCl$

The functional group in acyl chlorides is the COCl group:

$$-C{\overset{\displaystyle O}{\underset{Cl}{}}}$$

Acyl chlorides are named from the parent carboxylic acid.

The examples below show that the suffix '-*oic acid*' is changed to '-*oyl chloride*' in the corresponding acyl chloride.

$H_3C-C{\overset{O}{\underset{OH}{}}}$	$H_3C-C{\overset{O}{\underset{Cl}{}}}$	$CH_3CH_2-C{\overset{O}{\underset{OH}{}}}$	$CH_3CH_2-C{\overset{O}{\underset{Cl}{}}}$
ethanoic acid	ethanoyl chloride	propanoic acid	propanoyl chloride

Properties

Acyl chlorides are the most reactive organic compounds that are commonly used. Unlike most other organic functional groups, acyl chlorides do **not** occur naturally because of their high reactivity with water.

Acyl chlorides are the only common organic compounds that react violently with water.

$$R-\overset{\delta+}{C}{\overset{\displaystyle \overset{\delta-}{O}}{\underset{\underset{\delta-}{Cl}}{}}}$$

The relatively large $\delta+$ charge attracts the lone pair of a nucleophile

The electron-withdrawing effects of both the **chlorine** atom and carbonyl **oxygen** atom produces a relatively large $\delta+$ charge on the carbonyl carbon atom. Nucleophiles are strongly attracted to the electron-deficient carbon atom, increasing the reactivity.

Acyl chlorides in organic synthesis

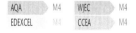

The high reactivity of acyl chlorides compared with carboxylic acids makes them particularly useful in the organic synthesis of related compounds.

Advantages of acyl chlorides over carboxylic acids:

- A good yield of product – reactions go to completion.
- Reactions often occur quickly and at lower temperatures.

Using acyl chlorides in synthesis (not AQA)

The high reactivity of an acyl chloride means that it is usually prepared from the parent carboxylic acid '*in situ*', i.e. when it is needed.

Acyl chlorides are prepared by reacting a carboxylic acid with:

*With acyl chlorides, anhydrous conditions are **essential** – acyl chlorides react with water.*

- phosphorus pentachloride
$$RCOOH + PCl_5 \longrightarrow RCOCl + POCl_3 + HCl$$

- or sulfur dichloride oxide, $SOCl_2$
$$RCOOH + SOCl_2 \longrightarrow RCOCl + SO_2 + HCl$$

- The required nucleophile is then added to the acyl chloride.

Reactions of acyl chlorides with nucleophiles

AQA	M4	SALTERS	M4
EDEXCEL	M4	WJEC	M4
		CCEA	M4

Nucleophiles of the type H–Y: react readily with acyl chlorides.

This is an **addition-elimination** reaction involving:
- addition of HY across the C=O double bond followed by
- elimination of HCl.

Addition-elimination reactions of acyl chlorides

Acyl chlorides can be reacted with different nucleophiles to produce a range of related functional groups. In each addition-elimination reaction, hydrogen chloride is eliminated as the second product.

The reaction scheme below shows the reactions of an acyl chloride RCOCl with the nucleophiles water, H_2O, methanol, CH_3OH, ammonia, NH_3 and methylamine, CH_3NH_2.

The high reactivity of an acyl chloride means that these reactions take place at room temperature.

(WJEC and Salters only)
Acyl chlorides also react with phenols to form esters.

Ammonia and methylamine are bases – they react with the HCl eliminated:

$$NH_3 + HCl \longrightarrow NH_4^+Cl^-$$
$$CH_3NH_2 + HCl \longrightarrow CH_3NH_3^+Cl^-$$

Mechanism (AQA only)

The mechanism below shows the **nucleophilic addition-elimination** reaction of ethanoyl chloride with methanol to form an ester. Each nucleophile in the reaction scheme above will react in a similar manner.

Acid anhydrides

AQA ▶ M4   WJEC ▶ CH4
OCR ▶ M4

Acyl chlorides are ideal for small-scale preparations in the laboratory. However, they are too expensive and too reactive for large-scale preparations, for which they are replaced by acid anhydrides.

The equation below shows the reaction between ethanoic anhydride and methanol:

> Notice the molecular structure of ethanoic anhydride – two ethanoic acid molecules have been condensed together with loss of H_2O. Hence the 'anhydride' from ethanoic acid.

acid anhydride

carboxylic acid as second product

- The reaction is slower and easier to control on a large scale.
- This is the same essential reaction as with an acyl chloride but RCOOH forms instead of HCl as the second product.

Synthesis of aspirin

Aspirin can be synthesised by acylation of 2-hydroxybenzenecarboxylic acid (*salicylic acid*) using ethanoic anhydride.

aspirin

ethanoic acid as second product

Progress check

1 Write down the structural formula of the organic product formed from the reaction of propanoyl chloride with:
   (a) water
   (b) ammonia.

2 What could you react together to make N-ethylbutanamide, $CH_3CH_2CH_2CONHCH_2CH_3$?

3 Write an equation for the preparation of aspirin from 2-hydroxybenzoic acid using an acyl chloride.

1 (a) CH_3CH_2COOH
  (b) $CH_3CH_2CONH_2$
2 $CH_3CH_2CH_2COCl$ and $CH_3CH_2NH_2$
3

4.6 Arenes

After studying this section you should be able to:

- apply rules for naming simple aromatic compounds
- understand the delocalised model of benzene
- explain the resistance to addition of benzene compared with alkenes

LEARNING SUMMARY

Aromatic organic compounds

AQA	M4	SALTERS	M5
EDEXCEL	M5	WJEC	M4
OCR	M4	CCEA	M5

Organic compounds with pleasant smells were originally classified as aromatic compounds. Many of these contain a benzene ring in their structure and nowadays an aromatic compound is one structurally derived from benzene, C_6H_6.

The diagrams below show different representations of a benzene molecule. It is usual practice to omit the carbon and hydrogen labels.

Arenes burn with a smoky flame. This reflects the relatively low hydrogen to carbon ratio compared to alkanes and alkenes.

Arenes

An **arene** is an aromatic hydrocarbon containing a benzene ring. Benzene is the simplest arene and substituted arenes have alkyl groups attached to the benzene ring. Examples of arenes are shown below.

benzene   methylbenzene   ethylbenzene

Functional groups

Arenes can have a functional group next to an **aryl** group (a group containing a benzene ring):

Aryl group   *Functional group*

or C_6H_5X

An aryl group contains a benzene ring.

- An aryl group is often represented simply as Ar—.
- The simplest aryl group is the **phenyl** group, C_6H_5, derived from benzene, C_6H_6.

Naming of aromatic organic compounds

The benzene ring of an aromatic compound is numbered from the carbon atom attached to a functional group or alkyl side-chain.
The names can be derived in two ways:

- Some compounds (e.g. hydrocarbons, chloroarenes and nitroarenes) are regarded as substituted benzene rings.

chlorobenzene   1,3-dimethylbenzene   2,4-dinitromethylbenzene

You should be able to suggest names for simple arenes.

- Other compounds (e.g. phenols and amines) are considered as phenyl compounds of a functional group.

phenol
not hydroxybenzene

phenylamine
not aminobenzene

2,4,6-trichlorophenol

The stability of benzene

AQA	M4	SALTERS	M5
EDEXCEL	M5	WJEC	M4
OCR	M4	CCEA	M5

Two structures are used to represent benzene:

- the **Kekulé** structure, developed between 1865 and 1872
- the modern **delocalised**, structure developed in the 1930s.

The Kekulé structure of benzene

The Kekulé model of a benzene molecule has alternate double and single bonds making up the ring.

The chemical name for the Kekulé structure of benzene is **cyclohexa-1,3,5-triene**, after the positions of the double bonds in the ring.

The original Kekulé structure showed benzene as a hexagonal molecule with alternate double and single bonds. Each carbon atom is attached to one hydrogen atom. This model was later modified to one with two isomers, rapidly interconverting into one another.

The Kekulé structure of benzene

The delocalised structure of benzene

The delocalised model of a benzene molecule has identical carbon–carbon bonds making up the ring.

The delocalised structure of benzene shows a benzene molecule as a hybrid state between Kekulé's two isomers with no separate single and double bonds.

In this hybrid state:

- each carbon atom contributes one electron from its p-orbital to form π-bonds
- the π-bonds are spread out or **delocalised** over the whole ring.

Key point from AS

- **Alkenes**
  Revise AS page 129

This model helps to explain the low reactivity of benzene compared with alkenes (see also page 131–132).

The double bond in an alkene is a **localised** π-bond between two carbon atoms. In the delocalised structure of benzene, each bond is identical, with electron density spread out to encompass and stabilise the whole ring. The diagram below shows one of the π-bonds formed by delocalisation of electrons in a benzene molecule.

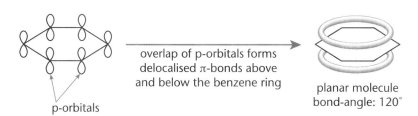

overlap of p-orbitals forms delocalised π-bonds above and below the benzene ring

p-orbitals

planar molecule
bond-angle: 120°

Although the Kekulé structure is used for some purposes, the delocalised structure is a better representation of benzene. You will find both representations in books.

The delocalised structure of a benzene molecule is shown as a hexagon to represent the carbon skeleton and a circle to represent the six delocalised electrons.

delocalised electrons
– all carbon-carbon
bonds identical

The shape of a benzene molecule

Key points from AS

- **Electron-pair repulsion theory**
  Revise AS pages 60–61

Using electron-pair repulsion theory:

- there are **three** centres of electron density surrounding each carbon atom
- the shape around each carbon atom is trigonal planar with bond angles of 120°.

This results in a benzene molecule that is **planar**.

3 electron centres surround each carbon atom

Experimental evidence for delocalisation

AQA	M4	SALTERS	M5
EDEXCEL	M5	WJEC	M4
OCR	M4	CCEA	M5

Bond length data

The Kekulé structure of benzene as cyclohexa-1,3,5-triene suggests two bond lengths for the separate single and double bonds:

- C—C bond length = 0.154 nm
- C=C bond length = 0.134 nm

Experiment shows only one C—C bond length of 0.139 nm, between the bond lengths for single and double carbon-carbon bonds.

0.154 nm   0.134 nm
0.134 nm   0.154 nm
0.154 nm   0.134 nm

all C–C bonds are the same length: 0.139 nm

> This shows that each carbon-carbon bond in the benzene ring is intermediate between a single and a double bond.

KEY POINT

Thermochemical evidence

Hydrogenation of cyclohexene

Each molecule of cyclohexene has **one** C=C double bond. The enthalpy change for the reaction of cyclohexene with hydrogen is shown below:

 + H$_2$ ⟶

$\Delta H^\ominus = -120$ kJ mol^{-1}

Hydrogenation of benzene

The Kekulé structure of benzene as cyclohexa-1,3,5-triene has **three** double C=C bonds. It would be expected that the enthalpy change for the hydrogenation of this structure would be three times the enthalpy change for the **one** C=C bond in cyclohexene.

The stability of benzene can also be demonstrated using thermochemical data for the reactions of bromine with cyclohexene and benzene.

 + 3 H$_2$ $\longrightarrow$

predicted enthalpy change:
$\Delta H^{\ominus} = 3 \times -120 = -360$ kJ mol^{-1}

- When benzene is reacted with hydrogen, the enthalpy change obtained is far less exothermic, $\Delta H^{\ominus} = -208$ kJ mol^{-1}.

KEY POINT

- The difference between the thermochemical data for cyclohexa-1,3,5-triene and benzene suggests that benzene has **more stable bonding** than the Kekulé structure.
- The delocalisation or resonance energy of benzene of -152 kJ mol^{-1} is the difference between the two enthalpy changes above. This is extra energy that must be provided to break the delocalised benzene ring.

Progress check

1 Name the **three** aromatic isomers of C$_6$H$_4$Br$_2$.

2 State two pieces of experimental evidence that support the delocalised structure of benzene.

3 Explain why alkenes, such as cyclohexene, are so much more reactive with electrophiles than arenes such as benzene.

3 Alkenes have localised π-bonds with a larger electron density than the delocalised π-bonds in benzene. The greater electron density is able to attract electrophiles more strongly. Also, the stability of the benzene ring must be disrupted if benzene is to react.

2 The enthalpy change of hydrogenation of benzene is less exothermic than three times the enthalpy change of hydrogenation of cyclohexene.
All the carbon-carbon bonds in the benzene ring are the same length.

1 1,2-dibromobenzene; 1,3-dibromobenzene; 1,4-dibromobenzene.

4.7 Reactions of arenes

After studying this section you should be able to:

- describe electrophilic substitution of arenes: nitration, halogenation, alkylation, acylation and sulfonation
- describe the mechanism of electrophilic substitution in arenes
- understand the importance of reactions of arenes in the synthesis of commercially important materials

LEARNING SUMMARY

Electrophilic substitution reactions of arenes

AQA	M4	SALTERS	M5
EDEXCEL	M5	WJEC	M4
OCR	M4	CCEA	M5

Many electrophiles react with alkenes by **addition**. However, electrophiles react with arenes by **substitution**, replacing a hydrogen atom on the ring. The difference in behaviour results from the high stability of the **delocalised** π-bonds in benzene compared with the **localised** π-bonds in alkenes (see pages 128–129).

> - Benzene reacts with only very reactive electrophiles.
> - The typical reaction of an arene is **electrophilic substitution**.
>
> **KEY POINT**

In this section, reactions of benzene are discussed to illustrate electrophilic substitution reactions of arenes.

Nitration of arenes

AQA	M4	SALTERS	M5
EDEXCEL	M5	WJEC	M4
OCR	M4	CCEA	M5

The nitration of arenes produces aromatic nitro compounds, which are important for the synthesis of many important products including explosives and dyes (see page 140).

Nitration of benzene

Benzene is nitrated by concentrated nitric acid at 55°C in the presence of concentrated sulfuric acid, which acts as a catalyst.

Although some heat is required (55°C), too much may give further nitration of the benzene ring forming 1,3-dinitrobenzene.

$$\text{benzene} + HNO_3 \xrightarrow[55°C]{H_2SO_4} \text{nitrobenzene} + H_2O$$

Mechanism

- The role of the concentrated sulfuric acid is to generate the **nitronium ion**, NO_2^+, as the active electrophile:

*The nitronium ion is also called a **nitryl cation**.*

$$HNO_3 + H_2SO_4 \longrightarrow H_2NO_3^+ + HSO_4^-$$
$$H_2NO_3^+ \longrightarrow NO_2^+ + H_2O$$

- The powerful NO_2^+ electrophile then reacts with benzene.

*This is **electrophilic substitution**.*

attack of NO_2^+ electrophile   proton loss

- The H^+ formed regenerates a molecule of H_2SO_4.

$$H^+ + HSO_4^- \longrightarrow H_2SO_4$$

*The H_2SO_4 therefore acts as a **catalyst**.*

- The H_2SO_4 molecule reacts with more nitric acid to form more nitronium ions.

Comparing halogenation of arenes and alkenes

EDEXCEL	M5	SALTERS	M5
OCR	M4	WJEC	M4
		CCEA	M5

Key points from AS

- Alkenes
  Revise AS page 129
- Addition reactions of alkenes
  Revise AS pages 130–132

Bromination of alkenes

Alkenes and cycloalkenes react with bromine by **electrophilic addition**. The reaction takes place in the dark at room temperature. The bromination of cyclohexene is shown below:

- Using the same conditions with benzene, there is **no reaction**.
- The stability of the delocalised system resists addition that would disrupt this stability (see pages 128–129).

Bromination of arenes

Arenes react with bromine by **electrophilic substitution**. The reaction takes place only in the presence of a **halogen carrier**, which acts as a catalyst.

Suitable halogen carriers include:

- iron
- aluminium halides, e.g. $AlCl_3$ for chlorination; $AlBr_3$ for bromination.

The bromination of benzene is shown below.

> **KEY POINT**
>
> Bromination of alkenes and arenes are different types of reaction:
>
> alkenes: electrophilic **addition**
>
> arenes: electrophilic **substitution**

Mechanism (not Salters)

- Iron first reacts with bromine forming iron(III) bromide, $FeBr_3$.
- $FeBr_3$ acts as a halogen carrier, polarising the Br—Br bond.

$$Br_2 + FeBr_3 \longrightarrow Br^{\delta+}—Br^{\delta-} \; FeBr_3$$

- This electrophile reacts with benzene.

This is a similar principle to the nitration of the benzene ring.

- The H^+ formed generates $FeBr_3$.

$$H^+ + FeBr_4^- \longrightarrow FeBr_3 + HBr$$

*The $FeBr_3$ therefore acts as a **catalyst**.*

- $FeBr_3$ can now polarise more bromine molecules.

Alkylation of arenes

Alkylation reactions of arenes are commonly known as *Friedel–Crafts* reactions. These are very important reactions industrially as they provide a means of introducing an alkyl group onto the benzene ring.

Alkylation of benzene

This is a similar principle to the halogenation of the benzene ring.

As with halogenation, a halogen carrier is required to generate a more reactive electrophile. With chloroethane, ethylbenzene is formed.

Mechanism (Edexcel only)

Using a tertiary halogenoalkane, RCl, in the presence of $AlCl_3$ as a halogen carrier, a **carbocation** is generated.

$$RCl + AlCl_3 \longrightarrow R^+ + AlCl_4^-$$

By using different halogenoalkanes, different alkyl groups can be substituted onto the benzene ring.

• This carbocation acts as a powerful electrophile, which reacts with benzene.

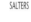

attack of R^+ electrophile   proton loss

• The H^+ formed generates more $AlCl_3$.

$$H^+ + AlCl_4^- \longrightarrow AlCl_3 + HCl$$

• The $AlCl_3$ reacts with more of the tertiary halogenoalkane RCl to generate more carbonium ions. The **$AlCl_3$** acts as a **catalyst**.

Acylation of arenes

The mechanism for acylation is similar to that of alkylation.

Acylation introduces an acyl group such as $CH_3C=O$ onto the benzene ring. An acyl chloride is used in the presence of a halogen carrier.

$$R-C\overset{O}{\underset{Cl}{\diagup\diagdown}} + AlCl_3 \longrightarrow R-\overset{+}{C}=O + AlCl_4^-$$

acylium ion

The acylium ion is then able to react with the benzene ring, introducing an acyl group onto the ring.

Acylation of benzene

The acylation of benzene is another example of a Friedel–Crafts reaction and is also important industrially.

Using ethanoyl chloride, CH_3COCl, the ethanoyl group $CH_3C=O$ can be introduced onto the benzene ring.

This reaction takes place by a similar mechanism to the alkylation reaction above (Edexcel only).

Sulfonation of arenes

Sulfonation of arenes produces sulfonic acids. Sodium salts of sulfonic acids are used for detergents and fabric conditioners.

133

Sulfonation of benzene

Benzene reacts with fuming sulfuric acid (concentrated sulfuric acid saturated with sulfur trioxide) forming a sulfonic acid.

The reaction takes place between benzene and sulfur trioxide.

$$\text{benzene} + SO_3 \xrightarrow{80\,°C} \text{benzenesulfonic acid (SO}_3\text{H)}$$

benzenesulfonic acid

Oxidation of substituted arenes

 WJEC ▶ M4

The side chains of substituted arenes can be readily oxidised by prolonged reflux with a strong oxidising agent.

Oxidation of methylbenzene

Using alkaline potassium manganate(VII), followed by addition of dilute aqueous acid, methylbenzene is oxidised to benzenecarboxylic acid (**benzoic acid**).

$$\text{CH}_3\text{-benzene} + 3[O] \xrightarrow{reflux} \text{COOH-benzene} + H_2O$$

benzenecarboxylic acid

Comparing hydrolysis of chloroarenes and chloroalkanes

Chloroalkanes, such as chloroethane, are hydrolysed by heating with aqueous alkali.

$$CH_3CH_2Cl + OH^- \longrightarrow CH_3CH_2OH + Cl^-$$

Using the same conditions, chloroarenes, such as chlorobenzene, are **not** hydrolysed.

- The aryl ring interacts with a p-orbital on the attached chlorine atom, drawing electron density towards the delocalised ring structure.

C–Cl bond
strengthened

- This results in the strengthening of the C–Cl bond, increasing the stability of chloroarenes compared with chloroalkanes.

Progress check

1 What is the common type of reaction of arenes?

2 Benzene reacts with bromine, nitric acid and chloroethane.
(a) Using C_6H_6 to represent benzene and C_6H_5 to represent the phenyl group, write balanced equations for each of these reactions.
(b) State the essential conditions that are needed for each of these reactions.

1 Electrophilic substitution.

2 (a) $C_6H_6 + Br_2 \longrightarrow C_6H_5Br + HBr$
$C_6H_6 + HNO_3 \longrightarrow C_6H_5NO_2 + H_2O$
$C_6H_6 + C_2H_5Br \longrightarrow C_6H_5C_2H_5 + HBr$
(b) With bromine, a halogen carrier such as Fe.
With nitric acid, concentrated sulfuric acid at 55°C.
With chloroethane, a halogen carrier such as Fe.

4.8 Phenols and alcohols

After studying this section you should be able to:

- state the uses of phenols in antiseptics and disinfectants
- describe the reactions of phenol with sodium and bases
- explain the ease of bromination of phenol compared with benzene

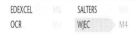 LEARNING SUMMARY

Phenols

General structure: ArOH (Ar is an aromatic ring)

 EDEXCEL | SALTERS
OCR | WJEC M4

Phenols and alcohols both have a hydroxyl functional group. In alcohols, the hydroxyl group is bonded to a carbon chain but phenols have an aromatic ring bonded directly to a hydroxyl (–OH) group. Some phenols are shown below.

| phenol, C_6H_5OH | 2,4,6-trichlorophenol (TCP) | 4-hexyl-3-hydroxyphenol (in cough drops) | thymol (in thyme) |

Uses

With $FeCl_3$ solutions, phenols produce a violet colour. This colour change is often used as a test for the phenol group.

Dilute solutions of phenol are used as disinfectants and phenol was probably the first antiseptic. However, phenol is toxic and can cause burns to the skin. Less toxic phenols are used nowadays in antiseptics.

Phenols are also used in the production of plastics, such as the widely used phenol-formaldehyde resins. More complex phenols, such as thymol shown above, can be used as flavourings and aromas and these are obtained from essential oils of plants.

Properties

Like alcohols, a phenol has a **hydroxyl group** which is able to form intermolecular **hydrogen bonds**. The resulting properties include:

- higher melting and boiling points than hydrocarbons with similar relative molecular masses
- some solubility in water.

Reactions of phenol as an acid

OCR | SALTERS
| WJEC M4

Unlike the neutrality shown by alcohols, phenols are very weak acids. The hydroxyl group donates a proton, H^+, by breaking the O–H bond.

$$C_6H_5OH \rightleftharpoons H^+ + C_6H_5O^-$$

Phenols take part in some typical acid reactions, reacting with sodium and bases to form ionic salts called **phenoxides**. However, the acidity is too weak to release carbon dioxide from carbonates and hydrogencarbonates (see carboxylic acids, page 119).

Key points from AS

- **Reaction with sodium**
  Revise AS page 138

Reaction with sodium

As with the hydroxyl groups in water and ethanol, phenol reacts with sodium, releasing hydrogen gas and forming a solution of sodium phenoxide, C_6H_5ONa.

$$2C_6H_5OH + 2Na \longrightarrow 2C_6H_5ONa + H_2$$

Compare these reactions with those of carboxylic acids page 119.

Reaction with an alkali

Phenol is a weak acid and is neutralised by aqueous alkalis. For example, aqueous sodium hydroxide forms the salt sodium phenoxide. Alcohols and water are not acidic enough to neutralise alkalis.

$$C_6H_5OH + NaOH \longrightarrow C_6H_5ONa + H_2O$$

Electrophilic substitution of the aryl ring of phenol

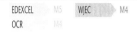

The aryl ring of phenol has a greater electron density than benzene because the adjacent oxygen reinforces the electron density in the ring. A p-orbital on the oxygen donates an electron pair to the benzene ring.

This results in:

• a greater electron density in the aryl ring, activating the ring
• greater reactivity of the ring towards electrophiles.

Bromination of phenol

• Phenol reacts directly with bromine. However, Benzene reacts with bromine only in the presence of a halogen carrier.
• Phenol undergoes multiple substitution with bromine. Benzene is monosubstituted only.

• The organic product 2,4,6-tribromophenol separates as a white solid.
• The bromine is decolourised.

Nitration of phenol (Edexcel only)

• Phenol reacts with dilute nitric acid in the cold to form a mixture of 2-nitrophenol and 4-nitrophenol.

• Benzene needs much stronger conditions – concentrated nitric acid and concentrated sulfuric acid at 55°C.

• This reaction again shows the increased reactivity of the aryl ring of phenol compared with benzene.

Reaction with iron(III) chloride, FeCl₃

Phenols form a violet coloration with $FeCl_3$ solution.

• This reaction is used as a test for a phenol.

• This allows phenols to be distinguished from alcohols.

Formation of esters

SALTERS   M4   WJEC   M4

Esters can be formed from an **alcohol** with a **carboxylic acid** in the presence of an acid catalyst (see page 120).

In phenols, the aryl ring reduces the electron density on the –OH group, reducing its effectiveness as a nucleophile. This means that phenols do **not** form esters in the same way as alcohols.

To form an ester from a phenol:

- an acyl chloride is used rather than the less reactive carboxylic acid (see page 125)
- alkaline conditions are used.

$$CH_3COCl + C_6H_5OH + NaOH \longrightarrow CH_3COOC_6H_5 + NaCl + H_2O$$

Reactions of the –OH group in alcohols

WJEC   M4

Substitution reactions of alcohols with hydrogen bromide

Alcohols react with hydrogen bromide by **nucleophilic substitution**.

The alcohol is mixed with NaBr and H_2SO_4. The reaction mixture generates hydrogen bromide as H^+ and Br^- ions.

You should also be aware of the use of ethanol as a biofuel.

$$C_2H_5OH + H^+ + Br^- \longrightarrow C_2H_5Br + H_2O$$

- The polarity of the C–OH bond allows a reaction with nucleophiles such as Br^-, which attack the electron-deficient carbon atom.

Dehydration of alcohols

When an alcohol is refluxed with a concentrated acid catalyst such as sulfuric acid H_2SO_4, or phosphoric acid H_3PO_4, an elimination reaction takes place: water is eliminated with formation of an alkene.

$$C_2H_5OH \longrightarrow C_2H_4 + H_2O$$

Progress check

1 Phenols are easily nitrated with nitric acid in the cold. With concentrated HNO_3, multiple substitution takes place.
   Suggest an equation for the reaction that takes place.

4.9 Amines

After studying this section you should be able to:

- explain the relative basicities of ethylamine and phenylamine
- describe the reactions of primary amines with acids to form salts
- describe the formation of phenylamine by reduction of nitrobenzene
- describe the synthesis of an azo dye from phenylamine
- understand the origin of colour in azo dyes
- describe the reactions of amines with halogenoalkanes and acyl chlorides
- describe the preparation and reaction of amides

LEARNING SUMMARY

Aliphatic and aromatic amines

Amines are organic compounds containing nitrogen derived from ammonia, NH_3. Amines are classified as primary, secondary or tertiary amines depending on how many of the hydrogen atoms in ammonia have been replaced by organic groups. The diagram below shows examples of aliphatic and aromatic amines.

Aliphatic amines   **Aromatic amine**

1 hydrogen replaced	2 hydrogens replaced	3 hydrogens replaced
primary amine	*secondary amine*	*tertiary amine*

NH_3 $H_3C\!-\!NH_2$ $\begin{array}{c} H_3C \\ \!\!\!\!\diagdown \\ H_3C \end{array}\!\!NH$ $\begin{array}{c} H_3C \\ H_3C\!-\!N \\ H_3C \end{array}$ $\langle\bigcirc\rangle\!-\!NH_2$

ammonia   methylamine   dimethylamine   trimethylamine   phenylamine

Amines in nature

Amines are found commonly in nature. They are weak bases and the shortest chain amines, e.g. ethylamine $C_2H_5NH_2$, smell of fish. Some diamines, such as putrescine $H_2N(CH_2)_4NH_2$, and cadaverine $H_2N(CH_2)_5NH_2$, are found in decaying flesh.

Polarity

Notice the similarity with ammonia.

The presence of an electronegative nitrogen atom in amines results in polar molecules. The diagram shows the polarity of the primary amine methylamine.

$$\begin{array}{c} \overset{\bullet\bullet\,\delta-}{N} \\ H_3C\cdots\!\!\diagup\!\!\diagdown H\,\delta+ \\ H \\ \delta+ \end{array}$$

Physical properties of amines

As with other polar functional groups, the solubility of amines in water decreases with increasing carbon chain length as the non-polar contribution to the molecule becomes more important.

The amino group dominates the physical properties of short-chain amines.

Hydrogen bonding takes place between amine molecules, resulting in:

- higher melting and boiling points than alkanes of comparable relative molecular mass
- solubility in water – amines with fewer than six carbons mix with water in all proportions.

Amines as bases

AQA	M4	SALTERS	M4
EDEXCEL	M5	WJEC	M4
OCR	M4	CCEA	M5

Amines are the organic bases.

Amines are weak bases because they only partially associate with protons:

$$RNH_2 + H^+ \rightleftharpoons RNH_3^+$$

The basic strength of an amine measures its ability to **accept** a proton, H^+. This depends upon:

- the size of the $\delta-$ charge on the amino nitrogen atom
- the availability of the nitrogen lone pair.

The basicity of aliphatic and aromatic amines

Aliphatic amines are stronger bases than ammonia; aromatic amines are substantially weaker.

Amines can also act as ligands with transition metal ions forming complex ions.

- Electron-donating groups, e.g. alkyl groups, **increase** the basic strength.
- Electron-withdrawing groups, e.g. C_6H_5, **decrease** the basic strength.

KEY POINT

inductive effect causes electron flow **away from** nitrogen

inductive effect causes electron flow **towards** nitrogen

decrease in electron density of nitrogen lone pair

increase in electron density of nitrogen lone pair

basicity increases

Amines react by the usual 'base reactions' producing organic ammonium salts.

Organic ammonium salts

Amines are **neutralised by acids** forming salts.

In the example below, methylamine is neutralised by hydrochloric acid forming a primary **ammonium salt**.

A proton, H^+, is added to the amino nitrogen atom.

$$CH_3NH_2 + HCl \longrightarrow CH_3NH_3^+Cl^-$$
methylammonium chloride

On evaporation of water, the primary ammonium salt crystallises out as an ionic compound.

Preparation of amines

AQA	M4	WJEC	M4
EDEXCEL	M5	CCEA	M5
OCR	M4		

Other reducing agents can also be used:
- *H_2/Ni catalyst*
- *$LiAlH_4$ in dry ether*

$LiAlH_4$ is an almost universal reducing agent and works in most examples of organic reduction.

Preparation of aromatic amines from nitroarenes

Aromatic amines can be prepared by **reducing** a nitroarene. For example, nitrobenzene is reduced by Sn in concentrated HCl to form phenylamine.

- Using tin and hydrochloric acid, the salt $C_6H_5NH_3^+Cl^-$ is formed.
- The amine is obtained from this salt by adding aqueous alkali.

Preparation of aliphatic amines from nitriles (not OCR)

Aliphatic amines can be prepared by **reducing** a nitrile with a suitable reducing agent (e.g. Na in ethanol; $LiAlH_4$ in dry ether).

$$CH_3CH_2C{\equiv}N + 4[H] \longrightarrow CH_3CH_2CH_2NH_2$$

The preparation of dyes from aromatic amines

Aromatic amines, such as phenylamine, are important industrially for the production of dyes. Modern dyes are formed in a two-stage synthesis:

- the aromatic amine is converted into a **diazonium salt**
- the diazonium salt is **coupled** with an aromatic compound such as phenol, forming an **azo dye**.

Formation of diazonium salts

An aromatic amine, such as phenylamine, forms a diazonium salt in the presence of nitrous acid, HNO_2, and hydrochloric acid.

- HNO_2 is unstable and is prepared *in situ* from $NaNO_2$ and $HCl(aq)$.

$$NaNO_2 + HCl \longrightarrow HNO_2 + NaCl$$

- This mixture is then reacted with phenylamine. It is important to keep the temperature **below 10°C** because diazonium salts decompose above this temperature.

$$C_6H_5NH_2 + HNO_2 + HCl \longrightarrow C_6H_5N_2{}^+Cl^- + 2H_2O$$

Formation of azo dyes by coupling

The diazonium salt is coupled with a suitable aromatic compound **below 10°C** in **aqueous alkali** to produce an azo dye.

E.g. coupling of benzenediazonium chloride with phenol:

- The azo dye is coloured.
- Coupling of a diazonium salt with different aromatic compounds forms dyes with different colours.

Using benzene for the synthesis of dyes

Benzene is used as the raw material for the synthesis of dyes.
A four-stage synthesis is shown below:

- benzene is nitrated to nitrobenzene (see page 131)
- nitrobenzene is reduced to phenylamine (see above)
- phenylamine is converted into a **diazonium salt**
- the diazonium salt is **coupled** with an aromatic compound such as phenol, forming an **azo dye**.

Diazonium salts in synthesis (CCEA only)

In addition to azo dyes, diazonium salts are useful for the synthesis of a variety of aromatic organic compounds.

- The diazonium group, $-N_2^+$, can be replaced by nucleophilic reagents.
- The N_2^+ group is lost as N_2 gas.

Preparation of phenol

The solution of the diazonium salt is simply warmed. Diazonium salts are unstable above 10°C, and hydrolysis with water takes place.

$$C_6H_5-N^+\equiv N\ Cl^- + H_2O \longrightarrow C_6H_5-OH + N_2 + HCl$$

Preparation of iodobenzene

The solution of the diazonium salt is warmed with aqueous potassium iodide.

$$C_6H_5-N^+\equiv N\ Cl^- + KI \longrightarrow C_6H_5-I + N_2 + KCl$$

Colour in azo dyes

SALTERS   M5   WJEC   M4

CCEA   M5

When energy is absorbed, the electron moves between electronic energy levels.

$\Delta E = hf$

Where ΔE = energy gap,

h = the Planck constant

f = frequency

(See page 90 for more details)

On pages 90–91, the origin of colour in transition elements is discussed. For organic compounds, the principles are the same.

When visible light strikes an organic compound that is coloured:

- an electron is promoted from a lower to a higher energy level
- **energy** of certain wavelengths is **absorbed**
- part of the visible spectrum has been removed and the **remaining wavelengths** are **reflected as colour**.

> **KEY POINT**
>
> A **chromophore** is part of an organic molecule responsible for its colour.

- Common chromophores include C=C, C=O, N=N, an aromatic ring, and features with lone pairs such as OH, NH_2.
- Colour is produced when the energy gap between energy levels matches the energy from part of the visible spectrum.
- For many organic molecules containing chromophores, this energy gap is large and matches energies in the ultraviolet region of the spectrum.
- Delocalisation reduces this energy gap. When the delocalisation is such that the energy gap matches energies in the visible part of the spectrum, the organic compound is coloured.

The ultraviolet region of the spectrum has more energy than the visible region – the gaps between its energy levels are greater.

The colour of an organic compound results from the **transfer** of an **electron** between energy levels in a delocalised system.

Colourless

ultraviolet

Kekulé structure of benzene
Some delocalisation but electron energy gap is large and matches ultraviolet region of spectrum

energy levels

Coloured

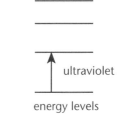

visible

energy levels

Azo dye has more delocalisation
(note the alternating double and single bonds)
Energy gap is smaller and matches visible part of spectrum

The key to colour in organic compounds is a delocalised arrangement of alternating double and single bonds, such as that in an azo dye.

Attaching dyes to fibres (Salters only)

Dyes attach themselves to fibres by:

- intermolecular bonds
- ionic attractions
- covalent bonding.

Modifying dyes (Salters only)

Chemists modify dyes by introducing different functional groups.

These may:

- modify the chromophore
- affect the solubility of the dye
- allow the dye to bond to fibres.

Complex ion formation with aqueous copper(II) ions

EDEXCEL M5

This reaction is similar to that of ammonia, NH_3 with transition metal ions.

However, the number of amine ligands that substitute may be different from NH_3.

An aqueous solution of copper(II) sulfate contains $[Cu(H_2O)_6]^{2+}$ complex ions.

- The lone pair of the amino group is able to form a coordinate bond to the Cu^{2+} ion.
- A ligand substitution reaction takes place (see page 93–94).
- This shows as a change in colour.

$$[Cu(H_2O)_6]^{2+} + 2C_6H_5NH_2 \longrightarrow [Cu(C_6H_5NH_2)_2(H_2O)_4]^{2+} + 2H_2O$$

Reactions of amines with halogenoalkanes

AQA M4 WJEC M4
OCR M4 CCEA M5

Key points from AS

- Nucleophilic substitution reactions of halogenoalkanes
  Revise AS pages 139–141

Formation of a primary amine by nucleophilic substitution

Amines act as **nucleophiles** with organic compounds such as halogenoalkanes.

Halogenoalkanes react with **excess ammonia** in hot ethanol, forming a primary amine. The reaction with bromomethane is shown below:

$$CH_3Br + 2NH_3 \longrightarrow CH_3NH_2 + NH_4^+Br^-$$

The reaction takes place in two steps:

- Nucleophilic substitution with ammonia forms an alkylammonium ion:

$$CH_3Br + NH_3 \longrightarrow CH_3NH_3^+ + Br^-$$

- Proton transfer with ammonia forms a primary amine:

$$CH_3NH_3^+Br^- + NH_3 \rightleftharpoons CH_3NH_2 + NH_4^+Br^-$$

Mechanism and further substitution (AQA only)

The mechanism for the formation of a primary amine from bromoethane is shown below.

With an **excess** of **ammonia**, the reaction essentially stops at this stage.

With an **excess** of the **halogenoalkane**, further substitution takes place, which continues until a quaternary ammonium salt is obtained.

$$RNH_2 \xrightarrow{RX} R_2NH \xrightarrow{RX} R_3N \xrightarrow{RX} R_4N^+X^-$$

primary	*secondary*	*tertiary*	*quaternary*
amine	*amine*	*amine*	*ammonium salt*

Each reaction stage proceeds by **nucleophilic substitution**.

The mechanism for the final stage in the formation of the quaternary salt tetramethylammonium bromide is shown below.

Quaternary ammonium salts are used as cationic surfactants in fabric softeners.

- This mechanism stops at the quaternary salt as there are no further protons to be exchanged.

Amides

Amides

Primary amides have the general formula $RCONH_2$ – R is an alkyl group.

The first six members of the **amides** homologous series are shown below.

structural formula	molecular formula	name
$HCONH_2$	CH_3NO	methanamide
CH_3CONH_2	C_2H_5NO	ethanamide
$CH_3CH_2CONH_2$	C_3H_7NO	propanamide
$CH_3CH_2CH_2CONH_2$	C_4H_9NO	butanamide
$CH_3CH_2CH_2CH_2CONH_2$	$C_5H_{11}NO$	pentanamide
$CH_3CH_2CH_2CH_2CH_2CONH_2$	$C_6H_{13}NO$	hexanamide

Types of amide

primary amide   secondary amide   tertiary amide

Preparation of amides

The amide group is stable and relatively unreactive. Preparation directly from a carboxylic acid is difficult (although this type of reaction is essentially what happens when peptides form from amino acids – see pages 150 and 157).

Amides are generally prepared by the reaction of ammonia and amines with **acyl chlorides** or with acid anhydrides – these are much more reactive than carboxylic acids (see pages 124–125).

- Reaction with ammonia produces a **primary amide**.

    $$CH_3COCl + NH_3 \longrightarrow + CH_3CONH_2 + HCl$$

    (Then $NH_3 + HCl \longrightarrow NH_4Cl$)

The acyl chloride is often prepared initially from a carboxylic acid. The acyl chloride can then be reacted with ammonia or amines to form amides.

- Reaction with a primary amine produces a **secondary amide**.

$$CH_3COCl + CH_3NH_2 \longrightarrow + \mathbf{CH_3CONHCH_3} + HCl$$

$$(Then\ NH_3 + HCl \longrightarrow NH_4Cl)$$

An amide can be prepared from a carboxylic acid by first converting the carboxylic acid into an acyl chloride (see pages 124–125 for more details).

Hydrolysis of amides (Salters and CCEA only)

Amides are stable organic compounds. They can be hydrolysed but they require much stronger conditions than esters. Alkaline hydrolysis requires prolonged refluxing with 6 mol dm^{-3} NaOH(aq).

Acid hydrolysis

Acid hydrolysis of amides form a carboxylic acid and an ammonium salt:

$$CH_3CONH_2 + H_2O + HCl \longrightarrow CH_3COOH + NH_4^+Cl^-$$

Alkaline hydrolysis

Alkaline hydrolysis of amides form an acid salt and ammonia:

$$CH_3CONH_2 + NaOH \longrightarrow CH_3COO^-Na^+ + NH_3$$

Dehydration of amides (CCEA only)

Amides are **dehydrated** by phosphorus pentoxide, P_2O_5, to form nitriles:

$$CH_3CONH_2 \longrightarrow CH_3C{\equiv}N + H_2O$$

Basicity of amides (CCEA only)

Compared to amines, amides are very weak bases. This lack of basicity is explained by delocalisation of the lone pair of electrons on the nitrogen atom by resonance.

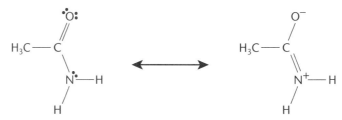

lone pair on nitrogen atom is delocalised

bond angles around N are close to 120° the molecule has a near planar shape about C–N

The delocalisation reduces the electron density on the nitrogen lone pair, reducing its ability to donate its electron pair.

Progress check

1 Explain why ethylamine is a stronger base than phenylamine.

2 (a) Write equations for the conversion of:
   (i) benzene to nitrobenzene
   (ii) nitrobenzene to phenylamine
   (iii) phenylamine to benzenediazonium chloride.

3 (a) Write down the formula of the organic product formed between:
   (i) bromoethane and excess ammonia
   (ii) ammonia and excess bromoethane.

Sample question and model answer

This question looks at the structure and reactions of benzene.

(a) The average enthalpy of hydrogenation of a single C–C bond is -120 kJ mol^{-1}.

(i) Assuming that benzene consists of a ring with three separate double bonds, predict the enthalpy change for the hydrogenation of benzene to form cyclohexane.

This is an easy mark.

Don't forget to include the sign!

3×-120 kJ mol$^{-1} = -360$ kJ mol^{-1} ✓

(ii) The actual enthalpy of hydrogenation of benzene is -205 kJ mol^{-1}. Using this information and your answer to (i), what conclusion can be drawn about the stability of the benzene ring? Use an enthalpy level diagram to illustrate your answer.

An enthalpy diagram is a good way of showing the delocalisation energy of delocalised benzene.

1 mark here is awarded for the relative positions of the delocalised and Kekulé structures of benzene.

Benzene has extra stability arising from delocalisation of π-electrons ✓

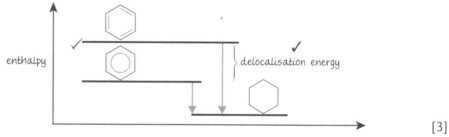

[3]

(b) Benzene can be nitrated to form nitrobenzene. Give the reagents, the equation and the conditions for this reaction.

This is standard bookwork that must be learnt.

Other reactions could have been chosen (e.g. bromination).

Remember that 'reagents' are the chemicals 'out of the bottle' that are reacted with the organic compound.

You must give the full name or formula for any 'reagent'.

Reagent(s):	concentrated HNO_3/H_2SO_4 ✓
Equation:	$C_6H_6 + HNO_3 \longrightarrow C_6H_5NO_2 + H_2O$ ✓
Conditions:	warm to 55°C ✓

[3]

(c) For the nitration of benzene:

(i) write the formula of the electrophile

NO_2^+ ✓

(ii) write the equation to show the formation of the electrophile

More standard bookwork.

An alternative response here would be:

$HNO_3 + 2H_2SO_4 \rightarrow NO_2^+ + 2HSO_4^- + 2H_2O$

$HNO_3 + H_2SO_4 \longrightarrow NO_2^+ + HSO_4^- + H_2O$ ✓

(iii) state the type of mechanism

electrophilic substitution ✓

(iv) outline the mechanism.

Notice how precise you need to be to score all three marks:

• arrow to electrophile ✓
• correct intermediate ✓
• arrow on C–H for loss of H$^+$ ✓.

The sulfuric acid catalyst is then regenerated:

$H^+ + HSO_4^- \longrightarrow H_2SO_4$ ✓

[7]

[Total:13]

Practice examination questions

1 Compounds **A**, **B** and **C** are structural isomers, each containing a carbonyl group. The compounds have a relative molecular mass of 72 and the following percentage composition by mass: C, 66.7%; H, 11.1%; O, 22.2%.

(a) Calculate the molecular formula of the isomers. [2]

(b) What are the structural formulae of **A**, **B** and **C**? [3]

(c) Name a reagent that reacts with **A**, **B** and **C**. State the observation. [2]

(d) Name a reagent that reacts with only two of the isomers **A**, **B** and **C**. Identify which isomers react and state the observation. [3]

(e) One of the isomers **A**, **B** and **C** is reacted with $NaBH_4$ to form a product that has optical isomers. Identify this product and the isomer that has been reacted with $NaBH_4$. [2]

[Total: 12]

2 (a) Compound **D**, $CH_3CH_2COOCH_2CH_3$, is an ester.

(i) Name compound **D**.

(ii) Compound **D** was hydrolysed by heating with aqueous sodium hydroxide. An alcohol **E** was formed, together with another organic product, **F**. Give the structural formulae of compounds **E** and **F**. [4]

(b) Compound **G** is a di-ester formed by the reaction of ethane-1,2-diol and methanoic acid.

(i) Draw the structural formula of compound **G**.

(ii) Identify a substance that could catalyse this reaction. [2]

(c) Compound **H** is a tri-ester. Complete and balance the equation below for the formation of the tri-ester **H** by esterification.

$$\longrightarrow \quad \begin{array}{l} CH_3COO-CH_2 \\ CH_3COO-CH \\ CH_3COO-CH_2 \end{array}$$

tri-ester H

[3]

[Total: 9]

3 An aromatic hydrocarbon has a relative molecular mass of 106 and has the following composition by mass: C, 90.56%; H, 9.44%.

(a) (i) Deduce the molecular formula of the aromatic hydrocarbon.

(ii) Draw structures for all the structural isomers of this aromatic hydrocarbon. [7]

(b) In the presence of a sulfuric acid, one of these isomers, **I**, reacts with nitric acid to give only one mononitro product **J**.

(i) Deduce which of the isomers in (a)(ii) is isomer **I**.

(ii) Draw the structure of **J**. [2]

[Total: 9]

4 Methylbenzene can be converted into an azo dye using the reaction scheme below.

(a) For Step 1:

(i) name the mechanism

(ii) state the reagents required. [2]

(b) For the reduction in Step 2:

(i) state the reagents required

(ii) write a balanced equation (use [H] to represent the reducing agent). [3]

(c) For Step 3:

(i) state the reagents required and the essential conditions

(ii) show the structure of the diazonium salt formed (the functional group should be displayed with charges clearly shown). [3]

(d) For Step 4:

(i) give the name and formula of a suitable organic compound for coupling with the diazonium salt

(ii) show the structure of the azo dye formed. [2]

[Total: 10]

5 Compound **C**, $CH_3CH_2CH_2COOCH(CH_3)_2$, can be prepared using the reaction scheme below.

$$CH_3CH_2CH_2CHO \xrightarrow{\text{Step 1}} \textbf{A} \ + \ \textbf{B} \xleftarrow{\text{Step 2}} CH_3COCH_3$$

$$\downarrow \text{Step 3}$$

$$CH_3CH_2CH_2COOCH(CH_3)_2$$

$$\textbf{C}$$

(a) Show the structural formulae of compounds **A** and **B**. [2]

(b) For each of the three steps, give the reagents and conditions and name the type of reaction. [9]

[Total: 11]

Chapter 5

Polymers, analysis and synthesis

The following topics are covered in this chapter:

- *Amino acids*
- *Polymers, proteins and nucleic acids*
- *Chromatography*

- *Analysis*
- *NMR spectroscopy*
- *Organic synthetic routes*

5.1 Amino acids

After studying this section you should be able to:

- describe the acid–base properties of amino acids and the formation of zwitterions
- explain the formation of polypeptides and proteins as condensation polymers of amino acids
- describe the acid hydrolysis of proteins and peptides

LEARNING SUMMARY

Amino acids

There are 22 naturally occurring amino acids.

A typical amino acid is an organic molecule with both acidic and basic properties comprising:

- a basic amino group, $-NH_2$
- an acidic carboxyl group, $-COOH$.

Each amino acid has a unique organic R group or side chain. The formula of a typical amino acid is shown below.

$$H_2N-\underset{\underset{H}{|}}{\overset{\overset{R}{|}}{C}}-COOH$$

Examples of some amino acids are shown below:

$$H_2N-\underset{\underset{H}{|}}{\overset{\overset{H}{|}}{C}}-COOH$$

R = H

glycine

$$H_2N-\underset{\underset{H}{|}}{\overset{\overset{CH_3}{|}}{C}}-COOH$$

R = CH_3

alanine

$$H_2N-\underset{\underset{H}{|}}{\overset{\overset{CH(CH_3)_2}{|}}{C}}-COOH$$

R = CH(CH_3)_2

valine

Acid–base properties of amino acids

Isoelectric points and zwitterions

Each amino acid has a particular pH called the **isoelectric point** at which the overall charge on an amino acid molecule is zero.

Examples of isoelectric points

amino acid	aspartic acid	glycine	histidine	arginine
isoelectric point	3.0	6.1	7.6	10.8

At the isoelectric point, the amino acid exists in equilibrium with its zwitterion form.

At the isoelectric point, an amino acid exists as a zwitterion:

- the **carboxyl** group **has donated** a proton to the **amino** group, which form a positive NH_3^+ ion.

A **zwitterion** is a dipolar ion with both positive and negative charges in different parts of the molecule.

$$H_3\overset{+}{N}-\underset{\underset{H}{|}}{\overset{\overset{R}{|}}{C}}-C\overset{\nearrow O}{\underset{\searrow O^-}{}}$$

zwitterion – two ions
in one molecule

Amino acids as bases

In strongly **acidic** conditions a **positive ion** forms:

- an amino acid behaves as a **base**
- the COO^- ion gains a proton.

$$H_3\overset{+}{N}-\underset{\underset{H}{|}}{\overset{\overset{R}{|}}{C}}-C\overset{\nearrow O}{\underset{\searrow O^-}{}} \quad + \quad H^+ \longrightarrow H_3\overset{+}{N}-\underset{\underset{H}{|}}{\overset{\overset{R}{|}}{C}}-C\overset{\nearrow O}{\underset{\searrow OH}{}}$$

positive ion

Because of their reactions with strong acids and strong bases, amino acids act as buffers and help to stabilise the pH of living systems.

Amino acids as acids

In strongly **alkaline** conditions a **negative ion** forms:

- an amino acid behaves as an **acid**
- the NH_3^+ ion loses a proton.

$$H_3\overset{+}{N}-\underset{\underset{H}{|}}{\overset{\overset{R}{|}}{C}}-C\overset{\nearrow O}{\underset{\searrow O^-}{}} \quad + \quad OH^- \longrightarrow H_2N-\underset{\underset{H}{|}}{\overset{\overset{R}{|}}{C}}-C\overset{\nearrow O}{\underset{\searrow O^-}{}} \quad + \quad H_2O$$

negative ion

> **KEY POINT**
> - At the **isoelectric point**, the amino acid is **neutral**.
> - At a pH more **acidic** than the isoelectric point, the amino acid forms a **positive ion**.
> - At a pH more **alkaline** than the isoelectric point, the amino acid forms a **negative ion**.

Physical properties of amino acids

AQA	M4	SALTERS	
EDEXCEL		WJEC	M4
OCR		CCEA	M5

Solid amino acids

Solid amino acids have higher melting points than expected from their molecular masses and structure. This suggests that amino acids crystallise in a giant lattice with strong electrostatic forces between the **zwitterions**.

Optical isomers

With the exception of aminoethanoic acid (*glycine*), H_2NCH_2COOH, all amino acids have a chiral centre and are optically active.

Other reactions of amino acids (CCEA only)

An amino acid has both amino and carboxyl groups. Each functional group can take part in their typical reactions.

Reaction with carbonates

With carbonates, $CO_2(g)$ is evolved and a salt is produced. The reaction of sodium carbonate with the amino acid glycine is shown below.

$$Na_2CO_3 + 2H_2NCH_2COOH \longrightarrow 2H_2NCH_2COO^-Na^+ + CO_2 + H_2O$$

This reaction is similar to that of ammonia, NH_3 with transition metal ions.

It is difficult to predict how many amino acid molecules have exchanged with water molecules.

Reaction with copper(II) sulfate

An aqueous solution of copper(II) sulfate contains $[Cu(H_2O)_6]^{2+}$ complex ions.

- The lone pair of the NH_2 part of the amino acid molecule is able to form a coordinate bond to the Cu^{2+} ion.
- A ligand substitution reaction takes place (see pages 93–94).
- This shows as a change in colour.

$$[Cu(H_2O)_6]^{2+} + x\ H_2NCH_2COOH \longrightarrow [Cu(H_2NCH_2COOH)_x(H_2O)_{6-x}]^{2+} + x\ H_2O$$

Reaction with nitrous acid

With nitrous acid, HNO_2, the amino part of the amino acid molecule reacts, producing a hydroxyl group and nitrogen gas:

$$HNO_2 + H_2NCH_2COOH \longrightarrow HOCH_2COOH + N_2 + H_2O$$

Polypeptides and proteins

A protein is formed by condensation polymerisation of amino acids. See page 157.

A polypeptide is the name given to a short chain of amino acids linked by peptide bonds.

A protein is simply the name given to a long-chain polypeptide.

In nature, individual amino acids are linked together in chains as **polypeptides** and **proteins** (see page 157).

The diagram below shows the condensation of the amino acids glycine and alanine to form a **dipeptide**. The amino acids are bonded together by a **peptide link**.

glycine   *alanine*   *peptide link*

Each peptide link forms:

- between the **carboxyl group** of glycine and the **amino group** of alanine
- with loss of a water molecule in a **condensation reaction**.

Further condensation reactions between amino acids build up a **polypeptide** or **protein**.

- For each amino acid added to a protein chain, one water molecule is lost.
- Most common proteins contain more than 100 amino acids.
- Each protein has a unique sequence of amino acids and a complex three-dimensional shape, held together by intermolecular bonds, including hydrogen bonds.

Hydrolysis of proteins

Proteins can also be hydrolysed with hot aqueous alkali.

Hydrolysis breaks down a protein into its separate amino acids.

- The protein is refluxed with 6 mol dm^{-3} HCl(aq) for 24 hours.
- The resulting solution is neutralised.

The equation below shows the hydrolysis of a dipeptide.

peptide link   *glycine*   *alanine*

- Hydrolysis of a protein is the reverse process to the condensation polymerisation that forms proteins.
- Biological systems use enzymes to catalyse this hydrolysis, which takes place at body temperature without the need for acid or alkali.

Identifying the amino acids in a protein

To determine the amino acids present in a protein, the protein is first boiled with 6 mol dm^{-3} hydrochloric acid. The amino acids formed are separated using paper chromatography and made visible by spraying the paper with ninhydrin.

Each amino acid moves a different distance on the chromatography paper, making for easy identification of the amino acids in the protein.

Progress check

1 The isoelectric point of serine (R = –CH$_2$OH) is 5.7. Draw the form of the molecule in aqueous solutions of pH 3.0, pH 5.7 and pH 10.0.

2 Draw the structure of the tripeptide with the sequence alanine–serine–aspartic acid (alanine: R = –CH$_3$; aspartic acid: R = –CH$_2$COOH).

The answers to questions 1 and 2 are shown (printed inverted) as chemical structural diagrams:

1. Three forms of serine:
   - pH = 3.0: $H_3\overset{+}{N}$–C–COOH with CH$_2$OH side chain
   - pH = 5.7: $H_3\overset{+}{N}$–C–COO$^-$ with CH$_2$OH side chain (zwitterion)
   - pH = 10.0: H_2N–C–COO$^-$ with CH$_2$OH side chain

2. Tripeptide structure:
   $H_2\overset{}{N}$–C–C–N–C–C–N–C–COOH
   with side chains CH$_3$, CH$_2$OH, CH$_2$COOH and peptide carbonyl (C=O) linkages.

5.2 Polymers, proteins and nucleic acids

After studying this section you should be able to:

- describe the characteristics of addition polymerisation
- describe the characteristics of condensation polymerisation in polypeptides, proteins, polyamides and polyesters
- discuss the disposal of polymers
- describe primary, secondary and tertiary structures of protein
- describe the characteristics of enzyme catalysis
- describe the structure and functions of DNA and its importance in directing protein synthesis

During the study of alkenes in AS Chemistry, you learnt about addition polymers. For A2 Chemistry, addition polymerisation is reviewed and compared with condensation polymerisation.

Monomers and polymers

AQA	M4	SALTERS	M4
EDEXCEL	M4 M5	WJEC	M4
OCR		CCEA	M5

A **polymer** is a compound comprising very large molecules that are multiples of simpler chemical units called **monomers**.

Monomers are small molecules that can combine together to form a single large molecule, called a **polymer**.

> **KEY POINT**
>
> Two processes lead to formation of a polymer.
> - **Addition polymerisation** – monomers react together forming the polymer **only**. There are no by-products.
> - **Condensation polymerisation** – monomers react together forming the polymer **and** a simple compound, usually water.

Addition polymerisation

AQA	M4	SALTERS	
EDEXCEL	M5	WJEC	M5
OCR		CCEA	M5

In addition polymerisation:

- the monomer is an **unsaturated** molecule with a double **C=C** bond
- the double bond is **lost** as the **saturated** polymer forms.

Many different addition polymers can be formed using different monomer units based upon alkenes.

Key points from AS

- **Addition polymerisation of alkenes**
  Revise AS pages 132–133

Make sure that you can show a repeat unit of a polymer from the monomer (and a monomer from a repeat unit).

Poly(propenamide) and poly(ethenol) are highly water-absorbent polymers because of the ability of the polar –OH and –CONH₂ groups to hydrogen bond with water.

phenylethene → poly(phenylethene) *polystyrene* — packaging as expanded foam

ethenol → poly(ethenol) — soluble laundry bags and liquid detergent capsules

propenamide → poly(propenamide) — water treatment to remove suspended matter, soft contact lenses

Properties of monomers and addition polymers

The monomers are volatile liquids or gases. Polymers are solids.
This difference can be explained in terms of van der Waals' forces.
- The van der Waals' forces acting between the large polymer molecules are much stronger than those acting between the much smaller monomer molecules.

Addition polymers of hydrocarbon monomers and of halogenated monomers are very stable and inert, e.g. poly(ethene), poly(propene) and poly(phenylethene) (polystyrene), poly(chloroethene) (PVC) and poly(tetrafluoroethene) (PTFE). These polymers are insoluble in solvents and are non-biodegradable.

Low density poly(ethene) (LDPE) and high density poly(ethene) (HDPE) (CCEA only)

Production

	LD poly(ethene)	HD poly(ethene)
Temperature	150–300 °C	40–90 °C
Pressure	(High) 1000–3000 atm	(Low) 1–50 atm
Catalyst/Initiator	O_2 initiator (>0.1%)	Ziegler-Natta catalysts
Branching	branched chains	straight chains
Crystallinity	50 – 75%	95%

- LDPE is relatively soft and flexible. It is used for plastic bags and flexible sheeting.
- HDPE is relatively hard and rigid. It is used to make bottles, washing up bowls and plastic pipes.

Branching, crystallinity and properties (Salters and CCEA only)

The flexibility and softening of polymers such as LDPE and HDPE depend upon temperature, branching and crystallinity.

Crystallinity

Crystallinity is the regular packing of the chains, due to the regular structure of the polymer. The lower the temperature, the more effective the intermolecular bonds.
- A polymer softens and loses its crystallinity above its 'crystalline melting temperature', T_m
- A polymer becomes brittle below its glass transition temperature, T_g.

Chain length and branching

- As chain length increases, there are more intermolecular bonds, leading to greater strength.
- With branched chains, the polymer chains cannot get as close together as in non-branched chains and intermolecular bonds will be weaker.
- Flexibility depends on the ability of the polymer chains to slide over each other and this is more evident in polymers with branched chains.

Modifying properties

Chemists can modify the properties of a polymer to meet particular needs by:
- cold-drawing, to make the structure more crystalline
- copolymerisation – producing mixed polymers by using more than one monomer
- use of plasticisers, which change T_g and increase flexibility.

HDPE is more crystalline than LDPE.

The polymer chains in HDPE can align with one another more easily.

HDPE has non-branched chains.

LDPE has branched chains.

HDPE has stronger intermolecular forces and has more strength than LDPE.

LDPE is flexible.

HDPE is rigid.

The properties of all materials depend on their structure and bonding.

Condensation polymerisation

AQA	M4	SALTERS	M4
EDEXCEL	M4, M5	WJEC	M4
OCR	M4	CCEA	M5

In condensation polymerisation, the formation of a bond between monomer units also produces a small molecule such as H_2O or HCl.

Condensation polymers can be divided into natural polymers and man-made (synthetic) polymers.

- Natural polymers in living organisms include proteins, cellulose, rayon and DNA.
- Man-made polymers include synthetic fibres such as polyamides (e.g. *nylon*) and polyesters (e.g. *terylene*).

Polyamides

Nylon, proteins and polypeptides are all polyamides.

Proteins and polypeptides are natural condensation polymers. The link between the amino acid monomer units is usually described as a **peptide link** but chemically this is identical to an **amide** group. Hence polypeptides and proteins are **natural polyamides**.

Nylon-6,6 was the first man-made condensation polymer and was synthesised as an artificial alternative to natural protein fibres such as wool and silk.

Compare the use of two monomers in the production of synthetic nylon-6,6 with the natural condensation polymerisation using amino acids only.

The principle used was to mimic the natural polymerisation process above but, instead of using an amino acid monomer with two different functional groups, **two** chemically different monomers are usually used:

- a dicarboxylic acid **A**
- a diamine **B**

Many different polyamides can be made using different carbon chains or rings which bridge the double functional group.

Each diamine molecule bonds to a dicarboxylic acid molecule with loss of water molecule.

- the two monomers **A** and **B** join alternately: –A–B–A–B–A–B–A–B–

Polyamides can also be made using a diacyl chloride instead of a dicarboxylic acid.

- The greater reactivity of an acyl chloride results in easier polymerisation.
- Hydrogen chloride is lost instead of water.

Formation of nylon-6,6 from its monomers

Nylon-6,6 gets its name from the number of carbon atoms in each monomer (diamine first).

The diamine has 6 carbon atoms.

The dicarboxylic acid has 6 carbon atoms.

Hence: nylon-6,6.

The diacyl chloride for preparing nylon-6,6 would be $ClOC(CH_2)_4COCl$.

hexane-1,6-dioic acid

1,6-diaminohexane

$H_2N—(CH_2)_6—NH_2$

n molecules of a dicarboxylic acid

n molecules of a diamine

polymerisation

amide link

repeat unit

nylon-6,6

$+ \, n \, H_2O$

One water molecule forms for each amide bond formed

Formation of Kevlar from its monomers

Kevlar

Other polyamides include Kevlar, one of the hardest materials known. Kevlar is used for bulletproof vests, belts for radial tyres, cables and reinforced panels in aircraft and boats.

Polyesters

Polyesters are polymers made by a condensation reaction between monomers with formation of an ester group as the linkage between the molecules.

The first man-made polyester produced was *Terylene*. As with polyamides, polyesters are used as fibres for clothing.

Learn the principle behind condensation polymerisation.

In exams, you may be required to predict structures from unfamiliar monomers.

However, the principle is the same.

As with artificial polyamides, man-made polyesters are usually made from **two** different monomers:

- a dicarboxylic acid
- a diol

carbon chain
or
ring structure

Each diol molecule bonds to a dicarboxylic acid molecule with loss of a water molecule.

This is the same basic principle as for polyamides.

Many different polyesters can be made by using different carbon chains or rings bridging the double functional group.

Formation of Terylene from its monomers

'Terylene' is a trade name for the polymer 'polyethylane terephthelate' (PET). PET is used to make the plastic for many drink bottles.

n molecules of a dicarboxylic acid

n molecules of a diol

benzene-1,4-dicarboxylic acid

monomer A

ethane-1,2-diol

monomer B

polymerisation

ester link

repeat unit

Terylene

One water molecule forms for each ester bond formed

155

Disposal of polymers

AQA ▶ M4   SALTERS ▶ M4
EDEXCEL ▶ M5   CCEA ▶ M4, M5
OCR ▶ M4

Solid domestic and industrial waste contains a high percentage of polymers.

Disposal requires waste management strategies such as:

- incineration – to reduce waste bulk and to generate energy
- recycling – to preserve natural oil-produced finite resources produced from oil.

Problems with addition polymers (polyalkenes)

Addition polymers are non-polar and chemically inert.
This creates potential environmental problems during disposal of polymers.
Disposal by landfill causes long-term problems.

- Addition polymers are **non-biodegradable** and take many years to break down.

Disposal by burning can produce toxic fumes.

- Depolymerisation produces poisonous monomers.
- Disposal of poly(chloroethene) (*pvc*) by incineration can lead to the formation of very toxic dioxins if the temperature is too low.

Condensation polymers

Condensation polymers are polar.

- Condensation polymers are broken down naturally by acid and alkaline **hydrolysis** into their monomer units. They are therefore biodegradable and easier to dispose of than addition polymers.
- They may be photodegradable as the C=O bond absorbs radiation.

Chemists are minimising environmental waste by developing degradable polymers, similar in structure to poly(lactic acid).

lactic acid   **repeat unit**   poly(lactic acid)

Poly(lactic acid) is used for waste sacks, packaging, disposable eating utensils and medical applications such as internal dissolvable stitches.

'Green polymers' (Salters only)

The principles of green chemistry are important in the manufacture, use, recycling and eventual disposal of polymers. They involve:

- minimising any hazardous waste during the production of raw materials and their resulting polymers to reduce any negative impact on the environment
- reducing carbon emissions resulting from the 'life cycle' of a polymer
- recycling to produce energy and chemical feedstocks.

Progress check

1 (a) Draw the structure of the monomer needed to make poly(tetrafluoroethene).
(b) Draw the repeat unit of the polymer perspex, made from the monomer methyl 2-methylpropenoate, shown below.

2 (a) Show the structures of the two monomers needed to make nylon-4,6.
(b) Draw a short section of nylon 4,6 and show its repeat unit.

3 Draw the structures of the two monomers needed to make the polyester shown below.

(Answers, printed inverted at the foot of the Progress check box:)

3 HO—⬡—OH and HOOC—⬡—COOH

2 (a) H₂N—(CH₂)₄—NH₂ and HOOC—(CH₂)₄—COOH
(b) repeat unit of nylon 4,6

1 (a) F₂C=CF₂
(b) repeat unit of perspex

Proteins and polypeptides

AQA	M4	SALTERS	M4
EDEXCEL	M5	WJEC	M4
OCR	M4	CCEA	M5

Proteins and polypeptides

A protein is a chain of many amino acids linked together with 'peptide bonds' formed by **condensation polymerisation**.

Each amino acid molecule has both **amino** and **carboxyl** groups.
A **peptide** bond forms, with loss of a water molecule, between:

- the **amino** group of one amino acid molecule and
- the **carboxyl** group of another amino acid molecule.

The diagram below shows how amino acid molecules are linked together during condensation polymerisation.

Although a 'repeat unit' is shown on the diagram, the R group may be different in each unit. There are over 20 different amino acids, each having a different R group.

- Most common proteins contain more than 100 amino acids.
- Each protein has a unique sequence of amino acids and a complex three-dimensional shape, held together by intermolecular bonds including hydrogen bonds.

Just think of how many
100 letter words you
could make out of 20
letters of the alphabet!

Primary structure (Salters and CCEA only)

The **primary** structure of a protein is the actual **sequence** of **amino acids** joined by **peptide** links.

e.g. ....ala–gly–lys–cys–asn–arg–gly–leu–try–val–............

- There are countless possible sequences, with over 20 different amino acids and sequences of 50–2000 amino acids in different proteins. Each amino acid has its own shorthand code (e.g. ala). You do not need to know these.
- The amino acid sequence determines the properties of proteins and accounts for the diversity of proteins in living things. But each protein has its own unique primary structure.

For CCEA, you should
know the R– groups in 2
amino acids:

glycine (gly): R = H

alanine (ala): R = CH_3

All other R– groups
would be provided.

Secondary structure (Salters and CCEA only)

The **secondary** structure of a protein is a region in which there are regular structures.

There are two types, an α-**helix** and a β-**sheet**, each held together by **hydrogen bonds** between C=O and N–H:

C=O ···· H–N

α-helix   β-sheet

Tertiary structure (Salters and CCEA only)

The **tertiary** structure of a protein is the **three-dimensional** shape. The tertiary structure will contain regions of secondary structures and will always have the sequence of amino acids in the primary structure.

The tertiary structure is held together by intermolecular bonds that act between R– groups of different amino acids.

Possible R– group interactions are:

The tertiary structure
determines the
properties of proteins.

- **hydrogen bonding** – between polar side chains such as CH_2OH and CH_2CONH_2
- **covalent bonding** caused by SH groups forming –S–S– linkages
- **ionic attractions** – between ions on side chains such as –CH_2COO^- and NH_3^+
- **instantaneous dipole–induced dipole attractive forces** (van der Waals' forces) – between non-polar hydrocarbon chains.

Enzyme catalysis

Specificity (Salters and CCEA only)

An **enzyme** is:

- a **biological catalyst**
- a protein with a definite **three-dimensional** tertiary structure.

An enzyme is **specific** about the reactions that can be catalysed. Within the tertiary structure, there is an **active site**.

The substrate has to have the correct shape but the intermolecular bonds must also fit. Otherwise there will be nothing to hold the substrate to the active site.

- The active site has a definite shape that will only accept a molecule if it is the correct shape – the 'Lock and Key' analogy.
- The molecule that fits the active site is called the **substrate**.
- The substrate **bonds** to the active site by **intermolecular** bonds.

By bonding to the active site, a reaction is able to take place with a pathway of lower activation energy.

The effect of pH on enzyme activity (Salters and CCEA only)

Enzymes work at a particular **optimum pH**. Changes in pH will affect the ability of the substrate to bond by intermolecular bonds to the active site.

Extremes of pH will interfere with the intermolecular bonds within the tertiary structure. This can change the shape of the active site so that the substrate no longer fits.

The effect of temperature on enzyme activity (Salters and CCEA only)

Higher **temperatures increase** the **rate** of enzyme-catalysed reactions at the active site. However, if the **temperature** is **too high**,

- the **intermolecular bonds** within the **tertiary structure** may start to **break**,
- the **shape** of the **active site is changed** and the substrate no longer fits.

This change is irreversible and the enzyme will no longer function – it is **denatured**.

Inhibition (Salters only)

If you study Biology, you may have also come across non-competitive inhibition in which an inhibitor binds to another site on the enzyme, changing the shape of the active site.

An **inhibitor** is a molecule with a **similar shape** and with a similar arrangement of **binding sites** as the substrate.

- The inhibitor **competes** with the substrate for the active site and may block the site for access to a substrate molecule.
- The process is **reversible** as the inhibitor will also leave the active site.
- The rate of the enzyme-catalysed reaction slows down.

This is known as **competitive inhibition**.

Rates of enzyme-catalysed reactions (Salters only)

The graph below shows how the rate of an enzyme-catalysed reaction changes with increasing substrate concentration.

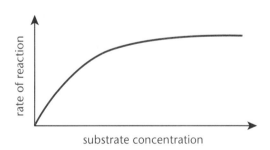

At **low substrate concentrations**:

See page 21 for shapes of rate/concentration graphs for different orders of reaction.

- the rate of reaction is **proportional** to the substrate concentration – the order with respect to substrate molecules is **one**
- active sites are **available** for any increase in substrate molecules.

As substrate **concentration increases**:

- there is progressively less effect until the rate levels off.
- the active sites become progressively **occupied**
- eventually all active sites are occupied and increasing the substrate concentration makes no difference to the rate – the order with respect to substrate molecules is **zero**.

Nucleic acids: DNA and RNA

SALTERS M4

DNA is a condensation polymer formed from monomer units called **nucleotides**, having three components: **phosphate**, **sugar** and **base**.
• The sugar is deoxribose in **DNA** (deoxyribonucleic acid).
• The sugar is ribose in **RNA** (ribonucleic acid).

Phosphate and sugars

Various models for DNA were devised before the currently accepted version was formulated.

phosphate   ribose   deoxyribose

Bases

DNA has thymine.

RNA has uracil.

The other three bases are the same in DNA and RNA.

Each **nucleotide** can have one of **four** possible bases:
• DNA has: adenine;   **thymine**;   cytosine;   guanine.
• RNA has: adenine;   **uracil**;   cytosine;   guanine.

uracil   cytosine   adenine   guanine

(thymine has a –CH₃ at position *)

The nucleotides

Each nucleotide has this sequence: **phosphate–deoxyribose–base**.

The phosphate unit and base are joined by condensation with deoxyribose, as shown below.

phosphate   base (cytosine)

condensation

deoxyribose

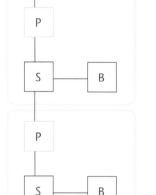

Simplified section of a DNA strand
• P = phosphate
• S = sugar
• B = base

Condensation polymerisation of nucleotides

The nucleotide units polymerise through the sugar–phosphate backbone as shown in the simplified diagram in the margin.

Base pairing in DNA

Two separate **strands** are linked together in the **double helix** structure of DNA. In DNA,
• Adenine and Thymine pair up with **two** hydrogen bonds.
• Cytosine and Guanine pair up with **three** hydrogen bonds.

The geometry of each base ensures that the correct bases pair up.

DNA analysis can be used for genetic fingerprinting. There are discussions about developing a national DNA database of every person in the UK but this carries with it ethical issues and concern about storage of personal information.

• A and T 2 H bonds	• C and G 3 H bonds
Adenine Thymine	Guanine Cytosine

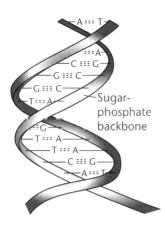

Sugar-phosphate backbone

Each strand of DNA contains a sequence of bases in a definite order. The **complimentary strand** must contain only the correct sequence of the complimentary bases (see the margin diagram).

DNA can **replicate** by first 'unzipping' each strand:

- Each separate strand attracts nucleotides.
- By base pairing, the nucleotides are attached in the correct order.
- The result is **two** double helices of DNA, made from one original double helix.
- Each new double helix is identical to the original.

Every person's DNA is unique but sections of DNA are identifiable as **genes**. Some genes are inherited and are handed down through generations. These genes give family characteristics. There are also other genes that are common to every person's DNA and these carry the necessary information for the body to create proteins.

DNA and proteins

DNA encodes for an amino acid sequence in a protein by using a base sequence within a '**protein gene**' section of DNA.

The DNA double helix can make a strand of RNA by base pairing.

RNA has some important differences:
- A strand contains four bases but with the base **uracil** instead of **thymine**.
- **Ribose** is the sugar instead of deoxyribose.
- The strand is formed from just a section of the DNA double helix; it is much shorter than a DNA strand.
- RNA exists as a single strand.

The protein code

The strand of RNA formed is called **messenger RNA, m-RNA**.

- The sequence of bases in m-RNA contains **triplet codes**, each with **three** bases.
- Each amino acid can bond to another type of RNA strand called **transfer RNA, t-RNA** within the ribosome region of a cell.
- Each **amino acid** has its own **special t-RNA** with a special sequence of bases that have triplet codes – these are **complimentary** to the code in m-RNA.
- The **m-RNA** strand **attracts t-RNA** molecules in sequence, each carrying an amino acid. The triplet codes ensure that the correct t-RNA molecule is attached.

The diagram below shows a section of the m-RNA strand with 12 bases: **four triplet codes**.

Diagram to illustrate protein synthesis

The amino acids join together in the correct sequence for a protein. As soon as two amino acid molecules join by a peptide link, the t-RNA molecule leaves to collect another amino acid molecule.

In the left-hand side of the diagram above, the t-RNA molecule that codes for Leu (leucine) is leaving. Note the correct base pairing within each matching of triplets.

Ala (alanine) is the next amino acid that needs to be added and you can see this on the right-hand side of the 'conveyor belt'.

5.3 Chromatography

After studying this section you should be able to:

- understand what is meant by chromatography
- know that components are separated by absorption and partition
- understand what is meant by R_f value and retention time
- know that chromatography can be combined with mass spectrometry

LEARNING
SUMMARY

Types of chromatography

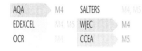

Chromatography is an analytical technique that separates components in a mixture between a **mobile phase** and a **stationary phase**. The technique is especially useful for the purification of an organic substance.

- The **mobile phase** moves in a definite direction and may be a liquid (as in thin-layer chromatography, TLC) or a gas (as in gas chromatography, GC).
- The **stationary phase** is the substance that is fixed in place during the chromatography. It may be a solid (as in TLC) or either a liquid or solid on a solid support (as in GC).
- A solid stationary phase separates by **adsorption** of the components in the mixture on the surface.
- A liquid stationary phase separates by **partition** between the liquid and the mobile phase.

Thin-layer and paper chromatography

Thin-layer chromatography (TLC) uses a stationary phase in the form of a thin layer of an adsorbent such as silica gel or alumina on a flat, inert support.

Paper chromatography and TLC use a similar technique.

- A small dot of sample solution is placed on the TLC plate or paper.
- The paper or plate is placed in a jar containing a shallow layer of solvent and sealed.
- As the solvent rises, it meets the sample of the mixture. Different compounds in the mixture travel different distances depending on how strongly they interact with the paper and their solubility in the solvent.

This allows the calculation of a **Retardation factor**, R_f for each component (see below).

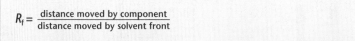

$$R_f = \frac{\text{distance moved by component}}{\text{distance moved by solvent front}}$$

KEY POINT

The R_f value can be compared to standard compounds to aid in the identification of an unknown substance. The diagram below shows some R_f values.

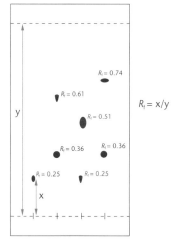

$R_f = x/y$

Using TLC, the separated components can be located as spots using iodine or ultraviolet radiation. Components can be identified by comparing the R_f values with known compounds.

The amino acids in a mixture can be separated and identified by paper chromatography and located using ninhydrin.

Two-way chromatography (CCEA only)

Sometimes more than one component in a mixture may have the same R_f value in a particular solvent. The chromatogram can be run a second time, at right-angles using a second solvent. This can result in separation of the components with the same R_f value in the first solvent. This is called **two-way chromatography**.

This is a useful technique for separating the components in complex mixtures of similar compounds, such as amino acids. The diagram below shows the result of carrying out two-way chromatography on a mixture of amino acids.

For amino acids, the spots are usually developed as purple spots using ninhydrin.

You can see which amino acids have been separated by each solvent by comparing the R_f values vertically and horizontally.

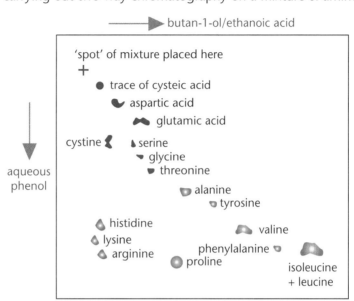

In this chromatogram:

- some amino acids have been separated by butan-1-ol/ethanoic acid and some by aqueous phenol
- 15 out of 17 amino acids have been separated, with only isoleucine and leucine unseparated.

AQA — M4
EDEXCEL — M3
WJEC — M4

Liquid chromatography

Liquid chromatography uses a stationary phase within a column. The liquid mobile phase is passed through the column and components are separated depending on their solubility in the mobile phase and retention in the stationary phase. Liquid chromatography can be run under pressure using very small packing particles. This is referred to as **high pressure liquid chromatography (HPLC)**.

AQA — M4   SALTERS — M5
EDEXCEL — M4   WJEC — M4
OCR — M4   CCEA — M5

Gas Chromatography GC

Gas chromatography (GC) or Gas-Liquid chromatography (GLC) is a separation technique in which the mobile phase is a gas. Gas chromatography is always carried out in a column. The column contains the stationary phase, which is either a high-boiling liquid or a solid on a solid, porous support.

- The sample is injected into the column, which is heated to vaporise the components in the mixture.
- The gas mobile phase flushes the mixture along the column.
- As the mixture moves through the column, components are slowed down as they interact with the stationary phase. Different components are slowed down by different amounts, which causes them to separate.
- Each component leaves the column at a different time.

This allows the calculation of the retention time for each component.

> **KEY POINT**
>
> **Retention time** is the time it takes for a component to pass from the column inlet to the detector.

The retention time can be compared to standard compounds to aid in the identification of an unknown substance. The peak area also gives the approximate percentage composition of a mixture. The diagram below shows a GC chromatogram with some retention times.

Limitations

Analysis by gas chromatography has limitations:

- Similar compounds often have similar retention times.
- Unknown compounds have no reference retention times for comparison.

OCR
SALTERS
CCEA   M5

Combining mass spectrometry with chromatography

Mass spectrometry can be combined with chromatography to provide a far more powerful analytical tool than chromatography alone. For example, mass spectrometry is used with GC as GC-MS and with high pressure liquid chromatography as HPLC-MS.

In GC-MS:

- the components in a mixture are first separated using GC
- the separated components are directed into the mass spectrometer
- the generated mass spectra can be analysed or compared with a spectral database by computer for positive identification of each component.

GC-MS is extensively used in modern analysis, e.g. in forensics, environmental analysis and airport security.

Progress check

1 What is meant by the following terms?
   (a) R_f value
   (b) retention time
2 How does GC-MS enable components in a mixture to be identified?

1 (a) $R_f = \dfrac{\text{distance moved by component}}{\text{distance moved by solvent front}}$
   (b) The time it takes for a component to pass from the column inlet to the detector.
2 GC separates the components. Each component then passes into a mass spectrometer. The resulting mass spectrum is analysed against a spectral database to identify the component.

5.4 Analysis

After studying this section you should be able to:

- review from AS Chemistry the use of infra-red spectroscopy and mass spectrometry in organic analysis

LEARNING SUMMARY

Infra-red spectroscopy and mass spectrometry in organic analysis

AQA	M4	SALTERS	M4, M5
EDEXCEL	M4 M5	WJEC	M4
OCR	M4		

Key points from AS

- **Analysis**
  Revise AS pages 143–146

In AS Chemistry, you were introduced to the use of two important instrumentation techniques in organic analysis:

- Infra-red spectroscopy.
- Mass spectrometry.

For A2 chemistry, you study a further important instrumentation technique called nuclear magnetic resonance, NMR. Analytical chemists often combine the best features of each spectral technique to identify an unknown compound. This section starts with a review of the key features of IR and MS. For full details, you will need to refer to *Revise AS Chemistry*, Chapter 4.7.

Using infra-red (IR) spectroscopy in organic analysis

IR spectroscopy is useful for identifying the functional groups in a molecule.

An infra-red spectrum is particularly useful for identifying:

IR spectroscopy is most useful for identifying C=O and O–H bonds.

Look for the distinctive patterns.

- an alcohol from absorption of the O–H bond (3230–3550 cm^{-1})
- a carbonyl compound from absorption of the C=O bond (1680–1750 cm^{-1})
- a carboxylic acid from absorption of the C=O bond (1680–1750 cm^{-1}) and broad absorption of the O–H bond (2500–3300 cm^{-1} (broad)).

Fingerprint region

The fingerprint region is unique for a particular compound.

- Between 1000 and 1550 cm^{-1}

Many spectra show a complex pattern of absorption in this range.

- This pattern can allow the compound to be identified by comparing with spectra of known compounds.

Using mass spectrometry in organic analysis

Mass spectrometry of organic compounds is useful for identifying:

- the relative molecular mass from the molecular ion peak, *M*
- parts of the skeletal formula by the fragmentation pattern.

The fragmentation pattern provides clues about the molecular structure of the compound. The table below shows the identity of the main peaks in the mass spectrum of butanone, $CH_3COCH_2CH_3$.

Note that only ions are detected in the mass spectrum.

Uncharged species such as the radical cannot be deflected within the mass spectrometer.

The skill in interpreting a mass spectrum is in searching for known patterns, evaluating all the evidence and reassembling all the information into a molecular structure.

m/z	*ion*	*fragment lost*
72	$CH_3COCH_2CH_3^+$	–
57	$CH_3CH_2CO^+$	$CH_3\bullet$
43	CH_3CO^+	$CH_3CH_2\bullet$
29	$CH_3CH_2^+$	$CH_3CO\bullet$
15	CH_3^+	$CH_3CH_2CO\bullet$

5.5 NMR spectroscopy

After studying this section you should be able to:

- understand that NMR spectroscopy is carried out with ^1H and ^{13}C
- predict the different types of proton and carbon present in a molecule from chemical shift values
- predict the relative numbers of each type of proton present from an integration trace
- predict the number of protons adjacent to a given proton from the spin-spin coupling pattern
- predict possible structures for a molecule from ^1H and ^{13}C spectra
- predict the chemical shifts and splitting patterns of the protons in a given molecule
- describe the use of D_2O in NMR spectroscopy

LEARNING SUMMARY

Nuclear magnetic resonance

AQA	M4	SALTERS	M5
EDEXCEL	M4, M5	WJEC	M4
OCR	M4	CCEA	M5

Nuclear Magnetic Resonance (NMR) spectroscopy is an extremely important modern method of analysis. It is used extensively in the pharmaceutical industry and in universities for identification and purity checking of organic compounds.

Key principles

The nucleus of an atom of hydrogen (i.e. a proton) has a magnetic spin.
When placed in a strong electromagnetic field:

- the nucleus can absorb energy from the low energy **radio-frequency** region of the spectrum to move to a higher energy state
- **nuclear magnetic resonance** occurs as protons resonate between their spin energy states.

> Only nuclei with an odd number of nucleons (neutrons + protons) possess a magnetic spin: e.g. ^1H, ^{13}C. Proton NMR spectroscopy is the most useful general purpose technique.

> The energy gap is equal to that provided by radio waves.

The different nuclear spin states in an applied magnetic field

An NMR spectrum shows **absorption peaks** corresponding to the radio-frequency absorbed.

Chemical shift, δ

Electrons around the nucleus **shield** the nucleus from the applied field.

- The magnetic field at the nucleus of a particular proton is different from the applied magnetic field.

> NMR spectroscopy is the same technology as used in MRI (magnetic resonance imaging) in medical body scanners. It is a non-invasive technique.

applied magnetic field

magnetic field produced from orbiting electron

> Protons in different environments absorb at different chemical shifts.

- Different radio-frequencies are absorbed, depending on the **environment** of the proton.
- **Chemical shift** is a measure of the magnetic field experienced by protons in different environments resulting from nuclear shielding.

> Chemical shift, δ, is measured relative to a standard: tetramethylsilane (TMS), $Si(CH_3)_4$.
> • The chemical shift of TMS is defined as $\delta = 0$ ppm.

Carbon-13 NMR spectroscopy

AQA   M4
OCR

Carbon-13 NMR spectroscopy allows the identification of carbon atoms in an organic molecule and is an important tool in the determination of structure in organic chemistry.

Typical chemical shifts

Chemical shifts indicate the chemical environment of the **carbon atoms** present. The table below shows typical chemical shift values for different types of carbon atom.

You don't need to learn these chemical shifts - the data is provided on exam papers.

The presence of an electronegative atom or group causes chemical shift 'downfield'. This is called 'deshielding'.

Notice the chemical shift caused by the carbonyl group, oxygen, halogens and a benzene ring.

type of carbon		chemical shift, δ/ppm
C–C		5–55
C–Cl or C–Br		30–70
C–N (amines)		35–60
C–O		50–70
C=C (alkenes)		115–140
aromatic		110–165
carbonyl (ester, carboxylic acid, amide)		160–185
carbonyl (aldehyde, ketone)		190–220

• The actual chemical shift may be slightly different depending upon the actual environment of the carbon.

In a carbon–13 NMR spectrum, the size of the peak tells us nothing about the number of carbon atoms responsible. This is in contrast to proton NMR spectroscopy. However, carbon-13 spectrum is far more sensitive with a much larger range of chemical shift values. Unless there are carbon atoms present in an identical environment, each carbon atom is likely to show as a separate single signal.

Interpreting a carbon-13 NMR spectrum

A carbon-13 NMR spectrum of ethyl ethanoate, $CH_3COOCH_2CH_3$ shows absorptions at **four** chemical shifts showing carbon atoms in **four different environments**:

167

- The four chemical shifts can be matched to the table of chemical shifts on page 167 so that the carbon atom responsible for each absorption can be identified.

chemical shift/ppm	environment
171	H_3C-C double bond O, single bond $O-CH_2-CH_3$
60	H_3C-C double bond O, single bond $O-CH_2-CH_3$
21	H_3C-C double bond O, single bond $O-CH_2-CH_3$
14	H_3C-C double bond O, single bond $O-CH_2-CH_3$

> **KEY POINT**
>
> Carbon-13 NMR spectroscopy is useful for identifying the **type** of environment of each carbon – the chemical shift.

Progress check

1 For each structure, predict the chemical shift of each carbon atom.
   (a) CH_3CH_2OH; (b) CH_3CH_2CHO; (c) CH_3COCH_3.

1 (a) CH_3CH_2OH, δ = 5–55 ppm; CH_3CH_2OH, δ = 50–70 ppm
(b) CH_3CH_2CHO, δ = 5–55 ppm; CH_3CH_2CHO, δ = 5–55 ppm;
CH_3CH_2CHO, δ = 190–220 ppm
(c) CH_3COCH_3, δ = 5–55 ppm; CH_3COCH_3, δ = 190–220 ppm

Proton NMR spectroscopy

AQA	M4	SALTERS	M5
EDEXCEL	M3	WJEC	M4
OCR	M4	CCEA	M5

Proton NMR spectroscopy allows the identification of hydrogen atoms in an organic molecule and is an important tool in the determination of structure in organic chemistry.

Typical chemical shifts

Chemical shifts indicate the **types of protons** and **functional groups** present. The table below shows typical chemical shift values for protons in different chemical environments.

The presence of an electronegative atom or group causes chemical shift 'downfield'. This is called 'deshielding'.

Notice the chemical shift caused by the carbonyl group, oxygen, halogens and a benzene ring.

You don't need to learn these chemical shifts – the data is provided on exam papers.

type of proton	chemical shift, δ/ppm
R-CH$_3$	0.7–1.6
R-CH$_2$-R	1.2–1.4
$H_3C-C\overset{O}{\diagdown}$ $R-CH_2-C\overset{O}{\diagdown}$	2.0–2.9
⬡-CH$_3$ ⬡-CH$_2$-R	2.3–2.7
X–CH$_3$ X–CH$_2$–R (X = halogen)	3.2–3.7
–O–CH$_3$ –O–CH$_2$–R	3.3–4.3
R–O–H	3.5–5.5
⬡-H	7
$H_3C-C\overset{O}{\underset{H}{\diagdown}}$	9.5–10
$H_3C-C\overset{O}{\underset{OH}{\diagdown}}$	11.0–11.7

The only reliable means of identifying O–H in NMR is to use D$_2$O: see Identifying O–H protons, page 171.

- The actual chemical shift may be slightly different depending upon the actual environment of the proton.
- The chemical shift for O–**H** can vary considerably and depends upon concentration, solvent and other factors.

Interpreting low resolution NMR spectra

AQA	M4	SALTERS	M5
EDEXCEL	M4, M5	WJEC	M4
OCR	M4	CCEA	M5

Low resolution NMR spectrum of ethanol

A low resolution NMR spectrum of ethanol shows absorptions at **three** chemical shifts, showing the **three different types of proton**:

The relative areas of each peak are usually measured by running a second NMR spectrum as an *integration trace*.

An NMR spectrum is obtained in solution. The solvent must be proton-free, usually CCl$_4$ or CDCl$_3$ is used.

- The **area** under each peak is in direct proportion to the **number of protons** responsible for the absorption.
- The three chemical shifts can be matched to the table of chemical shifts on page 169 so that the protons responsible for each absorption can be identified.

chemical shift/ppm	no. of protons	environment	type of proton
$\delta = 1.0$	3H	CH_3CH_2OH	CH_3 adjacent to a carbon chain
$\delta = 3.5$	2H	CH_3CH_2OH	CH_2 adjacent to $-O$
$\delta = 4.9$	1H	CH_3CH_2OH	OH

> **KEY POINT**
>
> A low resolution NMR spectrum is useful for identifying:
> - the **number** of different types of proton from the number of peaks
> - the **type** of environment of each proton from the chemical shift
> - **how many** protons of each type from the integration trace.

Interpreting high resolution NMR spectra

AQA	M4	WJEC	M4
OCR	M4	CCEA	M5
EDEXCEL	M4, M5	SALTERS	M5

Different types of proton have different chemical shifts.

A high resolution NMR spectrum shows splitting of peaks into a pattern of sub-peaks.

Spin-spin coupling patterns:
- arise from interactions between protons on **adjacent** carbon atoms which have **different chemical shifts**
- indicate the number of **adjacent** protons.

Equivalent protons (i.e. protons with the same chemical shift) will **not** couple with one another.

The **spin-spin coupling** pattern shows as a **multiplet** – a doublet, triplet, quartet, etc.

A singlet is next to C.
A doublet is next to CH.
A triplet is next to CH_2.
A quadruplet is next to CH_3.

> **KEY POINT**
>
> We can interpret the spin-spin coupling pattern using the **n+1 rule**.
> For **n adjacent protons**, the number of peaks in a multiplet = **n+1**.

The high resolution NMR spectrum of ethanol

The high resolution NMR spectrum of ethanol shows spin-spin coupling patterns – some of the signals have been split into multiplets:

The multiplicity (doublet, triplet, quartet) does not indicate the number of protons on that carbon. The number of protons is given by the integration trace.

quartet
∴ from n + 1 rule, a CH_3 is adjacent: CH_2–CH_3

triplet
∴ from n + 1 rule, a CH_2 is adjacent: CH_2–CH_3

singlet of 1 peak
O–H splitting is not normally seen owing to H-bonding

Notice the different sizes of each sub-peak within the splitting pattern:
- doublet 1:1
- triplet 1:2:1
- quartet 1:3:3:1.

- The **chemical shift** identifies the **type of protons** responsible for each peak.
- The **n+1 rule** can be used to identify the **number of adjacent protons**.

chemical shift /ppm	environment	number of adjacent protons (n)	splitting pattern (n+1)
δ = 1.2	CH₃CH₂OH	2H	2+1 = 3: triplet
δ = 3.6	CH₃CH₂OH	3H	3+1 = 4: quartet
δ = 4.5	CH₃CH₂OH	–	singlet

- Note that O–H splitting is not normally seen owing to H-bonding or exchange with the solvent used.

Equivalent protons do not interact with each other.

- The three equivalent CH₃ protons in ethanol cause splitting of the adjacent CH₂ protons, but not amongst themselves.

> A high resolution NMR spectrum is useful for identifying the **number** of **adjacent** protons from the spin-spin coupling pattern.

Identifying O–H protons

The OH absorption peak at 4.5 δ is absent in D₂O.

O–H protons can absorb at different chemical shifts depending on the solvent used and the concentration. The signals are often broad and usually show no splitting pattern.

These factors can make it difficult to identify O–H protons. However, they can be identified by using deuterium oxide, D₂O.

- A second NMR spectrum is run with a small quantity of D₂O added to the solvent.
- The D₂O exchanges with O–H protons and any such peak will disappear.

Solvents

Deuterated solvents such as CDCl₃ are used in NMR spectroscopy. Any protons in a solvent such as CHCl₃ would produce a large proton absorption peak. By using CDCl₃, this absorption is absent and the spectrum is that of the organic compound alone.

Progress check

1 For each structure, predict the number of peaks in its low resolution NMR spectrum corresponding to the different types of proton and also the number of each different type of proton
(e.g. CH₃CH₂OH has 3 peaks in the ratio 3:2:1).
(a) CH₃OH (b) CH₃CH₂CHO (c) CH₃COCH₃ (d) (CH₃)₂CHOH.

2 The NMR spectrum of compound **X** (C₂H₄O) has a doublet at δ 2.1 and a quartet at δ 9.8.
(a) Identify compound **X**. (Use the table of chemical shifts on page 169).
(b) How many protons are responsible for each multiplet?

2 (a) ethanal, CH₃CHO.
(b) doublet at δ 2.1, 3H; quartet at δ 9.8, 1H.

1 (a) 2 peaks in the ratio 3:1 (b) 3 peaks in the ratio 3:2:1 (c) 1 peak
(d) 3 peaks in the ratio 6:1:1

5.6 Organic synthetic routes

There are many variations possible and the schemes below could not include all reactions without appearing more complicated.

Organic chemists are frequently required to synthesise an organic compound in a multistage process. This is fundamental to the design of new organic compounds such as those needed for modern drugs to combat disease, a new dye or a new fibre. With so many reactions to consider, it is essential to see how different functional groups can be interconverted and summary charts are useful for showing these links. The flow-charts on the next pages show reactions drawn from both the AS and A2 parts of A Level Chemistry.

Construct a scheme of your own using only those reactions in your course.

In revision it is useful to draw your own schemes from memory.

Aliphatic synthetic routes

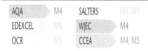

The scheme below is based upon two main sets of reactions:
- a set based around bromoalkanes
- a set based around carboxylic acids and their derivatives.

The two sets of reactions are linked via the oxidation of primary alcohols.
- A reaction scheme based upon a secondary alcohol would result in oxidation to a ketone only.

Radiation is usually linked with analysis but it is also used in some synthetic methods.

Microwaves are used for heating organic reagents in organic research. As with cooking, a reaction can take place in a fraction of the time using less vigorous conditions.

Ultraviolet radiation is used to initiate some reactions.

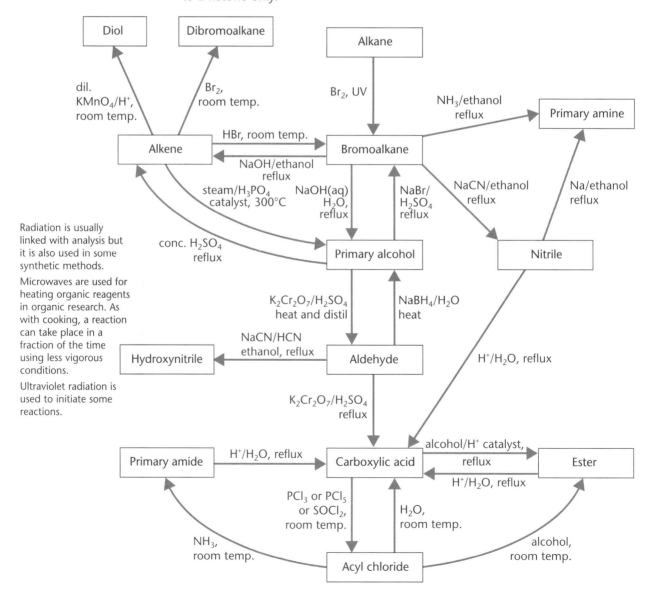

Aromatic synthetic routes

AQA	M4	SALTERS	M5
EDEXCEL	M5	WJEC	M4
OCR	M4	CCEA	M5

Compared with aliphatic organic chemistry at A Level, there are comparatively few aromatic reactions.

Benzene

- Reactions involving the benzene ring are mainly electrophilic substitution.

alkylation — RCl / AlCl₃ warm → R

acylation — RCOCl / AlCl₃ warm

nitration — H₂SO₄ / HNO₃ 55°C → NO₂

chlorination — Cl₂ / AlCl₃ warm → Cl

reduction — 1 Sn / conc HCl reflux 2 NaOH(aq) → NH₂

diazotisation — HNO₂/HCl <10°C → Cl⁻N⁺≡N

azo coupling — NaOH(aq) <10°C

It is essential, if you are to answer such questions, that you thoroughly learn suitable reagents and conditions for all the reactions in your syllabus.

- Questions are often set in exams asking for reagents, conditions or products.
- More searching problems may expect a synthetic route from a starting material to a final product. The synthetic route may include several stages.

Synthesis of medicines

Synthesis of chiral drugs (OCR and Edexcel only)

The structures of many drugs include a chiral carbon atom. Often the drug is **stereo-specific** with only one of the optical isomers having the desired effect.

Modern synthesis of a pharmaceutical is often carried out to produce a **single optical isomer**. This **reduces** any **risks** on the body from **side effects** of the 'other' optical isomer.

It is difficult to separate optical isomers produced by chemical synthesis and this increases costs. Various strategies have been developed to synthesise just the required optical isomer.

- **Enzymes** or bacteria could be used for any synthetic steps that introduce a chiral centre. Enzymes promote stereo-selectivity.

In the synthesis of stereo-specific drugs, it is important to understand the mechanism of the reaction and how this can help to plan the synthesis.

- Methods of chemical synthesis have been devised that use '**chiral catalysts**'.
- Natural chiral molecules, such as L-amino acids or D-sugars, can be used as starting materials for the synthesis. This makes use of the natural '**chiral pool**'.

Combinatorial chemistry (Edexcel and Salters only)

More effective medicines can be obtained by modifying the structure of existing medicines. **Combinatorial chemistry** is a modern technique that has been adopted by the pharmaceutical industry in drug research.

In combinatorial chemistry a **large number** of related compounds are made together so that their potential effectiveness as medicines can be assessed by **large-scale screening**.

This process often includes passing reactants over reagents on polymer supports.

Testing and identifying a new medicine (Salters only)

Testing a medicine involves clinical trials to assess a potential new drug:

- Step I – Is it safe?
- Step II – Does it work?
- Step III – Is it better than the standard treatment?

Chemists have a role in:

- designing and making new compounds for use as pharmaceuticals
- ethical testing
- using computer modelling in the design of medicines.

Pharmacologically active compounds (Salters only)

Many medicines and drugs are **pharmacologically active** – they induce a biological interaction of a living organism.

The structure and action of a pharmacologically active compound depends upon:

- the **pharmacophore** and groups that modify it
- its **interaction** with receptor sites
- the ways that species interact in three dimensions (size, shape, bond formation, orientation).

Identification of the pharmacophore is a key step in devising a synthesis for a new drug.

The pharmacophore is the part of a molecular structure that is responsible for a particular biological or pharmacological interaction.

Retrosynthesis

SALTERS   M5

Retrosynthesis is a modern method for devising an organic synthesis.

Conventional synthetic routes are developed by working forwards from starting materials to a target molecule.

- In retrosynthesis an **alternative synthesis** is worked out, **backwards** from a target molecule into simpler structures.
- This procedure is repeated until simple or commercially available starting materials are reached.

Retrosynthesis is useful for identifying **potential synthetic routes** in a logical and straightforward fashion.

The chemist will then **reverse** this process to develop a synthetic route towards the required target molecule.

Sample question and model answer

The structures of the four esters with the molecular formula $C_4H_8O_2$ are shown below.

$$CH_3CH_2COOCH_3 \quad CH_3COOCH_2CH_3 \quad HCOOCH_2CH_2CH_3 \quad HCOOCH(CH_3)_2$$
$$\textbf{A} \qquad\qquad \textbf{B} \qquad\qquad\quad \textbf{C} \qquad\qquad\qquad \textbf{D}$$

(a) For each structure predict the number of peaks in its NMR spectrum and the ratio of the number of protons responsible for each peak.

When we are predicting the number of peaks, we are identifying the number of different types of proton.

In compound **D**, $HCOOCH(CH_3)_2$, both the methyl groups (6H) are equivalent and will have the same chemical shift.

Ester **A**, $CH_3CH_2COOCH_3$, has 3 peaks ✓ in the ratio 3:2:3 ✓

Ester **B**, $CH_3COOCH_2CH_3$, has 3 peaks ✓ in the ratio 3:2:3 ✓

Ester **C**, $HCOOCH_2CH_2CH_3$, has 4 peaks ✓ in the ratio 1:2:2:3 ✓

Ester **D**, $HCOOCH(CH_3)_2$, has 3 peaks ✓ in the ratio 1:1:6 ✓   [8]

(b) The NMR spectrum of one of the four esters is shown below. Integration data is shown by each peak. Analyse the spectrum to find out which ester has produced it. Include all the supporting evidence from the spectrum.

Notice how the answer is methodical, going through each piece of evidence separately.

The combination of a triplet and quartet is a giveaway for a CH_3CH_2 combination.

The integration data and splitting patterns point to either structure **A** or **B**.

The chemical shifts are conclusive though.

The integration data supports either structure **A** or **B**:

There are three types of proton in the ratio 3:2:3 ✓

The splitting pattern supports either structure **A** or **B**:

The CH_3 group at $\delta = 1.14$ ppm has been split into a triplet ✓ by 2 protons on an adjacent carbon atom. ✓

The CH_2 group at $\delta = 2.29$ ppm has been split into a quartet ✓ by 3 protons on an adjacent carbon atom. ✓

The CH_3 group at $\delta = 3.68$ ppm is a singlet ✓ so there can be no protons on an adjacent carbon. ✓

From chemical shifts,

the CH_3 group at $\delta = 3.68$ ppm must be adjacent to an O atom ✓

the CH_2 group at $\delta = 2.29$ ppm must be adjacent to a C=O group. ✓

Taking all the evidence together:

the ester must be **A**, $CH_3CH_2COOCH_3$. ✓   [10]

(c) The IR and mass spectra can also be used to help confirm structures. What key IR absorptions and *m/z* values would you expect to see in the mass and IR spectra of ester **A**?

There are 4 marks here. The first is for the key information: the C=O absorption in the IR and the molecular ion peak in the mass spectrum. The second mark is for identification of another key feature.

IR: A C=O absorption at about 1700 cm^{-1}. ✓ A C–O absorption at 1000–1300 cm^{-1}. There should be no absorption above 3000 cm^{-1} as there is no –OH group present. ✓

Mass spectrum: A molecular ion peak at $m/z = 88$ confirming the molecular mass. ✓ Fragments ions at $m/z = 57$ from $CH_3CH_2CO^+$; also at $m/z = 31$ for CH_3O^+. ✓   [4]

Other fragment ions could have been chosen.

[Total: 22]

Practice examination questions

1 Valine, $(CH_3)_2CH(NH_2)COOH$, is an amino acid.

(a) What is the 'R' group in valine? [1]

(b) Valine reacts with an amino acid, **A**, to form the dipeptide below.

(i) Draw a circle around the peptide linkage.

(ii) Draw the structure of the amino acid, **A**.

(iii) Valine can react with the amino acid, **A**, to form a different dipeptide from that shown above. Draw the structural formula of this other dipeptide. [3]

(c) Valine has a chiral centre and can exist as two optical isomers.

(i) State what is meant by a *chiral centre* and explain how a chiral centre gives rise to optical isomerism.

(ii) Draw diagrams to show the two optical isomers of valine. State the bond angle around the chiral centre. [5]

(d) In aqueous solution, valine exists as different ions at different pH values. The zwitterion exists in the pH range 3–10. Draw the displayed formula of the ion of valine present at pH values of 2.0, 7.0 and 12.0. [3]

(e) Compound **B** is an amino acid with the molecular formula $C_3H_7NO_3$.

Draw the structure of amino acid **B**. [1]

[Total: 13]

2 Lactic acid has the structural formula $CH_3CH(OH)COOH$.

(a) What is the systematic name for lactic acid? [1]

(b) Lactic acid has optical isomers.

(i) What structural feature in lactic acid results in optical isomerism?

(ii) How does optical isomerism arise?

(iii) Draw a three-dimensional diagram to show the optical isomers of lactic acid. [4]

(c) Lactic acid can polymerise to form poly(lactic acid)

(i) What type of polymer is poly(lactic acid)?

(ii) Draw a short section of poly(lactic acid) showing two repeat units. [3]

[Total: 8]

Practice examination questions (continued)

3 The repeat units of two polymers **C** and **D** are shown below.

C   **D**

(a) (i) Draw the structure of the monomer of polymer **C**.

    (ii) Name the type of polymerisation. [2]

(b) (i) Draw the structure of the two monomers of polymer **D**.

    (ii) Name the type of polymerisation. [3]

(c) Suggest why polymer **D** would be more likely to be hydrolysed than polymer **C**. [2]

[Total: 7]

4 Compounds **E** and **F** are both diols with the molecular formula $C_4H_{10}O_2$.

(a) The proton NMR spectrum of a diol **E**, $C_4H_{10}O_2$, is shown below. The numbers by each peak represent the integration trace. The peak at $\delta = 3.65$ ppm disappears when the spectrum is run a second time with D_2O added.

    (i) How many different types of proton are present in the diol?

    (ii) The peaks at $\delta = 1.63$ ppm and $\delta = 0.90$ ppm result from a single alkyl group. Identify this group and explain the splitting pattern.

    (iii) What can be deduced from the single peak at $\delta = 1.32$ ppm?

    (iv) What can be concluded about the peak at $\delta = 3.65$ ppm?

    (v) Show the structure of the diol **E**. [9]

(b) The proton NMR spectrum diol **F**, run in D_2O, is shown below.

    (i) What can be concluded about the two peaks?

    (ii) Show the structure of the diol **F**. [5]

[Total: 14]

Chapter 6
Synoptic assessment

What is synoptic assessment?

Synoptic assessment emphasises your understanding and your application of the principles included in your chemistry course.

Part of your chemistry course is assessed using **synoptic questions**. These are written so that you can **draw together** knowledge, understanding and skills learned in **different parts** of AS and A2 Chemistry.

Synoptic assessment takes place mainly in **module exams** but part may be assessed in coursework.

What type of questions will be asked?

You will need to answer **two** main types of synoptic question.

You make the links between different areas of chemistry yourself. You choose and use the context for your answer.

1 **You make links** and **use connections** between different areas of chemistry. Some examples are given below.

   Using examples drawn from different parts of your chemistry course:
   • discuss the role of a lone pair in chemistry...
   • discuss the chemistry of water...
   • compare the common types of chemical bonding...

The context has been made for you. Often this will be a situation or will involve data that you will not have seen before.

2 **You use ideas and skills** which **permeate chemistry**.
   This type of question will be more structured and you are unlikely to have as much choice in how you construct your answer.

   Your task is to interpret any information using the 'big ideas' of chemistry.

   Examples of themes that permeate chemistry are shown below:
   • formulae, moles, equations and oxidation states
   • chemical bonding and structure
   • periodicity
   • reaction rates, chemical equilibrium and enthalpy changes.

How will you gain synoptic skills?

A synoptic question may present you with a new piece of information.

You may be expected to calculate formulae from data, write correct formulae, balance equations and perform quantitative calculations using the mole concept, use knowledge of the Periodic Table to predict reactions of unfamiliar elements or compounds, etc.

Luckily chemistry is very much a synoptic subject. Throughout your study of chemistry, you apply many of the ideas and skills learnt during AS Chemistry or your GCSE course.

In studying A2 Chemistry, you will have been using synoptic skills naturally, probably without realising it. You certainly can make little real progress in chemistry without a sound grasp of concepts such as equations, the mole, structure and bonding!

Worked synoptic question

Here, some of the big ideas of chemistry are being tested: formulae, bonding, redox, moles, equations. These will always appear on an exam paper with synoptic questions.

You have nothing to fear from this type of question provided that your chemistry is sound.

Structured type question

In its reactions, sulfuric acid, H_2SO_4, can behave as an acid, an oxidising agent and as a dehydrating agent.

The displayed formula of pure sulfuric acid is shown below.

(a) The boiling point of pure sulfuric acid, at 270°C, is higher than might be expected.

Suggest why the boiling point of sulfuric acid is higher than might be expected. Explain your reasoning. Show a diagram in your answer.

Sulfuric acid molecules form hydrogen bonds ✓

hydrogen bonds break (on boiling) ✓ [3]

Nice clues in the question.

You are given the structure of sulfuric acid for a reason. You are being tested on whether you can apply hydrogen bonding to a new situation.

This is not much different from water's H-bonding!

(b) Dilute sulfuric acid takes part in the typical acid reactions, reacting with metals, carbonates and bases.

Write balanced equations for reactions of sulfuric acid with

(i)   a metal $Mg + H_2SO_4 \longrightarrow MgSO_4 + H_2$

(ii) a carbonate $MgCO_3 + H_2SO_4 \longrightarrow MgSO_4 + CO_2 + H_2O$

(iii) a base $2NaOH + H_2SO_4 \longrightarrow Na_2SO_4 + 2H_2O$ [3]

Just keep your nerve when writing equations. You can choose any reasonable examples.

Do think about ionic charges. Many students blindly assume that a sulfate ion has a 1– charge but how can this be so if we have H_2SO_4 with $2H^+$ ions?

(c) Concentrated sulfuric acid will readily oxidise halide ions to the halogen.

Equation 5.1 represents the unbalanced equation for the oxidation of iodide ions by sulfuric acid.

$H^+ + SO_4^{2-} + I^- \longrightarrow I_2 + H_2S + H_2O$ Equation 5.1

(i) Determine the oxidation numbers of sulfur and iodine on either side of the equation.

$SO_4^{2-} : +6 \longrightarrow H_2S: -2$ ✓

$I^-: -1 \longrightarrow I_2: 0$ ✓   [2]

(ii) Balance Equation 5.1

$10H^+ + SO_4^{2-} + 8I^- \longrightarrow 4I_2 + H_2S + 4H_2O$ ✓   [1]

In (i), you are just using oxidation number rules.

– this should be 2 easy marks.

(ii) is far harder.

You have to make sure that there is the same oxidation number change up and down.

Here, each S goes down by 8 from +6 to – 2.

Each I goes up by 1 from –1 to 0 ... so we need 8 × I

Finally balance the H_2O and finally the H^+.

Worked synoptic question (continued)

Here you are provided with information and you must use your chemical knowledge and understanding to solve the problem.

Use the information – the whole question is about **dehydration** (loss of water) caused by sulfuric acid.

Good advice is to look for the obvious. If your answer looks like some chemistry that you have never seen before, then you have probably made a mistake!

Key points here:

A must be made out of H, C, or O.

B is black. What do you know that is black?

C is hard but is still based on dehydration.

(d) Concentrated sulfuric acid reacts with many organic compounds, forming water as one of the products.

For example, sulfuric acid dehydrates ethanol by eliminating water to form ethane.

$$C_2H_5OH \longrightarrow C_2H_4 + H_2O$$

Three other examples are shown below.

- Sulfuric acid dehydrates methanoic acid to form a gas, **A**, with the same molar mass as ethene.
- Sulfuric acid dehydrates sucrose, $C_{12}H_{22}O_{11}$, to form a black solid, **B**.
- Sulfuric acid dehydrates ethane-1,2-diol to form a compound **C** with a molar mass of 88 g mol^{-1}.

Suggest the identity of **A**, **B** and **C**. Write equations for each reaction and deduce the structural formula of compound **C**.

A: CO ✓ $HCOOH \longrightarrow CO + H_2O$ ✓

B: C ✓ $C_{12}H_{22}O_{11} \longrightarrow 12C + 11H_2O$ ✓

C: $C_4H_8O_2$ ✓ $2C_2H_6O_2 \longrightarrow C_4H_8O_2 + 2H_2O$ ✓

Structure:

✓

[7]

[Total: 16]

OCR Unifying concepts Q5 Jan 2007

Practice examination questions

1 Organic acids occur widely in nature.

(a) Butanoic acid, $CH_3(CH_2)_2COOH$, is a straight-chain organic acid, largely responsible for the odour of rancid butter.

Caprylic acid is a straight-chain organic acid. It is produced in the body in small amounts as an antifungal agent in human sweat.

(i) Some caprylic acid was isolated from human sweat and analysed. The sample of caprylic acid had the percentage composition by mass:

C, 66.7%; H, 11.1%; O, 22.2%. $M_r = 144$.

Calculate the molecular formula of caprylic acid and suggest its structural formula.

(ii) Tracking dogs are trained to follow odours such as the characteristic blend of organic acids in the sweat from a person's feet. A dog is able to detect extremely small quantities of these acids.

Sweat containing equal amounts of butanoic and caprylic acids produces more butanoic acid vapour than caprylic acid vapour.

Suggest a reason for this. Explain your answer. [6]

(b) Compound **A** is a straight chain organic acid. A chemist analysed a sample of acid **A** by the procedure below.

The chemist first prepared a 250 cm³ solution of **A** by dissolving 10.8 g of **A** in water.

In a titration, 25.00 cm³ 0.500 mol dm⁻³ NaOH were neutralised by exactly 21.40 cm³ of solution **A**.

(i) Calculate the pH of the NaOH(aq) used in the titration.
$K_w = 1.00 \times 10^{-14}$ mol² dm⁻⁶.

(ii) Use the results to calculate the molar mass of acid **A** and suggest its identity. [8]

[Total: 14]

OCR Unifying concepts Q4 Jan 2003

2 *Refer to data on Page 169 for this question.*

Compounds **A** to **G** are all isomers with the molecular formula $C_6H_{12}O_2$.

(a) Isomer **A**, $C_6H_{12}O_2$, is a neutral compound and is formed by the reaction between compounds **X** and **Y** in the presence of a small amount of concentrated sulfuric acid.

X and **Y** can both be formed from propanal by different redox reactions.

X has an absorption in its infra-red spectrum at 1750 cm⁻¹.

Deduce the structural formulae of **A**, **X** and **Y**. Give suitable reagents, in each case, for the formation of **X** and **Y** from propanal and state the role of concentrated sulfuric acid in the formation of **A**. [8]

(b) Isomers **B**, **C**, **D** and **E** all react with aqueous sodium carbonate to produce carbon dioxide.

Deduce the structural formulae of the three isomers that contain an asymmetric carbon atom.

The fourth isomer has only three singlet peaks in its proton NMR spectrum. Deduce the structural formula of this isomer and label it **E**. [4]

(c) Isomer **F**, $C_6H_{12}O_2$, has the structural formula shown below, on which some of the protons have been labelled.

$$\underset{a}{H_3C - CH_2} - O - CH_2 - \underset{b}{CH_2} - \overset{\overset{\textstyle O}{\|}}{C} - CH_3$$

A proton NMR spectrum is obtained for **F**. Predict a value of δ for the protons labelled *a* and also for those labelled *b*. State and account for the splitting patterns of the peaks assigned to the protons *a* and *b*. [6]

(d) Isomer **G**, $C_6H_{12}O_2$, contains six carbon atoms in a ring. It has an absorption in its infra-red spectrum at 3270 cm⁻¹ and shows only three different proton environments in its proton NMR spectrum. Deduce a structural formula for **G**. [2]

[Total: 20]

AQA Further Physical and Organic Chemistry Q6 Jun 2003

Practice examination answers

Chapter 1 Rates and equilibria

1 (a) (i) 1st order with respect to H_2O_2 ✓
Double concentration of H_2O_2, rate doubles ✓
1st order with respect to I^- ✓
7 times concentration of I^-, rate x 7 ✓
Zero order with respect to H^+ ✓
Double concentration of H^+, rate stays constant ✓

(ii) Rate = $k[H_2O_2][I^-]$ ✓
2.8×10^{-2} dm^3 mol^{-1} s^{-1} ✓✓ 　　　　　　　　　　[9]

(b) (i) The slowest step of a multi-step process ✓

(ii) $H_2O_2 + I^- \longrightarrow H_2O + IO^-$ ✓ 　　　　　　　　　　　　[2]

　　　　　　　　　　　　　　　　　　　　　　　　　　　　　　[Total: 11]

2 (a) (i) Rate = $k[A]^2[B]$ ✓

(ii) 3 ✓

(iii) 27 ✓

(iv) $k = 3.14 \times 10^{-3}$ dm^6 mol^{-2} s^{-1} ✓✓ 　　　　　　　　　[5]

(b) (i) H^+ is a catalyst ✓
H^+ appears in the rate equation but not the overall equation ✓

(ii) $CH_3COCH_3(aq) + H^+(aq) \longrightarrow [CH_3COHCH_3(aq)]^+$ ✓ 　　[3]

　　　　　　　　　　　　　　　　　　　　　　　　　　　　　　[Total: 8]

3 (a) $K_c = \dfrac{[CH_3COOH]\ [CH_3CH_2OH]}{[CH_3COOCH_2CH_3]\ [H_2O]}$ ✓

$K_c = 0.26$ ✓; No units ✓ 　　　　　　　　　　　　　　　　　　[3]

(b) (i) Equilibrium would move to the left ✓ to counteract the added
CH_3CH_2OH ✓

(ii) No effect ✓ K_c only changes with temperature ✓ 　　　　　　[4]

　　　　　　　　　　　　　　　　　　　　　　　　　　　　　　[Total: 7]

4 (a) (i) $[H_2(g)]$, 0.44 mol dm^{-3}; ✓ $[I_2(g)]$, 0.02 mol dm^{-3}; ✓
$[HI(g)]$, 0.32 mol dm^{-3} ✓

(ii) $K_c = \dfrac{[HI(g)]^2}{[H_2(g)]\ [I_2(g)]}$ ✓ $= \dfrac{0.32^2}{0.44 \times 0.02} = 11.6$ ✓ no units ✓ 　　[6]

(b) Equilibrium moves to left ✓; Forward reaction is exothermic (or
the reverse reaction is endothermic) ✓ 　　　　　　　　　　　　[2]

　　　　　　　　　　　　　　　　　　　　　　　　　　　　　　[Total: 8]

5 (a) Proton donor ✓ 　　　　　　　　　　　　　　　　　　　　　　[1]

(b) $HCOOH + HNO_3 \rightleftharpoons HCOOH_2^+ + NO_3^-(aq)$ ✓
HCOOH is a base because it accepts a proton ✓ 　　　　　　　　[2]

(c) (i) pH = 0.75 ✓

(ii) $K_w = [H^+(aq)][OH^-(aq)]$ ✓
$[H^+(aq)] = 1.0 \times 10^{-14}/0.372 = 2.69 \times 10^{-14}$ mol dm^{-3} ✓ pH = 13.57 ✓

(iii) $K_a = \dfrac{[H^+(aq)][HCOO^-(aq)]}{[HCOOH(aq)]} = \dfrac{[H^+(aq)]^2}{[HCOOH(aq)]}$ ✓

$[H^+(aq)] = \sqrt{(1.6 \times 10^{-4} \times 0.263)} = 6.49 \times 10^{-3}$ mol dm^{-3} ✓ pH = 2.19 ✓

(iv) $[H^+(aq)] = K_a \times [HCOOH(aq)]/[HCOO^-(aq)]$ ✓
$[H^+(aq)] = 1.6 \times 10^{-4} \times 0.255/0.325 = 1.26 \times 10^{-4}$ mol dm^{-3} ✓
pH = 3.90 ✓　　　　　　　　　　　　　　　　　　　　　　[10]

[Total: 13]

6 (a) (i) A strong acid completely dissociates to donate protons.
A weak acid partially dissociates to donate protons. ✓

(ii) Proton acceptor ✓　　　　　　　　　　　　　　　[2]

(b) (i) $[OH^-(aq)] = 1.50 \times 10^{-3} \times 1000/25 = 0.060$ mol dm^{-3} ✓
$K_w = [H^+(aq)][OH^-(aq)]$ ✓
$[H^+(aq)] = 1.0 \times 10^{-14}/0.060 = 1.67 \times 10^{-13}$ mol dm^{-3} ✓ pH = 12.78 ✓

(ii) Amount of HCl added = 0.0125 mol ✓
Amount of HCl remaining = $0.0125 - 1.50 \times 10^{-3} = 0.011$ mol ✓
$[HCl] = 0.011 \times 1000/75 = 0.147$ mol dm^{-3} ✓ pH = 0.83 ✓　[8]

(c) $[H^+(aq)] = 10^{-pH} = 0.0302$ mol dm^{-3} ✓
amount of HCl = $0.0302 \times 25/1000 = 7.55 \times 10^{-4}$ mol ✓
$[H^+(aq)]$ in diluted solution = $7.55 \times 10^{-4} \times 1000/40 = 0.0189$ mol dm^{-3} ✓
pH = 1.72 ✓　　　　　　　　　　　　　　　　　　　[4]

[Total: 14]

Chapter 2 Energy changes in chemistry

1 (a) (i)

definition	letter
1st ionisation energy of potassium	B
1st electron affinity of bromine	F ✓
the enthalpy of atomisation of potassium	C
the enthalpy of atomisation of bromine	A ✓
the lattice enthalpy of potassium bromide	E
the enthalpy of formation of potassium bromide	D ✓

(ii) $-394 - (89 + 419 + 112 - 325)$ ✓ $= -689$ kJ mol^{-1} ✓　　[5]

(b) (i) KI is less exothermic because I$^-$ has a smaller charge density ✓ and
attraction between K$^+$ and I$^-$ ions is weaker ✓

(ii) CaBr$_2$ is more exothermic because Ca^{2+} has a greater charge density ✓ and
attraction between Ca^{2+} and Br$^-$ ions is stronger ✓　　　[4]

[Total: 9]

2 (a) (i) $Mg^{2+}(g) + 2Cl^-(g) \longrightarrow MgCl_2(s)$ ✓

(ii) $Mg^{2+}(g) + aq \longrightarrow Mg^{2+}(aq)$ ✓

(iii) $MgCl_2(s) + aq \longrightarrow Mg^{2+}(aq) + 2Cl^-(aq)$ ✓　　　[3]

(b) $\Delta H(\text{solution}) = -(\text{lattice enthalpy}) + \Sigma(\text{enthalpy changes of hydration})$ ✓
$= -(-2526) + (-1891 + 2 \times -384)$ ✓ $= -133$ kJ mol^{-1} ✓　　[3]

[Total: 6]

3 $\Delta H_r = \Sigma\Delta H_f(\text{products}) - \Sigma\Delta H_f(\text{reactants})$

      $= (-1676) - (-602)$ ✓

      $= -1074 \text{ kJ mol}^{-1}$ ✓

      $\Delta S_r = \Sigma S \text{ (products)} - \Sigma S \text{ (reactants)}$

      $= [(2 \times 81) + 51] - [27 + (2 \times 28)]$ ✓

      $= 130 \text{ J K mol}^{-1}$ ✓ $= 0.13 \text{ kJ K}^{-1} \text{ mol}^{-1}$ ✓

      $\Delta G = \Delta H - T\Delta S$ ✓

      $\Delta G = -1074 - 298 \times 0.13 = -1113 \text{ kJ mol}^{-1}$ ✓

      Reaction is feasible at 298 K because $\Delta G < 0$ ✓

                                                    [Total: 8]

4   (a)   298 K; $[Ni^{2+}(aq)]$ 1 mol dm^{-3} ✓; $[Fe^{2+}(aq)]$ and $[Fe^{3+}(aq)]$ are both 1 mol dm^{-3} ✓   [2]

     (b)   It allows ions to flow between half-cells ✓   [1]

     (c)   (i)   1.02 V ✓

          (ii)   Electrons flow along wire from nickel half-cell ✓

          (iii)   Ni is – electrode; Pt is + electrode ✓   [3]

     (d)   (i) $Ni \longrightarrow Ni^{2+} + 2e^-$ ✓

                $Fe^{3+} + e^- \longrightarrow Fe^{2+}$ ✓

          (ii) $2Fe^{3+} + Ni \longrightarrow 2Fe^{2+} + Ni^{2+}$ ✓   [3]

     (e)   In Ni redox equilibrium, increase in $[Ni^{2+}(aq)]$ shifts equilibrium to the right. Electrons are less available and electrode potential becomes less negative. ✓

         Difference in cell potentials becomes less and cell potential will decrease. ✓   [2]

                                                  [Total: 11]

5   (a)   (i) $Mn^{3+}(aq)$ ✓

          (ii) $Mn^{3+}(aq)$ ✓ I^- must provide electrons and have the more negative $E^{\ominus}$ (+0.54). ✓ The more positive $E^{\ominus}$ system (+1.49 V) will gain electrons. ✓   [4]

     (b)   0.66 V ✓

         $4V^{2+}(aq) + O_2(g) + 2H_2O(l) \longrightarrow 4OH^-(aq) + 4V^{3+}(aq)$ ✓✓   [3]

                                                    [Total: 7]

Chapter 3 The Periodic Table

1   (a) MgO, Al_2O_3, SiO_2; ✓ P_4O_{10}, SO_3 ✓   [2]

     (b)   The oxidation number matches the number of outer shell electrons involved in bonding to oxygen. ✓   [1]

     (c) $MgO(s) + H_2O(l) \longrightarrow Mg(OH)_2(aq)$ ✓

         Al_2O_3 and SiO_2 are insoluble ✓

         $P_4O_{10}(s) + 6H_2O(l) \longrightarrow 4H_3PO_4(aq)$ ✓

         $SO_3(l) + H_2O(l) \longrightarrow H_2SO_4(aq)$ ✓   [4]

     (d)   Giant ionic oxides form alkaline solutions ✓

         Giant covalent oxides are insoluble ✓

         Simple molecular oxides form acidic solutions ✓   [3]

                                                   [Total: 10]

2   (a)   Ni: $1s^2 2s^2 2s^6 3s^2 3p^6 3d^8 4s^2$ ✓;   Ni^{2+}: $1s^2 2s^2 2s^6 3s^2 3p^6 3d^8$ ✓   [2]

     (b)   A d-block element has its highest energy electron in a d sub-shell. ✓

         A transition element has at least one ion with a partially filled d sub-shell. ✓   [2]

(c) (i) Lone pair of electrons on oxygen atom ✓

(ii) Ion $[Ni(H_2O)_6]^{2+}$ ✓

(iii) Covalent bonding in H_2O ✓ and coordinate bonding from water ligands to the Ni^{2+} ion. ✓

(iv) Octahedral ✓ ; 90° ✓ [6]

(d) (i) $Ni(OH)_2$ or $Ni(OH)_2(H_2O)_4$ ✓
$Ni^{2+}(aq) + 2OH^-(aq) \longrightarrow Ni(OH)_2(s)$ ✓
or $[Ni(H_2O)_6]^{2+}(aq) + 2OH^-(aq) \longrightarrow Ni(OH)_2(H_2O)_4(s) + 2H_2O(l)$

(ii) $[NiCl_4]^{2-}$ ✓

(iii) $[Ni(H_2O)_6]^{2+}(aq) + 6NH_3(aq) \longrightarrow [Ni(NH_3)_6]^{2+}(aq) + 6H_2O(l)$ ✓
ligand substitution ✓ [5]

[Total: 15]

3 (a) (i) +6 ✓

(ii) +1 ✓

(iii) +4 ✓ [3]

(b) $2Cr^{3+} + 10OH^- + 3H_2O_2 \longrightarrow 2CrO_4^{2-} + 8H_2O$ ✓ [1]

(c) Vanadium as V_2O_5 in the contact process for the manufacture SO_3 for H_2SO_4 ✓
Iron in the Haber process for the manufacture of NH_3 ✓ [2]

[Total: 6]

4 (a) (i) +7 ✓

(ii) $5H_2O_2 + 2MnO_4^- + 6H^+ \longrightarrow 5O_2 + 8H_2O + 2Mn^{2+}$ ✓ [2]

(b) Reducing agent ✓ [1]

(c) Amount of $KMnO_4 = 0.0150 \times \dfrac{23.80}{1000} = 3.57 \times 10^{-4}$ mol ✓

Amount of H_2O_2 that reacted $= 2.5 \times 3.57 \times 10^{-4} = 8.925 \times 10^{-4}$ mol ✓

Concentration of $H_2O_2 = \dfrac{1000}{25.0} \times 8.925 \times 10^{-4} = 0.0357$ mol ✓ [3]

[Total: 6]

5 (a) (i) A coordinate (dative covalent bond) forms ✓ between a lone pair on a ligand and the central metal ion. ✓

(ii) Bidentate ligand ✓ [3]

(b) (i) +3 ✓

(ii) $1s^2 2s^2 2s^6 3s^2 3p^6 3d^b$ ✓ [2]

(c) (i) 6 ✓

(ii) Octahedral ✓

(iii) Add an excess of 1,2-diaminoethane or $NH_2CH_2CH_2NH_2$ ✓
$[Ni(H_2O)_6]^{2+} + 3NH_2CH_2CH_2NH_2 \rightarrow [Ni(NH_2CH_2CH_2NH_2)_3]^{2+} + 6H_2O$ ✓ [4]

[Total: 9]

Chapter 4 Chemistry of organic functional groups

1 (a) $C : H : O = 66.7/12 : 11.1/1 : 22.2/16$ ✓
   Molecular formula = C_4H_8O ✓ [2]

(b) $CH_3CH_2COCH_3$ ✓; $CH_3CH_2CH_2CHO$ ✓; $(CH_3)_2CHCHO$ ✓ [3]

(c) 2,4-dinitrophenylhydrazine ✓ orange precipitate ✓ [2]

(d) Tollens' reagent ✓. Reacts with butanal and methylpropanal ✓
   Silver mirror forms ✓ (or $H_2SO_4/K_2Cr_2O_7$; from orange to green) [3]

(e) Product: $CH_3CH_2CH(OH)CH_3$ ✓ formed from $CH_3CH_2COCH_3$ ✓ [2]

[Total: 12]

2 (a) (i) Ethyl propanoate ✓

(ii) **E**: CH_3CH_2OH ✓
   F: $CH_3CH_2COO^-Na^+$ ✓✓ (1 mark if CH_3CH_2COOH) [4]

(b) (i) ✓

(ii) Concentrated H_2SO_4 ✓ [2]

(c)

$3CH_3COOH$ + HO—CH₂ ... → ... + $3H_2O$ ✓ [3]

[Total: 9]

3 (a) (i) Molar ratio $C:H = 90.56/12 : 9.44/1$ ✓
   Empirical formula = C_4H_5 ✓
   Molecular formula = C_8H_{10} ✓

(ii) [7]

(b) (i) 1,4-dimethylbenzene ✓

(ii) [2]

[Total: 9]

4 (a) (i) Electrophilic substitution ✓

(ii) Concentrated HNO_3 and concentrated H_2SO_4 ✓ [2]

(b) (i) Tin and concentrated HCl ✓

(ii)

+ 6 [H] → + $2H_2O$ ✓✓ [3]

(c) (i) Reagents NaNO$_2$/HCl(aq) ✓ Conditions: <10°C ✓

(ii)

[3]

(d) (i) A phenol, e.g. C$_6$H$_5$OH ✓

(ii)

[2]

[Total: 10]

5 (a) Compound **A**: CH$_3$CH$_2$CH$_2$COOH ✓
Compound **B**: (CH$_3$)$_2$CHOH ✓ [2]

(b) Step 1: H$_2$SO$_4$/K$_2$Cr$_2$O$_7$ ✓ reflux ✓ oxidation ✓
Step 2: NaBH$_4$ ✓ with H$_2$O or ethanol ✓ reduction ✓
Step 3: Conc. H$_2$SO$_4$ catalyst ✓ reflux ✓ esterification ✓ [9]

[Total: 11]

Chapter 5 Polymers, analysis and synthesis

1 (a) (CH$_3$)$_2$CH ✓ [1]

(b) (i)

(ii)

(iii)

[3]

(c) (i) A chiral carbon atom has **four** different groups attached to it. ✓ It provides two possible forms that are non-superimposable mirror images of one another. ✓

(ii)

Bond angle = 109.5° ✓ [5]

(d)

pH = 2   pH = 7   pH = 12   ✓✓✓ [3]

(e)

✓   [1]

[Total: 13]

2 (a) 2-hydroxypropanoic acid ✓   [1]

(b) (i) A chiral carbon with four different groups attached ✓

(ii) Optical isomers are non-superimposable mirror images ✓

(iii)

✓✓   [4]

(c) (i) Polyester or a condensation polymer ✓

(ii)

ester link ✓ structure ✓   [3]

[Total: 8]

3 (a) (i)

✓

(ii) Addition ✓   [2]

(b) (i)

✓✓

(ii) Condensation ✓   [3]

(c) Polymer **D** has a polar CONH group. Polymer **C** is non-polar. ✓

Polar group in polymer **D** will attract water or hydroxide ions. ✓   [2]

[Total: 7]

4 (a) (i) 4 ✓

(ii) C_2H_5 ✓

The CH_2 group at $\delta = 1.63$ ppm is split into a quartet ✓ by 3 protons on an adjacent carbon. ✓

The CH_3 group at $\delta = 0.90$ ppm is split into a triplet ✓ by 2 protons on an adjacent carbon. ✓

(iii) A CH_3 group with no protons on an adjacent carbon. ✓

(iv) Two equivalent OH protons as the absorption disappears with D_2O. ✓

(v) $CH_3CH_2C(OH)_2CH_3$ ✓   [9]

(b) (i) There are 8H atoms with a peak ratio of 1:3. The peak at δ = 3.64 ppm is from a CH_2–O ✓ with no protons on an adjacent carbon atom ✓
The peak at δ = 1.24 ppm is from two equivalent CH_3 groups ✓ with no protons on an adjacent carbon atom. ✓

(ii) $(CH_3)_2C(OH)CH_2OH$ ✓ [5]

[Total: 14]

Chapter 6 Synoptic assessment

1 (a) (i) $C : H : O = \dfrac{66.7}{12} / : \dfrac{11.1}{1} : \dfrac{22.2}{16}$ ✓

= 5.56 : 11.1 : 1.39 = 4 : 8 : 1
empirical formula = C_4H_8O ✓
48 + 8 + 16 = 72 which is half of M_r
Therefore molecular formula = $C_8H_{16}O_2$ ✓
Structural formula = $CH_3(CH_2)_6COOH$ ✓

(ii) caprylic acid has a longer carbon chain and contains more electrons ✓
caprylic acid has more van der Waals' forces ✓ [6]

(b) (i) $[H^+(aq)] = \dfrac{K_w}{[OH^-(aq)]}$ ✓ $= \dfrac{1.00 \times 10^{-14}}{0.500}$

= 2.00×10^{-14} mol dm^{-3} ✓
pH = $-\log[H^+(aq)]$ = $-\log(2.00 \times 10^{-14})$ = 13.70 ✓

(ii) moles NaOH in 25.00 cm^3 = moles NaOH = 0.0125 mol ✓
moles **A** in 21.40 cm^3 = moles NaOH = 0.0125 mol ✓
moles **A** in 250 cm^3 = $\dfrac{0.0125 \times 250}{21.40}$ = 0.146 mol ✓

molar mass of **A** = $\dfrac{10.8}{0.146}$ = 74 g mol^{-1} ✓

Therefore **A** is propanoic acid / CH_3CH_2COOH ✓ [8]

[Total: 14]

2 (a) **A** = $CH_3CH_2COOCH_2CH_2CH_3$ ✓
X = CH_3CH_2COOH ✓
Y = $CH_3CH_2CH_2OH$ ✓
For propanal ⟶ **X**, H_2SO_4 ✓ and $K_2Cr_2O_7$ ✓
For propanal ⟶ **Y**, $NaBH_4$ ✓ in ethanol or water ✓
Concentrated H_2SO_4 is a catalyst ✓ [8]

(b) 3 isomers with asymmetric carbon atom: **B**, **C** and **D**
$CH_3CH_2CH_2CH(CH_3)COOH$ ✓
$CH_3CH_2CH(CH_3)CH_2COOH$ ✓
$(CH_3)_2CHCH(CH_3)COOH$ ✓
E: $(CH_3)_3CCH_2COOH$ ✓ [4]

(c) *a*: δ = 3.3–4.3 ppm ✓; *b*: δ = 2.0–2.9 ppm ✓
a: splitting will be a quartet ✓ as there is an adjacent CH_3 ✓
b: splitting will be a triplet ✓ as there is an adjacent CH_2 ✓ [6]

(d)

✓ (structure) ✓ (2 x OH groups at correct positions) [2]

[Total: 20]

Periodic Table

The Periodic Table

Key:

relative atomic mass	1.0
atomic symbol	**H**
atomic number	1
name	Hydrogen

Group 1	Group 2												Group 3	Group 4	Group 5	Group 6	Group 7	Group 0
																		4.0 **He** 2 Helium
6.9 **Li** 3 Lithium	9.0 **Be** 4 Beryllium												10.8 **B** 5 Boron	12.0 **C** 6 Carbon	14.0 **N** 7 Nitrogen	16.0 **O** 8 Oxygen	19.0 **F** 9 Fluorine	20.2 **Ne** 10 Neon
23.0 **Na** 11 Sodium	24.3 **Mg** 12 Magnesium												27.0 **Al** 13 Aluminium	28.1 **Si** 14 Silicon	31.0 **P** 15 Phosphorus	32.1 **S** 16 Sulfur	35.5 **Cl** 17 Chlorine	39.9 **Ar** 18 Argon
39.1 **K** 19 Potassium	40.1 **Ca** 20 Calcium	45.0 **Sc** 21 Scandium	47.9 **Ti** 22 Titanium	50.9 **V** 23 Vanadium	52.0 **Cr** 24 Chromium	54.9 **Mn** 25 Manganese	55.8 **Fe** 26 Iron	58.9 **Co** 27 Cobalt	58.7 **Ni** 28 Nickel	63.5 **Cu** 29 Copper	65.4 **Zn** 30 Zinc		69.7 **Ga** 31 Gallium	72.6 **Ge** 32 Germanium	74.9 **As** 33 Arsenic	79.0 **Se** 34 Selenium	79.9 **Br** 35 Bromine	83.8 **Kr** 36 Krypton
85.5 **Rb** 37 Rubidium	87.6 **Sr** 38 Strontium	88.9 **Y** 39 Yttrium	91.2 **Zr** 40 Zirconium	92.9 **Nb** 41 Niobium	95.9 **Mo** 42 Molybdenum	[98] **Tc** 43 Technetium	101.1 **Ru** 44 Ruthenium	102.9 **Rh** 45 Rhodium	106.4 **Pd** 46 Palladium	107.9 **Ag** 47 Silver	112.4 **Cd** 48 Cadmium		114.8 **In** 49 Indium	118.7 **Sn** 50 Tin	121.8 **Sb** 51 Antimony	127.6 **Te** 52 Tellurium	126.9 **I** 53 Iodine	131.3 **Xe** 54 Xenon
132.9 **Cs** 55 Caesium	137.3 **Ba** 56 Barium	138.9 **La*** 57 Lanthanum	178.5 **Hf** 72 Hafnium	180.9 **Ta** 73 Tantalum	183.8 **W** 74 Tungsten	186.2 **Re** 75 Rhenium	190.2 **Os** 76 Osmium	192.2 **Ir** 77 Iridium	195.1 **Pt** 78 Platinum	197.0 **Au** 79 Gold	200.6 **Hg** 80 Mercury		204.4 **Tl** 81 Thallium	207.2 **Pb** 82 Lead	209.0 **Bi** 83 Bismuth	[209] **Po** 84 Polonium	[210] **At** 85 Astatine	[222] **Rn** 86 Radon
[223] **Fr** 87 Francium	[226] **Ra** 88 Radium	[227] **Ac*** 89 Actinium	[261] **Rf** 104 Rutherfordium	[262] **Db** 105 Dubnium	[266] **Sg** 106 Seaborgium	[264] **Bh** 107 Bohrium	[277] **Hs** 108 Hassium	[268] **Mt** 109 Meitnerium	[271] **Ds** 110 Darmstadtium	[272] **Rg** 111 Roentgenium								

Elements with atomic numbers 112–116 have been reported but not fully authenticated

lanthanides

140.1 **Ce** 58 Cerium	140.9 **Pr** 59 Praseodymium	144.2 **Nd** 60 Neodymium	144.9 **Pm** 61 Promethium	150.4 **Sm** 62 Samarium	152.0 **Eu** 63 Europium	157.2 **Gd** 64 Gadolinium	158.9 **Tb** 65 Terbium	162.5 **Dy** 66 Dysprosium	164.9 **Ho** 67 Holmium	167.3 **Er** 68 Erbium	168.9 **Tm** 69 Thulium	173.0 **Yb** 70 Ytterbium	175.0 **Lu** 71 Lutetium

actinides

232.0 **Th** 90 Thorium	[231] **Pa** 91 Protactinium	238.1 **U** 92 Uranium	[237] **Np** 93 Neptunium	[242] **Pu** 94 Plutonium	[243] **Am** 95 Americium	[247] **Cm** 96 Curium	[245] **Bk** 97 Berkelium	[251] **Cf** 98 Californium	[254] **Es** 99 Einsteinium	[253] **Fm** 100 Fermium	[256] **Md** 101 Mendelevium	[254] **No** 102 Nobelium	[257] **Lr** 103 Lawrencium

Index

Index